SSM 与 Spring Boot 开发实战

肖海鹏　牟东旭◎编著

人 民 邮 电 出 版 社
北 京

图书在版编目（CIP）数据

SSM与Spring Boot开发实战 / 肖海鹏，牟东旭编著
. -- 北京 : 人民邮电出版社，2020.8（2023.8重印）
ISBN 978-7-115-54001-0

Ⅰ. ①S… Ⅱ. ①肖… ②牟… Ⅲ. ①JAVA语言－程序
设计 Ⅳ. ①TP312.8

中国版本图书馆CIP数据核字(2020)第078334号

内容提要

本书以 Java EE 为主要开发平台，系统讲解了通过 Spring、Spring MVC 和 MyBatis（SSM）三大框架开发企业项目的方法、技术与实践。本书主要介绍了 Spring、Spring MVC 和 MyBatis 的基础知识，Spring 的资源管理，如何实现控制反转，如何通过 Spring 表达式语言简化代码，如何通过面向切面编程降低业务逻辑各部分之间的耦合度，如何整合数据层，并结合具体案例讲述了如何通过 SSM、Spring Boot 实现项目的整合。

本书适合 Java 程序员、SSM 开发人员、Spring Boot 开发人员阅读。

◆ 编　　著　肖海鹏　牟东旭
　责任编辑　谢晓芳
　责任印制　王　郁　焦志炜
◆ 人民邮电出版社出版发行　　北京市丰台区成寿寺路 11 号
　邮编　100164　　电子邮件　315@ptpress.com.cn
　网址　https://www.ptpress.com.cn
　北京七彩京通数码快印有限公司印刷
◆ 开本：800×1000　1/16
　印张：27.5　　2020 年 8月第 1 版
　字数：557 千字　　2023 年 8 月北京第 9 次印刷

定价：99.00 元

读者服务热线：(010)81055410　印装质量热线：(010)81055316
反盗版热线：(010)81055315
广告经营许可证：京东市监广登字 20170147 号

前　言

Java 平台已经连续多年在 TIOBE 排名中保持第一。在 2020 年 3 月的 TIOBE 排名中，Python 位列第三，C 仍然稳居第二名的位置。经常有人问："到底是学习 Python 还是学 Java？哪个更有前途？Java 平台过几年会不会被其他开发平台代替？"

要回答这些问题，就需要了解软件的应用市场以及 Python、C 和 Java 语言的应用方向。显然，软件最大的应用市场是企业应用——大中型企业要购买软件，不论是国内还是国外，企业应用永远是第一大市场。

Python 的主要应用是人工智能，C 的定位是系统级开发，Java 的明确目标是企业应用，这 3 种语言的定位不同，不是竞争者的关系。Java 多年来在 TIOBE 排名中独霸首位和它的定位是密不可分的，而且至少目前没有看到 Java 的挑战者。在企业级开发市场中，Java 一枝独秀的局面至少在 10 年内不会被打破。

作为成熟的企业开发平台，Java 是一个完全开放的、开源的平台，其官方架构不断面临着其他开源框架的冲击。其中最大的挑战就是 Spring 对企业 Java Bean（Enterprise Java Bean，EJB）发起的冲击。持久层框架也发生了变化，从 Hibernate 一家独大，变成了 Hibernate 与 MyBatis 两强并存的局面。视图层则逐渐从模型-视图-控制器（Model-View-Controller，MVC）转向了富客户端的 Web 2.0 技术。

基于 Java 平台的框架很多，目前流行的是 SSM（Spring+Spring MVC+MyBatis）+ Spring Boot+分布式的开发模式。为了学习分布式开发，如 Spring Cloud 或 Dubbo，要先掌握 SSM 与 Spring Boot。

学习 Java 开源框架是一个漫长的过程，一蹴而就是不可能的。尤其是开源框架与 Java EE 平台的关系、用 SSM 搭建高并发系统的方法等，需要长时间的学习。因此，本书不仅适合 Java 初学者阅读，即使有 3~5 年开发经验的 Java 与 Spring 开发人员，也会从本书受益很多。

关于 SSM，推荐图书与视频结合的学习模式。另外，一定要多动手练习、多思考。掌握框架的设计思想，远比简单掌握框架的应用更有意义。

本书有完整的配套教学视频，在 CSDN、51CTO、腾讯课堂等网站搜索“肖海鹏”，即可找到相关的视频。

服务与支持

本书由异步社区出品，异步社区（https://www.epubit.com/）为您提供相关后续服务。

提交勘误

作者和编辑尽最大努力来确保书中内容的准确性，但难免会存在疏漏。欢迎您将发现的问题反馈给我们，帮助我们提升图书的质量。

当您发现错误时，请登录异步社区，按书名搜索，进入本书页面，单击“提交勘误”，输入勘误信息，单击“提交”按钮即可（见下图）。本书的作者和编辑会对您提交的勘误进行审核，确认并接受后，您将获赠异步社区的100积分。积分可用于在异步社区兑换优惠券、样书或奖品。

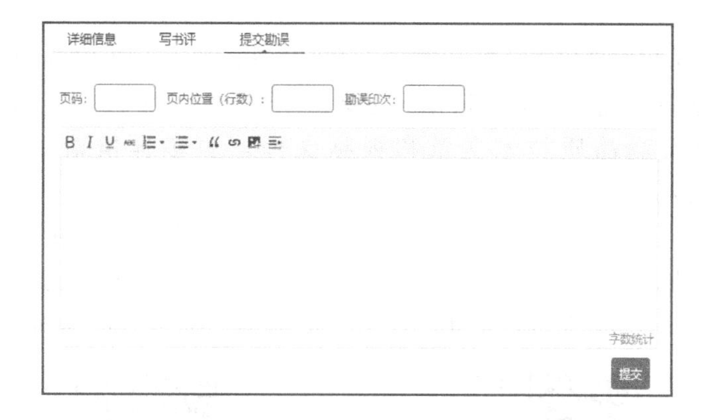

扫码关注本书

扫描下方二维码，您将会在异步社区微信服务号中看到本书信息及相关的服务提示。

与我们联系

我们的联系邮箱是 contact@epubit.com.cn。

如果您对本书有任何疑问或建议，请您发邮件给我们，并请在邮件标题中注明本书书名，以便我们更高效地做出反馈。

如果您有兴趣出版图书、录制教学视频，或者参与图书翻译、技术审校等工作，可以发邮件给我们；有意出版图书的作者也可以到异步社区在线投稿（直接访问 www.epubit.com/selfpublish/submission 即可）。

如果您所在学校、培训机构或企业想批量购买本书或异步社区出版的其他图书，也可以发邮件给我们。

如果您在网上发现有针对异步社区出品图书的各种形式的盗版行为，包括对图书全部或部分内容的非授权传播，请您将怀疑有侵权行为的链接通过邮件发送给我们。您的这一举动是对作者权益的保护，也是我们持续为您提供有价值的内容的动力之源。

关于异步社区和异步图书

"异步社区"是人民邮电出版社旗下 IT 专业图书社区，致力于出版精品 IT 技术图书和相关学习产品，为作译者提供优质出版服务。异步社区创办于 2015 年 8 月，提供大量精品 IT 技术图书和电子书，以及高品质技术文章和视频课程。更多详情请访问异步社区官网 https://www.epubit.com。

"异步图书"是由异步社区编辑团队策划出版的精品 IT 专业图书的品牌，依托于人民邮电出版社近 30 年的计算机图书出版积累和专业编辑团队，相关图书在封面上印有异步图书的 LOGO。异步图书的出版领域包括软件开发、大数据、AI、测试、前端、网络技术等。

异步社区

微信公众号

目　录

第 1 章 Spring 基础知识

Spring Framework 的创始人是著名学者 Rod Johnson。Rod Johnson 还是 JSR 154（Servlet 2.4）和 JDO 2.0 的规范专家、Java 社区过程（Java Community Process，JCP）的积极成员，是 Java 开发社区中的杰出人物。

Rod Johnson 的出名源于两本书，分别是 2002 年出版的 *Expert One-on-One J2EE Design and Development* 和 2004 年出版的 *Expert One-on-One J2EE Development without EJB*。这两本书的影响力非常深远，彻底打破了 Java EE 开发的传统模式，使 Java 平台的应用从大型企业快速转向中小型企业。

在最新的 TIOBE 排名中，Java 依然高居榜首。Rod Johnson 对 Java 平台的推广功不可没。

另外，要了解 Rod Johnson，一定要记住轮子理论。西方有一句著名谚语——不要重复发明轮子（Don't reinvent the wheel）。Rod Johnson 深受这句谚语的启发，励志独辟蹊径，不走常人之路。在其他计算机专家都想在某一领域独占鳌头时，Rod Johnson 想的是如何把五花八门的框架整合在一起。

Spring Framework 的开发思想首先基于轮子理论，通过提供统一的抽象接口，让各种框架可以优雅地协同工作。

1.1 Spring 与 Java EE

1.1.1 下载 Java EE 资源

Java EE（Java Enterprise Edition）的本质是一套完整的开发规范。Java 平台原来由 Sun 公司开发，后来 Oracle 公司兼并了 Sun 公司。因此，所有 Java 和 Java EE 的资源都要到 Oracle 官网下载。具体下载步骤如下。

（1）在 Oracle 官网中，选择首页最下面的 Resources for 中的 Developers（见图 1-1）。

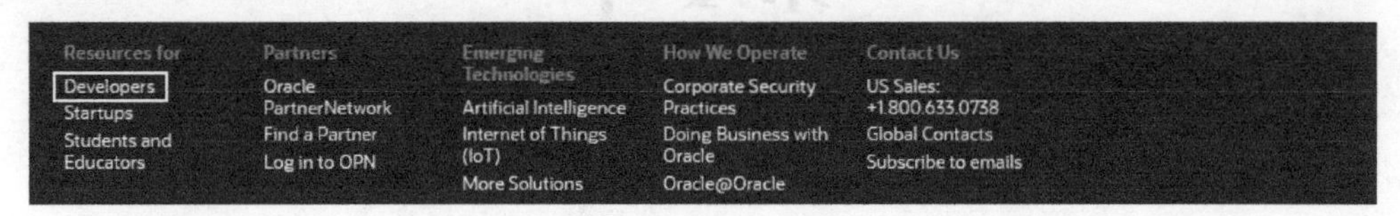

图 1-1 选择 Developers

（2）在弹出的页面中，选择上方的 Technologies，进入 Technologies 界面，选择 Java（见图 1-2）。

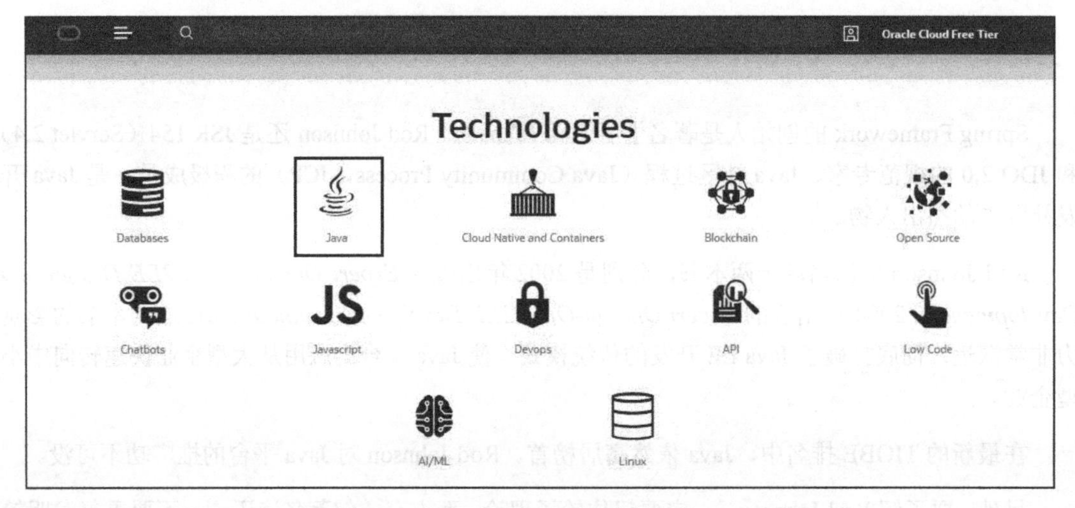

图 1-2 在 Technologies 界面中选择 Java

（3）单击 Java SE Download 链接（见图 1-3）。

（4）在弹出的页面中，选择左侧的 Java EE，进入 Java EE 界面，选择右侧的 Java EE 7 Technologies（见图 1-4）。

（5）在弹出的 Java EE 7 Technologies 界面中，选择要下载的资源，即可开始下载 Java EE 资源。

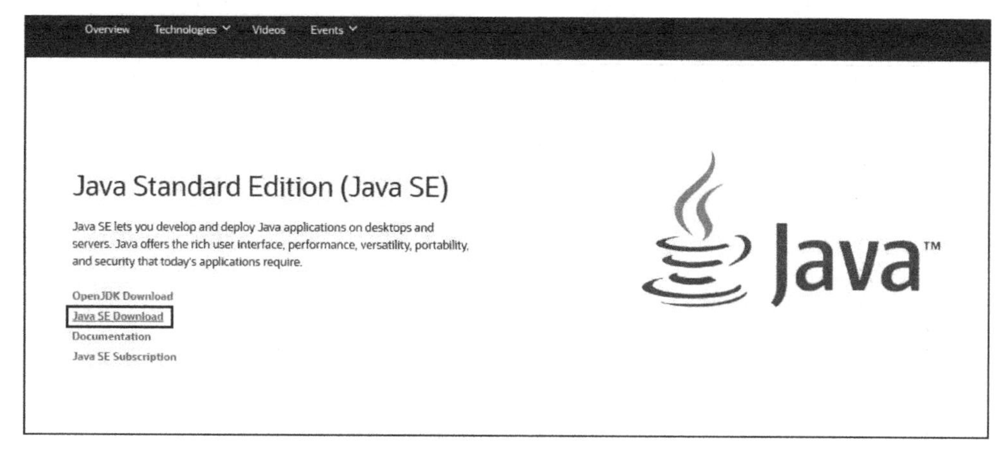

图 1-3　选择 Java SE Download 链接

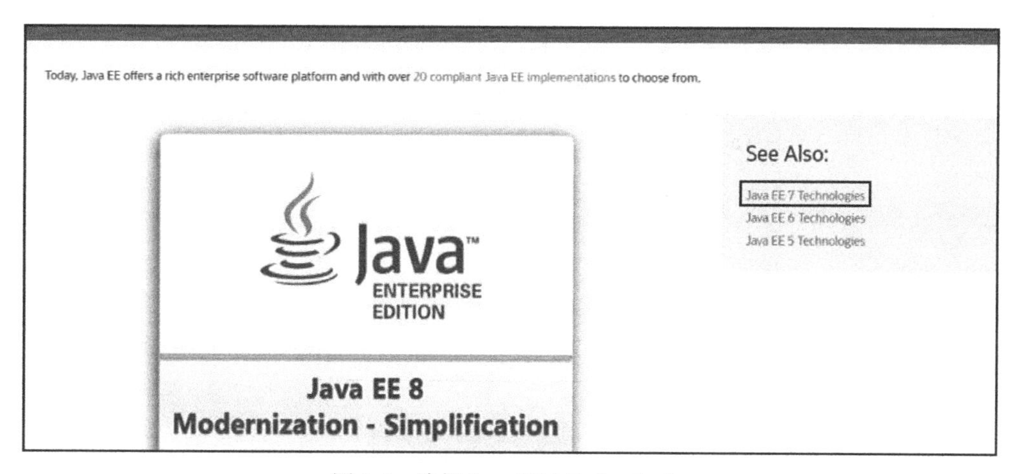

图 1-4　选择 Java EE 7 Technologies

Java EE 8 是当前的最新规范，但由于主流的开发规范是 Java EE 7，因此我们后面的开发以 Java EE 7 为基础。

1.1.2　Java EE

Java EE 是 Sun 公司制定的一套 Java 开发规范。

规范是什么？例如，冯·诺依曼定义了计算机架构，即计算机由输入设备、输出设备、运算器、控制器、内存、外存组成。这就可以看成一套高层的规范。

规范就是一组约定。规范相当于工业标准，协议也可以看成某种规范。

Java EE 的大量规范性文档是由一系列的 Java 规范提案（Java Specification Request，JSR）组成的。图 1-5（a）～（c）展示了 Java EE 7 的 JSR 列表。

Java EE Platform	
Java Platform, Enterprise Edition 7 (Java EE 7)	JSR 342
Web Application Technologies	
Java API for WebSocket	JSR 356
Java API for JSON Processing	JSR 353
Java Servlet 3.1	JSR 340
JavaServer Faces 2.2	JSR 344
Expression Language 3.0	JSR 341
JavaServer Pages 2.3	JSR 245
Standard Tag Library for JavaServer Pages (JSTL) 1.2	JSR 52
Enterprise Application Technologies	
Batch Applications for the Java Platform	JSR 352
Concurrency Utilities for Java EE 1.0	JSR 236
Contexts and Dependency Injection for Java 1.1	JSR 346
Dependency Injection for Java 1.0	JSR 330
Bean Validation 1.1	JSR 349
Enterprise JavaBeans 3.2	JSR 345
Interceptors 1.2 (Maintenance Release covered under JSR 318)	JSR 318

（a）

Java EE Connector Architecture 1.7	JSR 322
Java Persistence 2.1	JSR 338
Common Annotations for the Java Platform 1.2	JSR 250
Java Message Service API 2.0	JSR 343
Java Transaction API (JTA) 1.2	JSR 907
JavaMail 1.5	JSR 919
Web Services Technologies	
Java API for RESTful Web Services (JAX-RS) 2.0	JSR 339
Implementing Enterprise Web Services 1.3	JSR 109
Java API for XML-Based Web Services (JAX-WS) 2.2	JSR 224
Web Services Metadata for the Java Platform	JSR 181
Java API for XML-Based RPC (JAX-RPC) 1.1 (Optional)	JSR 101
Java APIs for XML Messaging 1.3	JSR 67
Java API for XML Registries (JAXR) 1.0	JSR 93
Management and Security Technologies	
Java Authentication Service Provider Interface for Containers 1.1	JSR 196
Java Authorization Contract for Containers 1.5	JSR 115
Java EE Application Deployment 1.2 (Optional)	JSR 88
J2EE Management 1.1	JSR 77
Debugging Support for Other Languages 1.0	JSR 45

（b）

Java EE-related Specs in Java SE	
Java Architecture for XML Binding (JAXB) 2.2	JSR 222
Java API for XML Processing (JAXP) 1.3	JSR 206
Java Database Connectivity 4.0	JSR 221
Java Management Extensions (JMX) 2.0	JSR 003
JavaBeans Activation Framework (JAF) 1.1	JSR 925
Streaming API for XML (StAX) 1.0	JSR 173

（c）

图 1-5　Java EE 7 的 JSR 列表

1.1.3　Java EE 7 的架构

图 1-6 为 Java EE 7 的架构，其中显示了 Java EE 7 平台支持的功能和主要运行模式。

Java EE 架构是分布式架构，其核心是容器与组件协同工作。

Java EE 容器包括 Applet 容器、Web 容器、应用程序客户端容器、EJB 容器。

Java EE 组件包括 Applet、JSP、Servlet、EJB 和 JavaBean。

理解 Java EE 中容器与组件的工作模式，对于学习 Spring Framework 非常重要，因为 Spring 也涉及容器与组件的工作模式。

组件简称 Bean。容器如何管理 Bean 对象、Bean 的生命周期、Bean 的状态管理等，都是重要的知识点。对比 Java EE 与 Spring 的 Bean 是非常重要的学习方法。

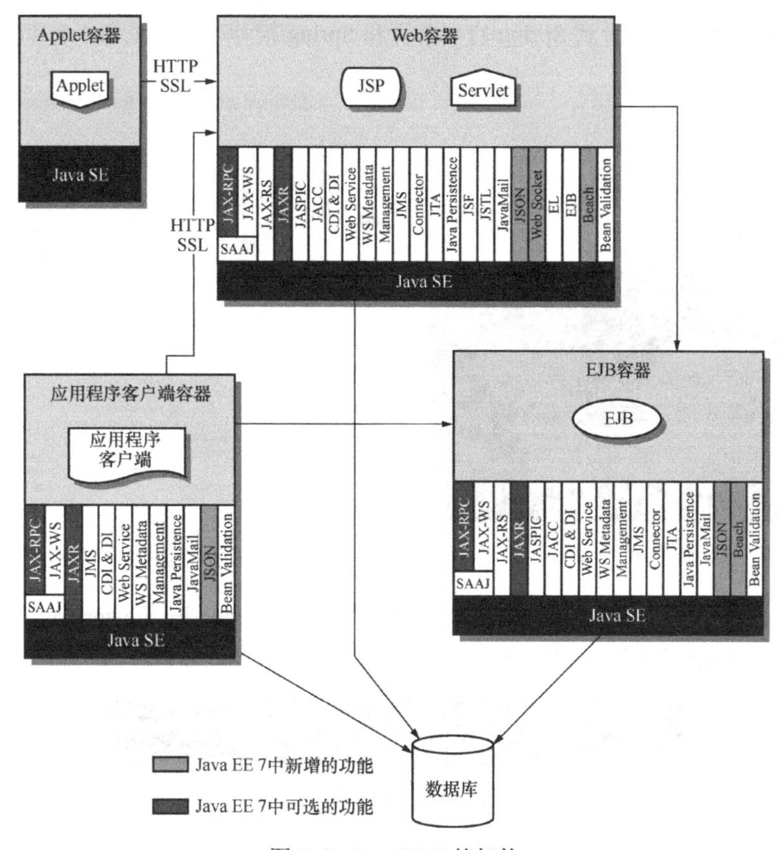

图 1-6　Java EE 7 的架构

Spring 容器与 Java EE 容器如何协同工作？性能如何？这些也是重要的话题。

1.1.4　Spring 与 Java EE 的关系

Spring 是轻量级框架，EJB 是重量级框架。Spring 的出发点是用声明性事务代替 EJB，因此 Spring 和 Java EE 是竞争关系。Spring 是第三方框架，Java EE 是规范，Spring 的所有开发必须满足 Java EE 平台的要求。Spring 用新的方案解决了 Java EE 中的问题，因此 Spring 和 Java EE 又是互补的关系。

例如，Spring 的面向切面编程（Aspect Oriented Programming，AOP）提出了中小企业的开发方案，替代了 EJB，Spring 支持 EJB 的调用，Spring 有很多自定义的注解，但是也支持 JSR 中的注解。

1.2　Spring 项目

从 Spring Framework 开始，Spring 现在已经开发了很多项目。“Spring 全家桶”是当前非常流行的开发模式。本节介绍 Spring 项目。

进入 Spring 官网，可以看到 Spring 官网首页和 Spring 推荐项目（见图 1-7 和图 1-8）。

图 1-7　Spring 官网首页　　　　图 1-8　Spring 推荐项目

现在 Spring 主推的项目有 Spring Boot、Spring Cloud 和 Spring Framework 5。当然，其他项目也非常受欢迎。Spring 是框架，它所属的公司是 Pivotal 公司（见图 1-9）。

图 1-9　Spring 所属公司

Spring Boot 的功能见图 1-10，它主要用于解决 Maven 资源下载的依赖问题，同时通过内嵌服务器提供了快速的开发、部署环境。

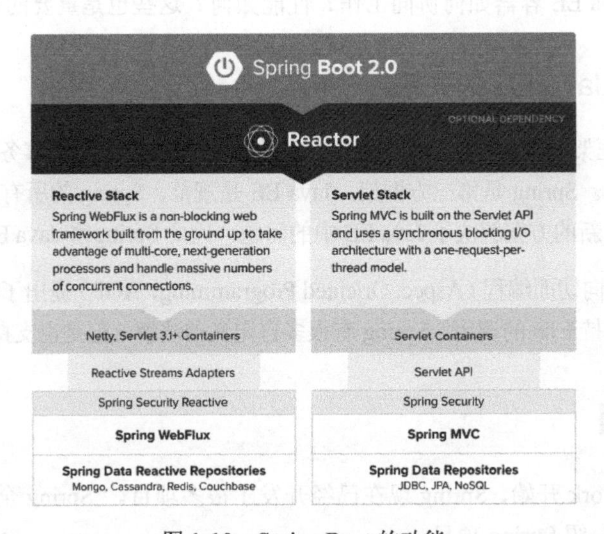

图 1-10　Spring Boot 的功能

Spring Cloud 的功能见图 1-11，它以 Netflix 公司的开源架构为基础，为中型网站提供了一套可以自由伸缩的分布式网络架构。

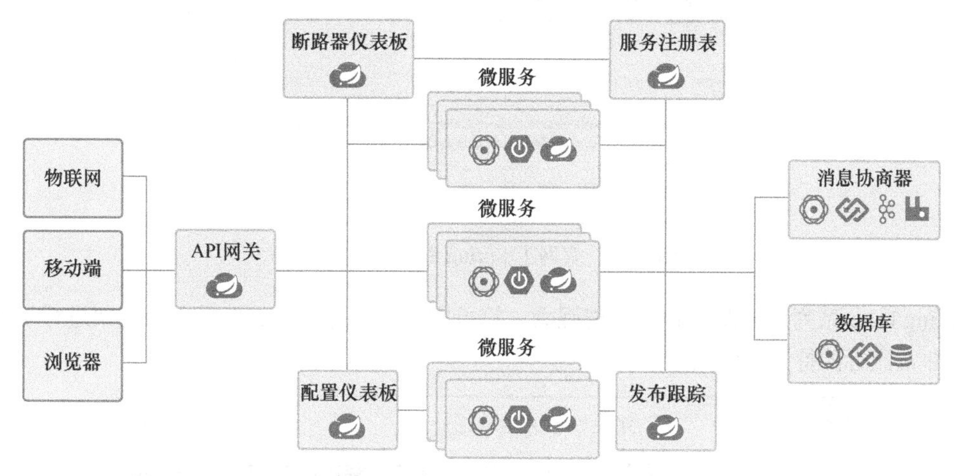

图 1-11　Spring Cloud 的功能

官网上 Spring Framework 的介绍见图 1-12，它强调了 Spring 作为基础架构的管道作用，可以让开发人员聚焦在业务逻辑上。

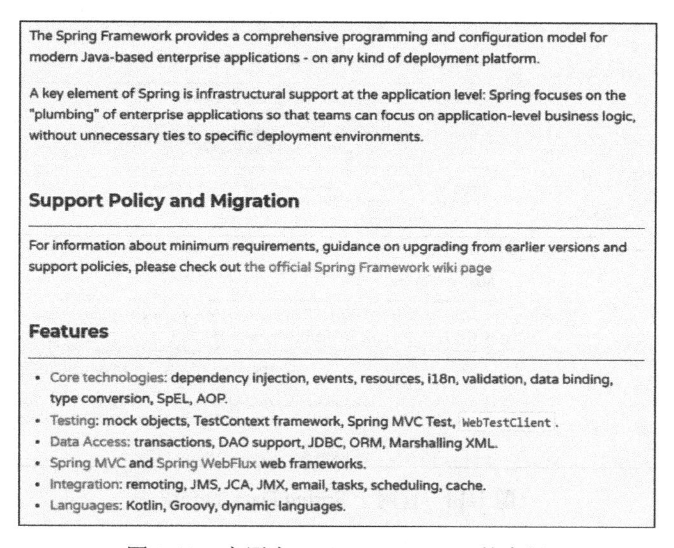

The Spring Framework provides a comprehensive programming and configuration model for modern Java-based enterprise applications - on any kind of deployment platform.

A key element of Spring is infrastructural support at the application level: Spring focuses on the "plumbing" of enterprise applications so that teams can focus on application-level business logic, without unnecessary ties to specific deployment environments.

Support Policy and Migration

For information about minimum requirements, guidance on upgrading from earlier versions and support policies, please check out the official Spring Framework wiki page

Features

- Core technologies: dependency injection, events, resources, i18n, validation, data binding, type conversion, SpEL, AOP.
- Testing: mock objects, TestContext framework, Spring MVC Test, `WebTestClient`.
- Data Access: transactions, DAO support, JDBC, ORM, Marshalling XML.
- Spring MVC and Spring WebFlux web frameworks.
- Integration: remoting, JMS, JCA, JMX, email, tasks, scheduling, cache.
- Languages: Kotlin, Groovy, dynamic languages.

图 1-12　官网上 Spring Framework 的介绍

官网上 Spring Security 的介绍见图 1-13，这是一套完善的安全框架，可以应用于大型企业和网站。Spring Security 采用 AOP 方式，提供了一套灵活的权限分配和校验机制，同时针对各种可能的网络安全隐患提出了相应的解决方案。

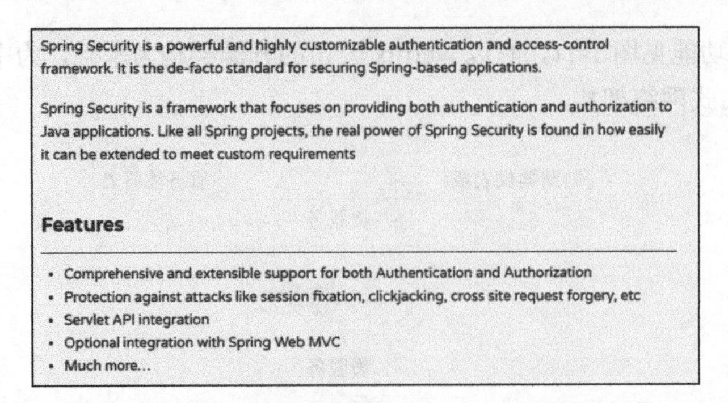
Spring Security is a powerful and highly customizable authentication and access-control framework. It is the de-facto standard for securing Spring-based applications.

Spring Security is a framework that focuses on providing both authentication and authorization to Java applications. Like all Spring projects, the real power of Spring Security is found in how easily it can be extended to meet custom requirements

Features

- Comprehensive and extensible support for both Authentication and Authorization
- Protection against attacks like session fixation, clickjacking, cross site request forgery, etc
- Servlet API integration
- Optional integration with Spring Web MVC
- Much more...

图 1-13 官网上 Spring Security 的介绍

Spring Data 系列（见图 1-14）功能强大，简单实用，对持久层的各种主流框架进行了整合，是现在非常流行的持久层解决方案。

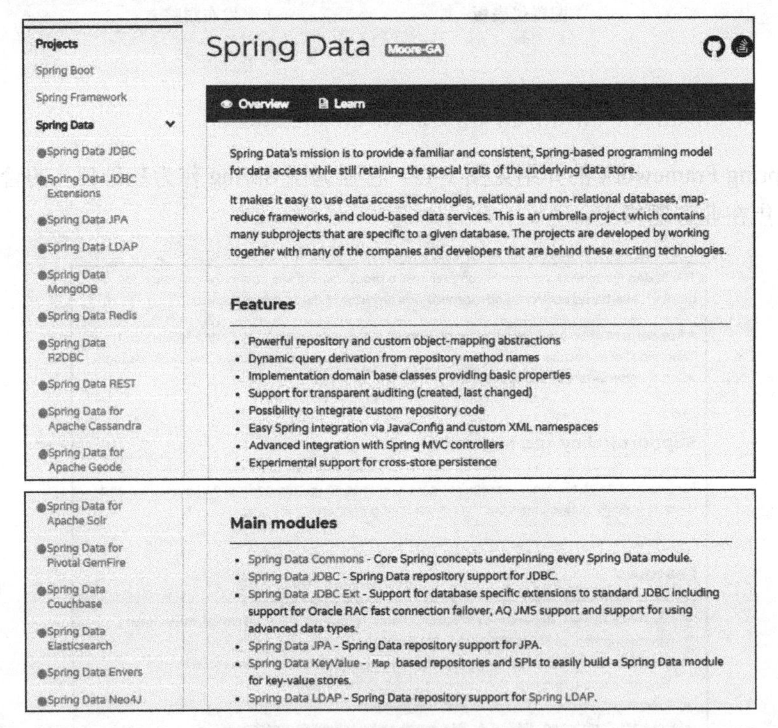

图 1-14 官网上 Spring Data 的介绍

1.3 比较 Spring Framework 历史版本

1.3.1 下载 Spring Framework 资源

Spring Framework 的更新信息已经移植到了 GitHub 上（在 GitHub 上搜索 Spring-Framework-

Versions 即可找到相关资源）。

Spring Framework 的主要版本如下。

- 5.2 版是 2019 年的最新版。
- 以 5.1 版为主线，从 2018 年 9 月支持到 2020 年第一季度。
- 4.3.*x* 系列在 2020 年之后不再更新，是现在主流企业使用的版本。
- 3.2.*x* 后期不再维护。

JDK 的依赖版本如下。

- JDK 8～14 依赖于 Spring Framework 5.2*x*。
- JDK 8～12 依赖于 Spring Framework 5.1*x*。
- JDK 8～10 依赖于 Spring Framework 5.0*x*。
- JDK 6～8 依赖于 Spring Framework 4.3*x*。

对于本书的所有案例，使用的测试版本是 Spring Framework 4.3.19、JDK 8、Java EE 7。

下载 Spring 开发文档的步骤如下。

（1）在图 1-15 所示的 Spring Framework 资源下载页中选择“Branch: 4.3.x”。

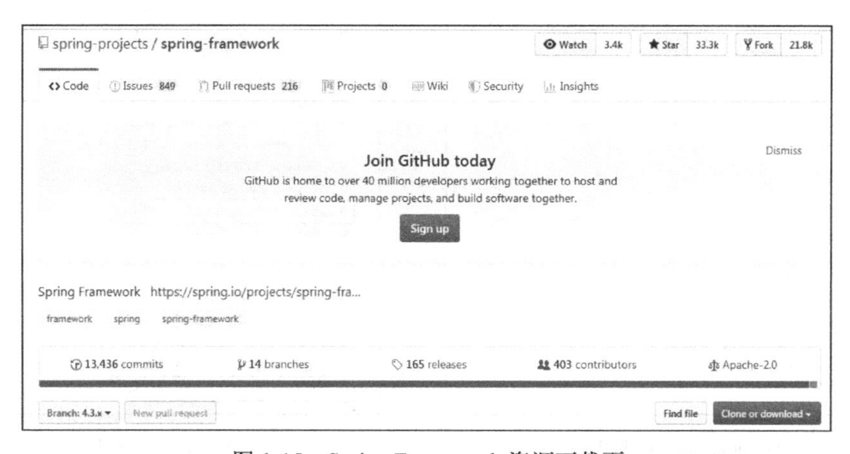

图 1-15　Spring Framework 资源下载页

（2）单击 Clone or download 按钮。

1.3.2　Spring 4.*x* 相对于 Spring 3.*x* 的变化

如图 1-16 所示， Spring 4.*x* 中模块的变化不是很大，在 Spring 3.*x* 中模块的基础上增加了发送消息的功能和 WebSocket 功能，不再支持 Struts。

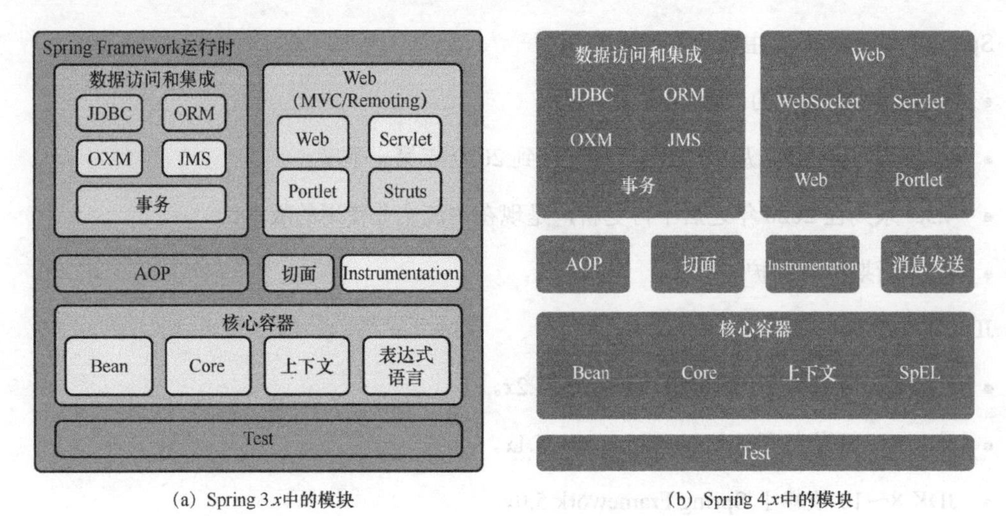

（a）Spring 3.x中的模块　　（b）Spring 4.x中的模块

图 1-16　Spring 3.x 与 Spring 4.x 中模块的对比

如图 1-17 所示，用方框标识了 Spring 3.2 和 Spring 4.0 中 JAR 包的变化。

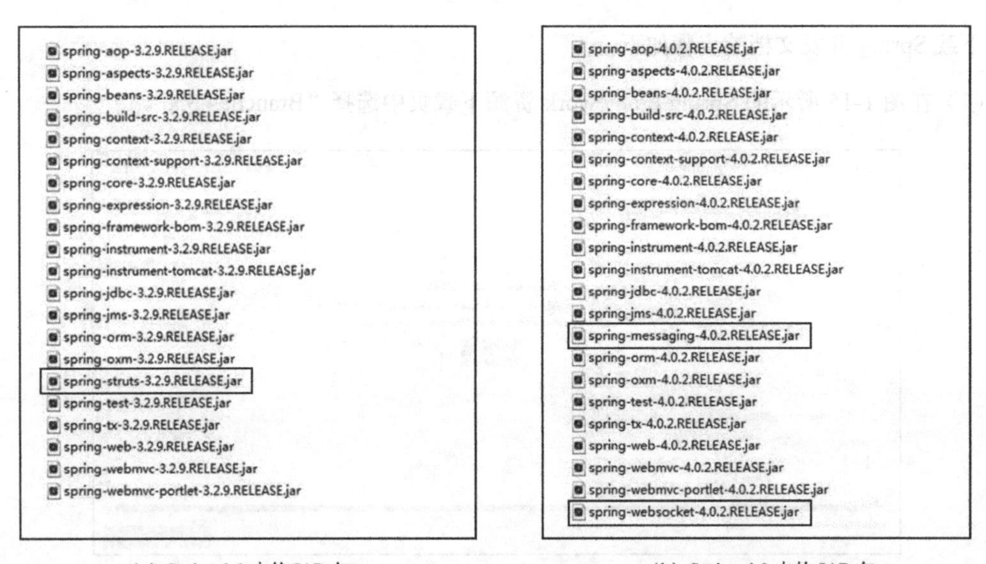

（a）Spring 3.2 中的 JAR 包　　（b）Spring 4.0 中的 JAR 包

图 1-17　Spring 3.2 与 Spring 4.0 中 JAR 包的变化

Spring 4.3.x 支持如下第三方库。

- Hibernate ORM 5.2（仍然支持 4.2/4.3 版本和 5.0/5.1 版本，3.6 版本已经废弃）。
- Hibernate Validator 5.3（仍然支持 4.3 版本）。

- Jackson 2.8（最低版本是 Jackson 2.6，和 Spring 4.3 一起使用）。
- OkHttp 3.x（仍然支持 OkHttp 2.x）。
- Tomcat 8.5 和 9.0。
- Netty 4.1。
- Undertow 1.4。
- WildFly 10.1。

另外，Spring Framework 4.3 在 spring-core.jar 中嵌入了 ASM 5.1 包、CGLIB 3.2.4 包、Objenesis 2.4 包。

需要注意的是 cglib.jar，这是用于动态代理的重要包，以前要从外部导入，从 Spring 4.3 之后无须再次导入这个包。

1.3.3 Spring 5.*x*的新增功能

Spring 5.*x* 支持 Java EE 8。从 Spring Framework 5.0 开始，Spring 开发环境要求最低为 Java EE 7、Servlet 3.1、JPA 2.1，同时支持所见即所得的 Java EE 8 新 API 集成。

Spring 5.*x* 支持如下 JSR 规范（注意，不是支持所有的 JSR）：

- Servlet API（JSR 340）；
- WebSocket API（JSR 356）；
- Concurrency Utilities（JSR 236）；
- JSON Binding API（JSR 367）；
- Bean Validation（JSR 303）；
- JPA（JSR 338）；
- JMS（JSR 914）。

Sping 5.*x* 的 Web 编程出现了很大变化，它提出了反应栈（reactive stack）模式。

Spring 5.*x* 同时支持 spring-webmvc 和 spring-webflux。

Spring Framework 的早期 Web 框架（即 Spring Web MVC），是基于 Servlet API 和 Servlet 容器进行构建的。Spring 5.*x* 的反应栈模式是完全非阻塞的、支持响应流的，可以运行在 Netty、Undertow 和 Servlet 3.1+容器中。

1.4 Spring Framework 技术

1.4.1 核心技术

简单来说，Spring Framework 的核心技术是 IoC 与 AOP，见图 1-18（a）与（b）。

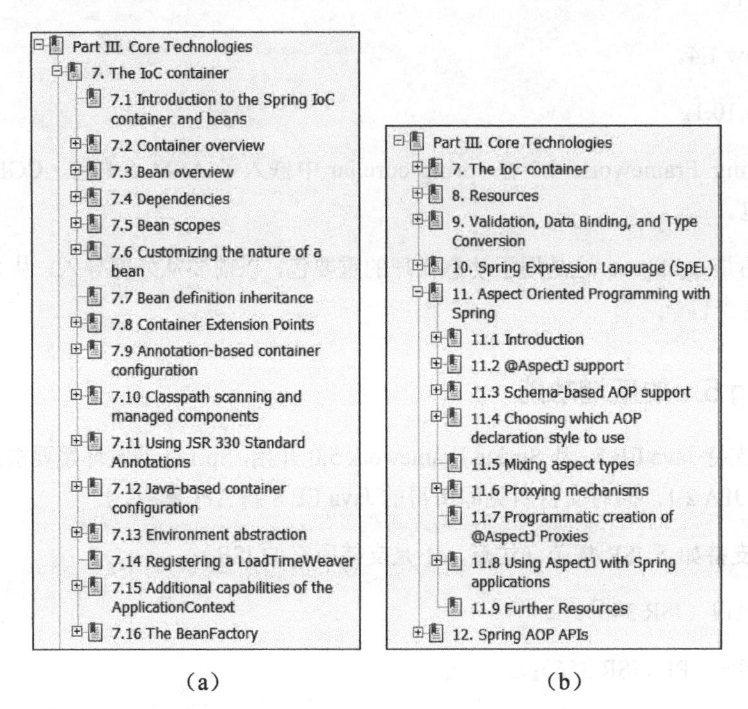

（a）　　　　　　　　（b）

图 1-18　Spring Framework 的核心技术

Spring Framework 的主要技术如下。

- IoC 容器：容器管理、Bean 管理。
- 资源：Spring 对资源的管理。
- 验证：Spring 数据校验。
- SpEL：Spring 表达式语言。
- AOP：分为@AspectJ 与基于 XML 的 AOP。

1.4.2 数据访问层的整合

Spring 通过提供一套统一的、抽象的接口整合数据访问层。这是 Spring 最重要的功能。通过 AOP 和动态代理技术，Spring 实现了声明性事务管理，这有效地减少了企业对 EJB 的依赖。

图 1-19（a）与（b）为数据访问层整合的主要功能。

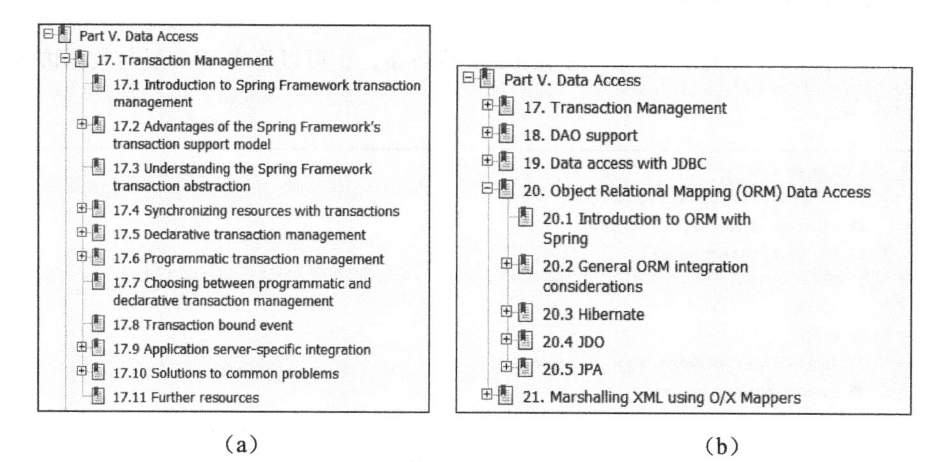

（a）　　（b）

图 1-19　数据访问层整合的主要功能

1.4.3　Web 层技术

Spring Framework 提供了 Web MVC 的实现，这是唯一打破常规的实现，理由是其他 Web MVC 框架确实有不足之处。

View Technologies 部分是 Spring Web MVC 对主流视图技术的支持，JSP 仅仅是常用视图技术之一。图 1-20（a）与（b）为 Web 层的主要功能。

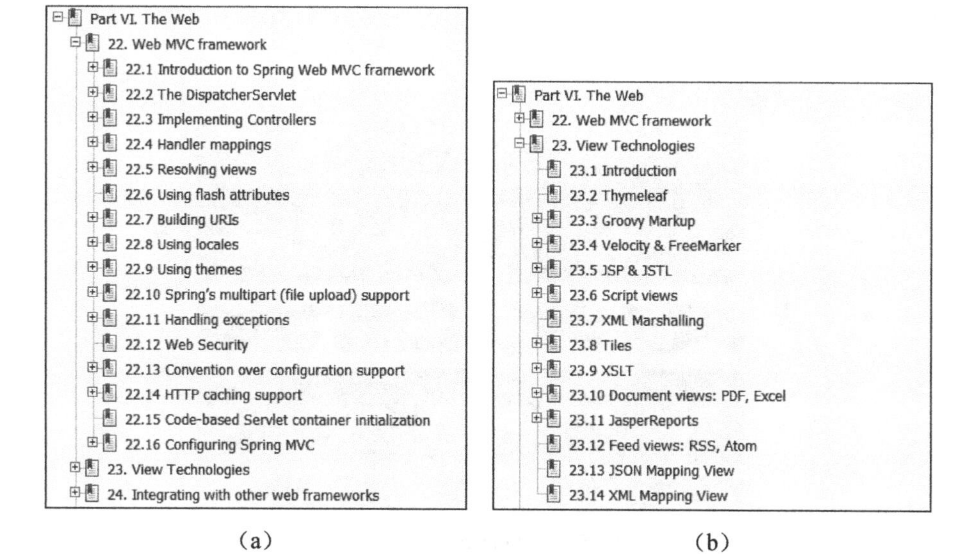

（a）　　（b）

图 1-20　Web 层的主要功能

1.4.4　与外部系统的集成

图 1-21 是 Spring Framework 与外部系统的集成方案，也可以看成 Web 层与数据访问层之外的其他内容汇总。

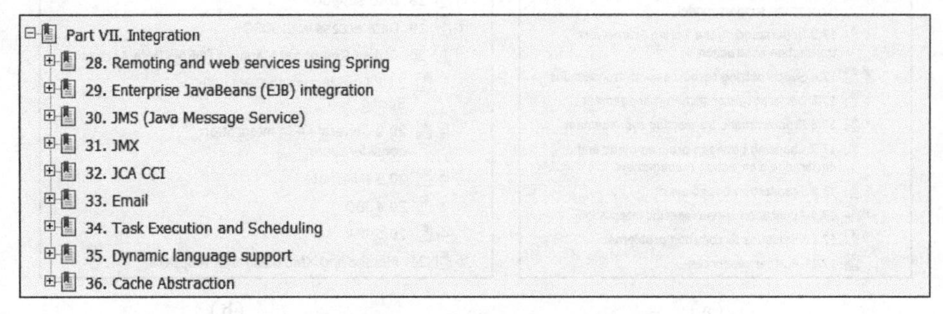
Part VII. Integration
28. Remoting and web services using Spring
29. Enterprise JavaBeans (EJB) integration
30. JMS (Java Message Service)
31. JMX
32. JCA CCI
33. Email
34. Task Execution and Scheduling
35. Dynamic language support
36. Cache Abstraction

图 1-21　Spring Framework 与外部系统的集成方案

1.5　Spring Framework 模块的组成

1.5.1　模块架构

图 1-22 是 Spring Framework 模块的组成。学习 Spring 开发，要求掌握 Spring Framework 有哪些模块，每个模块的主要用途是什么。Spring Framework 模块划分为核心容器、测试模块、数据访问和集成模块、Web 模块、AOP 模块、切面模块、Instrumentation 模块以及消息发送模块。

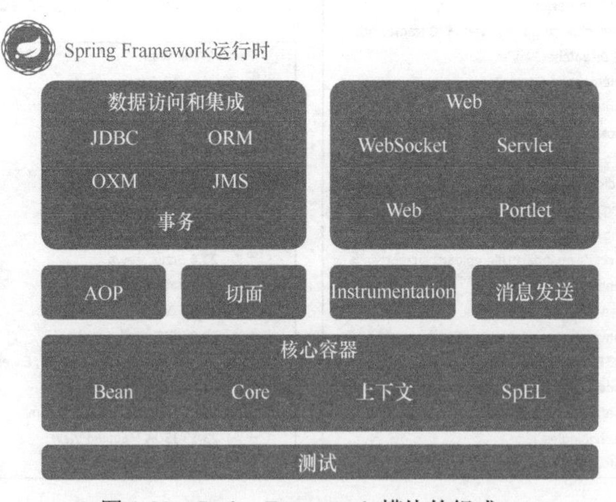

图 1-22　Spring Framework 模块的组成

1.5.2 模块与 JAR 包的对应关系

如图 1-23 所示，Spring Framework 的每个模块都有相应的 JAR 包。只有熟悉每一个 JAR 包的功能，才能在开发阶段的包导入过程中不至于混乱。

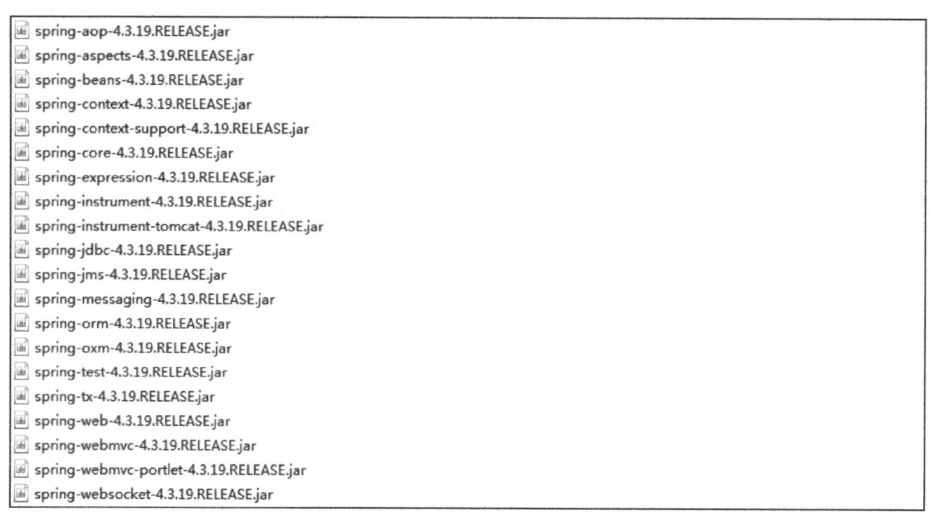

图 1-23 Spring Framework 中各模块的 JAR 包

1.5.3 模块的功能

在熟悉 Spring Framework 模块的基础上，要求掌握每个模块对应的 JAR 包，以及各个模块的含义（见表 1-1）。

表 1-1 Spring Framework 模块

模块（GroupId）	对应 JAR 包（ArtifactId）	功能描述
org.springframework	spring-aop	基于代理的 AOP 支持
org.springframework	spring-aspects	基于切面的 AspectJ
org.springframework	spring-beans	支持 Bean，包含 Groovy
org.springframework	spring-context	Spring 应用运行环境
org.springframework	spring-context-support	支持集成第三方库到 Spring 应用环境中
org.springframework	spring-core	核心工具集，可应用于其他各模块
org.springframework	spring-expression	SpEL
org.springframework	spring-instrument	JVM 启动的服务器设备代理
org.springframework	spring-instrument-tomcat	Tomcat 的设备代理
org.springframework	spring-jdbc	JDBC 支持包，包含建立数据源和访问 JDBC 数据
org.springframework	spring-jms	JMS 支持包
org.springframework	spring-messaging	支持消息架构和协议

续表

模块（GroupId）	对应 JAR 包（ArtifactId）	功能描述
org.springframework	spring-orm	支持 ORM，包含对 JPA 和 Hibernate 的支持
org.springframework	spring-oxm	对象/XML 映射
org.springframework	spring-test	支持单元测试与 Spring 组件集成测试
org.springframework	spring-tx	事务管理基础架构，包含 DAO 支持和 JCA 集成
org.springframework	spring-web	Web 支持基础包，包含 Web 客户端与基于 Web 的远程调用
org.springframework	spring-webmvc	基于 HTTP 的 MVC 与 REST 模型支持
org.springframework	spring-webmvc-portlet	在 Portlet 环境下的 MVC 支持
org.springframework	spring-websocket	WebSocket 与 SockJS 的基础

第 2 章
Spring 的资源管理

JDK 提供的标准 java.net.URL 类可以管理各种不同的资源，但是不能满足低层资源的管理需求。如从 classpath 读取资源，从 ServletContext 的相对路径读取资源等，都需要使用其他方式。JDK 中对 java.net.URL 的描述如下。

```
public final class URL
     extends Object  implements Serializable {}
```

类 URL 表示一种统一的资源定位符，在万维网上指向一种资源。资源可以简单看成文件、文件目录或复杂对象的引用（如查询数据库或搜索引擎）。

2.1 资源管理类

Spring Framework 使用如下资源管理类。

- UrlResource：包含 java.net.URL，如文件系统资源、HTTP 资源、FTP 资源。
- ClassPathResource：包含 classpath 下的资源。
- FileSystemResource：包含 ava.io.File。
- ServletContextResource：解析为基于 Web 应用根路径下的相对路径资源。

- InputStreamResource：使用字节流读取资源。
- ByteArrayResource：读取字节数组资源。

在非 Web 环境下，FileSystemResource 可以调用磁盘绝对路径，读取资源。ClassPathResource 从发布项目的 bin 文件夹下读取资源。

在 Web 环境下，使用 ServletContext 的 getRealPath( )读取 Web 资源路径，它可以把 HTTP 的虚拟路径转换为磁盘绝对路径。

2.2 资源

表 2-1 总结了把字符串转换为资源的方式。

表 2-1　把字符串转换为资源的方式

字符串	示例	说明
以“classpath:”开头的字符串	classpath:com/myapp/config.xml	从 classpath 下加载
以“file:”开头的字符串	file:///data/config.xml	从文件系统加载 URL 资源
以“http:”开头的字符串	http://myserver/logo.png	加载 URL 资源
以“(none)”开头的字符串	/data/config.xml	相对于 ApplicationContext 路径加载

org.springframework.util 下的 ResourceUtils 是一个工具类，用于在 Spring 框架内解析各种资源。这个抽象类中定义了 Spring 管理的资源类型（见表 2-2）。

表 2-2　Spring 管理的资源类型

资源类型	描述
classpath	伪 URL 前缀，用于从类路径下加载资源
file	URL 前缀，用于从文件系统加载资源
jar	URL 前缀，用于从 JAR 文件加载资源
war	URL 前缀，用于从 Tomcat 下的 war 加载资源
zip	使用 URL 协议访问 ZIP 文件
wsjar	使用 URL 协议访问 Websphere 的 JAR 文件
vfszip	使用 URL 协议访问 Jboss 的 JAR 文件
vfsfile	使用 URL 协议访问 Jboss 的文件系统
vfs	使用 URL 协议访问 Jboss 的 VFS 资源

2.3 资源访问接口

Spring 使用 Resource 接口访问底层资源。

```
package org.springframework.core.io;
public interface Resource extends InputStreamSource {
     boolean        exists();
     boolean        isOpen();
     boolean        isReadable();
     URL            getURL() throws IOException;
     File           getFile() throws IOException;
     Resource       createRelative(String relativePath) throws IOException;
     String         getFilename();
     String         getDescription();
}
public interface InputStreamSource {
     InputStream getInputStream() throws IOException;
}
```

实现 Resource 接口，可以管理各种资源。从 Spring 的 API 中可以看到 Resource 接口的各种已知实现类，不同实现类用于解析不同的资源类型。Resource 接口的定义如下。

```
org.springframework.core.io Interface Resource
```

- 父接口：包括 InputStreamSource。
- 子接口：包括 ContextResource、EncodedResource、VersionedResource、WritableResource。
- 所有已知实现类：包括 AbstractFileResolvingResource、AbstractResource、ByteArrayResource、ClassPathResource、DefaultResourceLoader.ClassPathContextResource、DescriptiveResource、FileSystemResource、InputStreamResource、PathResource、PortletContextResource、ServletContextResource、TransformedResource、UrlResource、VfsResource。

2.4 资源加载

使用 ResourceLoader 接口加载资源。

```
package org.springframework.core.io;
public interface ResourceLoader {
     Resource getResource(String location);
     ClassLoader   getClassLoader();
}
```

Spring 中所有的 ApplicationContext 都实现了 ResourceLoader 接口，因此在 Spring 环境下可以随时加载资源。

```
public interface ApplicationContext
               extends EnvironmentCapable, ListableBeanFactory,
                        HierarchicalBeanFactory,MessageSource,
                        ApplicationEventPublisher, ResourcePatternResolver {}

public interface ResourcePatternResolver extends ResourceLoader {}
```

抽象类 AbstractApplicationContext 的父类是 DefaultResourceLoader，它是 ResourceLoader 接口的重要实现，参见 API 中 DefaultResourceLoader 的描述信息。

```
org.springframework.core.io  Class  DefaultResourceLoader
java.lang.Object org.springframework.core.io.DefaultResourceLoader
```

- 所有实现的接口：包括 ResourceLoader。
- 所有已知子类：包括 AbstractApplicationContext、ClassRelativeResourceLoader、FileSystemResourceLoader、PortletContextResourceLoader、ServletContextResourceLoader。

示例 2-1：容器读取资源。

```
Resource template =
        ctx.getResource("some/resource/path/myTemplate.txt");
Resource template =
        ctx.getResource("classpath:some/resource/path/myTemplate.txt");
Resource template =
        ctx.getResource("file:///some/resource/path/myTemplate.txt");
Resource template =
        ctx.getResource("http://myhost.com/resource/path/myTemplate.txt");
Resource res  = new FileSystemResource("beans.xml");
Resource res  = new
        ServletContextResource(application,"/WEB-INF/classes/conf/file1.txt");
```

示例 2-2：从类路径下加载资源。

```
ApplicationContext context =
                new ClassPathXmlApplicationContext('beans.xml');
```

示例 2-3：在 Web 环境下，通过 ServletConfig 读取配置参数（/WEB-INF/spring-mvc.xml），然后根据 ServletContext 读取 Web 站点的绝对路径信息，结合上面配置好的相对路径，即可找到资源。

```
<servlet>
    <servlet-name>aa</servlet-name>
    <servlet-class>org.springframework.web.servlet.DispatcherServlet</servlet-class>
    <init-param>
        <param-name>contextConfigLocation</param-name>
        <param-value>/WEB-INF/spring-mvc.xml</param-value>
    </init-param>
    <load-on-startup>1</load-on-startup>
</servlet>
```

解析资源的核心代码如下。

```
WebApplicationContext rootContext =
        WebApplicationContextUtils.getWebApplicationContext(getServletContext());
ConfigurableWebApplicationContext wac = (ConfigurableWebApplicationContext)rootContext;
wac.setServletContext(getServletContext());
wac.setServletConfig(getServletConfig());
```

示例 2-4：通过监听器加载 ServletContextResource。

```
<context-param>
    <param-name>contextConfigLocation</param-name>
    <param-value>/WEB-INF/root-context.xml</param-value>
</context-param>
<servlet>
    <servlet-name>dispatcher</servlet-name>
    <servlet-class>org.springframework.web.servlet.DispatcherServlet</servlet-class>
    <init-param>
        <param-name>contextConfigLocation</param-name>
        <param-value></param-value>
    </init-param>
    <load-on-startup>1</load-on-startup>
</servlet>
<servlet-mapping>
    <servlet-name>dispatcher</servlet-name>
    <url-pattern>/*</url-pattern>
</servlet-mapping>
<listener>
    <listener-class>org.springframework.web.context.ContextLoaderListener
    </listener-class>
</listener>
```

解析资源的核心代码如下。

```
ServletContext sc;
ConfigurableWebApplicationContext wac;
String configLocationParam = sc.getInitParameter(CONFIG_LOCATION_PARAM);
if (configLocationParam != null) {
    wac.setConfigLocation(configLocationParam);
}
```

2.5 从配置中获取资源

从 Spring 的配置文件中加载资源。

```
<bean id="myBean" class="...">
    <property name="template" value="some/resource/path/myTemplate.txt"/>
</bean>
```

注意：此处的资源路径无前缀。在不同的 IoC 环境中可以分别使用 ClassPathResource、FileSystemResource 或 ServletContextResource 来接收资源。如果需要指明类型，则可以使用 “classpath: ” 或 “file:” 等前缀。

```
<property name="template" value="classpath:some/resource/path/myTemplate.txt">
<property name="template" value="file:///some/resource/path/myTemplate.txt"/>
```

2.6 应用上下文与资源

从 Spring 应用上下文中可以直接访问资源。

示例 2-5：通过 ClassPathXmlApplicationContext 从 classpath 下加载资源。

```
ApplicationContext ctx = new ClassPathXmlApplicationContext("conf/appContext.xml");
ApplicationContext ctx = new ClassPathXmlApplicationContext("beans.xml");
```

示例 2-6：通过 FileSystemXmlApplicationContext 加载 file 资源。

在当前运行程序的工作目录中，用相对路径加载资源。

```
ApplicationContext ctx
                = new FileSystemXmlApplicationContext("conf/appContext.xml");
```

使用指定前缀，从 classpath 下加载资源。

```
ApplicationContext ctx =
        new FileSystemXmlApplicationContext("classpath:conf/appContext.xml");
```

如果需要同时加载多个资源，则可以使用通配符“*”来加载资源。

```
/WEB-INF/*-context.xml
com/mycompany/**/applicationContext.xml
file:C:/some/path/*-context.xml
classpath:com/mycompany/**/applicationContext.xml
ApplicationContext ctx =
     new ClassPathXmlApplicationContext("classpath*:conf/appContext.xml");
ApplicationContext ctx =
     new ClassPathXmlApplicationContext("classpath*:META-INF/*-beans.xml");
```

注意：classpath 只加载指定的第一个路径下的资源，classpath*加载 classpath 下的所有资源。

2.7 MyBatis 的资源配置

当在 MyBatis 中配置映射文件时，使用的资源配置与 Spring 资源类似，参考配置信息如下。

```
<!-- 使用相对路径加载资源 -->
<mappers>
    <mapper resource="org/mybatis/builder/AuthorMapper.xml"/>
    <mapper resource="org/mybatis/builder/BlogMapper.xml"/>
    <mapper resource="org/mybatis/builder/PostMapper.xml"/>
</mappers>
<!-- 使用绝对路径加载资源 -->
<mappers>
    <mapper url="file:///var/mappers/AuthorMapper.xml"/>
    <mapper url="file:///var/mappers/BlogMapper.xml"/>
    <mapper url="file:///var/mappers/PostMapper.xml"/>
</mappers>
```

```xml
<!-- 使用 Mapper 接口和类描述资源 -->
<mappers>
    <mapper class="org.mybatis.builder.AuthorMapper"/>
    <mapper class="org.mybatis.builder.BlogMapper"/>
    <mapper class="org.mybatis.builder.PostMapper"/>
</mappers>
<!-- 注册包内的所有接口 -->
<mappers>
    <package name="org.mybatis.builder"/>
</mappers>
```

第 3 章 IoC

Spring 和 Java EE 一样基于“容器 + 组件”的运行模式。Spring 基于轻量级容器，EJB 基于重量级容器（如 Weblogic），Tomcat 基于中量级容器。

3.1 IoC 与 DI 的概念

控制反转（Inversion of Control，IoC）习惯上也称为依赖注入（Dependency Injection，DI）。当 IoC 容器创建 Bean 对象后，把它注入业务系统，同时建立 Bean 对象之间的依赖关系。这个处理过程是一个反转操作，即先创建对象后建立依赖，因此称为控制反转。

IoC 是一个复杂的概念，此处推荐阅读英文的定义。

The container then injects those dependencies when it creates the bean. This process is fundamentally inverse, hence the name Inversion of Control (IoC).

控制什么样的切面被反转了？马丁 · 福勒在 2004 年的 IoC 大会上提出这个问题，随后他建议重新将 IoC 命名为 DI。

3.2 IoC 容器与 ApplicationContext

接口 org.springframework.context.ApplicationContext 代表 IoC 容器，它同时负责配置、实例化

和装配 Bean。

Spring API 文档中的 ApplicationContext 接口定义如下。

```
org.springframework.context  Interface ApplicationContext
```

- 所有父接口：包括 ApplicationEventPublisher、BeanFactory、EnvironmentCapable、HierarchicalBeanFactory、ListableBeanFactory、MessageSource、ResourceLoader、ResourcePatternResolver。
- 所有已知子接口：包括 ConfigurableApplicationContext、ConfigurablePortletApplicationContext、ConfigurableWebApplicationContext、WebApplicationContext。
- 所有已知实现类：包括 AbstractApplicationContext、AbstractRefreshableApplicationContext、AbstractRefreshableConfigApplicationContext、AbstractRefreshablePortletApplicationContext、AbstractRefreshableWebApplicationContext、AbstractXmlApplicationContext、AnnotationConfigApplicationContext、AnnotationConfigWebApplicationContext、ClassPathXmlApplicationContext、FileSystemXmlApplicationContext、GenericApplicationContext、GenericGroovyApplicationContext、GenericWebApplicationContext、GenericXmlApplicationContext、GroovyWebApplicationContext、ResourceAdapterApplicationContext、StaticApplicationContext、StaticPortletApplicationContext、StaticWebApplicationContext、XmlPortletApplicationContext、XmlWebApplicationContext。

Spring Framework 的 IoC 容器操作基于两个包——org.springframework.beans 和 org.springframework.context。

BeanFactory 接口提供了管理各种 Bean 的高级配置机制。

ApplicationContext 是 BeanFactory 的子接口。ApplicationContext 提供了集成 AOP、管理消息资源、发布事件、在应用层注入上下文等功能。简单来说，BeanFactory 提供了配置框架和管理 Bean 的基本功能，ApplicationContext 添加了更多的企业应用功能。

3.3 容器的创建与使用

3.3.1 创建 IoC 容器

1. 容器与配置

一旦创建了 ApplicationContext 接口的实现类对象，就创建了 IoC 容器。在创建 IoC 容器时，习惯上要将 Spring 的配置文件传送给 IoC 容器的构造函数。

示例 3-1：使用 ClassPathXmlApplicationContext 创建 IoC 容器。

```
ApplicationContext context =
        new ClassPathXmlApplicationContext("services.xml", "daos.xml");
```

示例 3-2：使用 FileSystemXmlApplicationContext 创建 IoC 容器。

```
ApplicationContext ctx =
        new FileSystemXmlApplicationContext("classpath:conf/appContext.xml");
```

示例 3-3：使用 XmlWebApplicationContext 创建 IoC 容器。

```
XmlWebApplicationContext appContext = new XmlWebApplicationContext();
appContext.setConfigLocation("/WEB-INF/spring/dispatcher-config.xml");
```

注意：XmlWebApplicationContext 只有无参构造函数，这里创建 IoC 容器的方式与创建其他容器的方式不同。

如图 3-1 所示，IoC 容器读取配置信息，注入业务对象（包含 POJO 对象）。配置信息包含了容器要创建、配置、装配的对象信息，以及元数据信息。一般情况下，不仅使用简单直观的 XML 配置元数据的描述格式，还可以使用注解的方式进行配置。

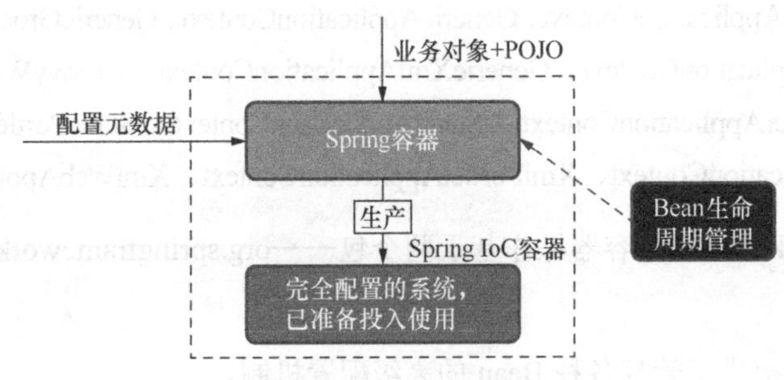

图 3-1　IoC 容器的架构

示例 3-4：Bean 的配置。

```
<?xml version="1.0" encoding="UTF-8"?>
<beans xmlns="http://www.springframework.org/schema/beans"
      xmlns:xsi="http://www.w3.org/2001/XMLSchema-instance"
      xsi:schemaLocation="http://www.springframework.org/schema/beans
      http://www.springframework.org/schema/beans/spring-beans.xsd">
      <bean id="..." class="...">
            <!--在此处装配并配置 Bean -->
      </bean>
      <bean id="..." class="...">
            <!--在此处装配并配置 Bean -->
      </bean>
      <!-- 更多 Bean 的定义 -->
</beans>
```

2. 从多个配置文件实例化容器

服务层和持久层的业务对象在 Spring 框架中都按 Bean 对象管理，在配置 Bean 时分别放在

services.xml 和 daos.xml 中。

（1）services.xml 的配置如下（petStore 为服务层对象）。

```
<?xml version="1.0" encoding="UTF-8"?>
<beans xmlns="http://www.springframework.org/schema/beans"
     xmlns:xsi="http://www.w3.org/2001/XMLSchema-instance"
     xsi:schemaLocation="http://www.springframework.org/schema/beans
     http://www.springframework.org/schema/beans/spring-beans.xsd">
     <!-- 服务对象 -->
     <bean id="petStore" class="org.springframework.samples.jpetstore.services.PetStore
          ServiceImpl">
          <property name="accountDao" ref="accountDao"/>
          <property name="itemDao" ref="itemDao"/>
          <!-- 对 Bean 的附加装配和配置信息 -->
     </bean>
          <!-- 对服务的更多 Bean 定义 -->
</beans>
```

（2）daos.xml 的配置如下（accountDao 和 itemDao 为持久层对象）。

```
<?xml version="1.0" encoding="UTF-8"?>
<beans xmlns="http://www.springframework.org/schema/beans"
     xmlns:xsi="http://www.w3.org/2001/XMLSchema-instance"
     xsi:schemaLocation="http://www.springframework.org/schema/beans
     http://www.springframework.org/schema/beans/spring-beans.xsd">
     <bean id="accountDao"
          class="org.springframework.samples.jpetstore.dao.jpa.JpaAccountDao">
               <!-- 对 Bean 的附加装配和配置信息 -->
     </bean>
     <bean id="itemDao"
           class="org.springframework.samples.jpetstore.dao.jpa.JpaItemDao">
               <!-- 对 Bean 的附加装配和配置信息 -->
     </bean>
     <!-- 对数据访问的更多 Bean 定义 -->
</beans>
```

（3）读取配置文件，生成 IoC 容器。

```
ApplicationContext context =
               new ClassPathXmlApplicationContext("services.xml", "daos.xml");
```

3. 组合配置文件

首先定义多个配置文件，然后组成一个配置文件，这在大型项目实践中是非常有用的一种配置方法。例如，在 spring-mvc.xml 中导入 services.xml 和 daos.xml。

```
<beans>
     <import resource="services.xml"/>
     <import resource="daos.xml"/>
     <import resource="/resources/themeSource.xml"/>
     <bean id="bean1" class="..."/>
```

```
    <bean id="bean2" class="..."/>
</beans>
```

Tomcat 启动时，会读取 spring-mvc.xml，同时加载 services.xml 和 daos.xml。

```
XmlWebApplicationContext appContext = new XmlWebApplicationContext();
appContext.setConfigLocation("/WEB-INF/spring-mvc.xml");
```

注意资源的路径，此处使用的是 Web 站点的相对路径，还可以使用其他资源的相对路径。

3.3.2 从容器读取 Bean 对象

要从 IoC 容器中读取 Bean 对象，操作步骤如下。

（1）创建 IoC 容器。

```
ApplicationContext context =
    new ClassPathXmlApplicationContext("services.xml", "daos.xml");
```

（2）根据类型找到 Bean 对象。

```
PetStoreService service = context.getBean(PetStoreService.class);
```

（3）根据 Bean 的名字找到 Bean 对象。

```
PetStoreService service = context.getBean("petStore");
```

（4）使用业务对象。

```
List<String> userList = service.getUsernameList();
```

3.3.3 “Hello,Spring”示例

前面我们学习了创建 IoC 容器，从容器中读取 Bean，下面介绍一个完整的入门示例。

操作步骤如下。

（1）导入 Spring 核心包与依赖包 commons-logging.jar，见图 3-2。

Referenced Libraries
- spring-beans-4.3.19.RELEASE.jar
- spring-context-4.3.19.RELEASE.jar
- spring-context-support-4.3.19.RELEASE.jar
- spring-core-4.3.19.RELEASE.jar
- spring-expression-4.3.19.RELEASE.jar
- commons.logging-1.1.1.jar

图 3-2　导入 Spring 核心包与依赖包 commons-logging.jar

（2）配置 XML。

```
<?xml version="1.0" encoding="UTF-8"?>
<beans xmlns="http://www.springframework.org/schema/beans"
    xmlns:xsi="http://www.w3.org/2001/XMLSchema-instance"
```

```
    xsi:schemaLocation="http://www.springframework.org/schema/beans
    http://www.springframework.org/schema/beans/spring-beans-4.3.xsd">
        <bean id="helloBean" class="com.icss.biz.HelloBiz" />
</beans>
```

（3）定义 Bean。

```
public class HelloBiz {
     public String sayHello(String name) {
          return "hello: Mr. " + name;
     }
}
```

（4）创建容器，读取 Bean。

```
public static void main(String[] args) {
     // 创建 Spring 容器，解析 XML 文件
     ApplicationContext context = new ClassPathXmlApplicationContext("beans.xml");
     // 根据 Bean 的 id，查找 Bean 对象
     HelloBiz hi = (HelloBiz) context.getBean("helloBean");
     String hello = hi.sayHello("tom");
     System.out.println(hello);
}
```

测试结果如下。

```
org.springframework.beans.factory.xml.XmlBeanDefinitionReader loadBeanDefinitions
信息: Loading XML bean definitions from class path resource [beans.xml]
hello: Mr. tom
```

3.4 Bean 对象的管理

Spring 的 IoC 容器用于管理一个或多个 Bean 对象。IoC 容器读取配置元数据，创建 Bean 对象。由于 Bean 对象的作用域有多种情况，不能直接管理 Bean 对象的引用，因此 IoC 容器是通过 BeanDefinition 对象来管理 Bean 的。

3.4.1 BeanDefinition

在 Sping API 文档中，有关 BeanDefinition 的描述如下。

```
org.springframework.beans.factory.config  Interface  BeanDefinition
```

- 所有父接口：包括 AttributeAccessor、BeanMetadataElement。
- 所有已知子接口：包括 AnnotatedBeanDefinition。
- 所有已知实现类：包括 AbstractBeanDefinition、AnnotatedGenericBeanDefinition、ChildBean Definition、GenericBeanDefinition、RootBeanDefinition、ScannedGenericBeanDefinition。

一个 BeanDefinition 描述一个 Bean 实例，Bean 可以有属性值、构造参数值等，更多信息由 BeanDefinition 的实现类提供。

BeanDefinition 定义了如下信息。

- 包名、类名：创建 Bean 实例时使用的包名、类名。
- Bean 行为：决定 Bean 在容器中的行为（如作用域、生命周期回调等）。
- 对其他 Bean 的引用：在 Bean 的协作和依赖操作中有用。
- 其他配置信息：如使用 Bean 来定义数据库连接池，可以通过属性或者构造参数指定连接数及连接池大小等。

表 3-1 展示了 Bean 的属性信息。

表 3-1　　Bean 的属性信息

属性名称	说明
class	包名、类名
name	可以使用名字查找 Bean 对象
scope	Bean 的作用域
constructor arguments	在依赖注入中通过构造函数注入
properties	通过属性配置依赖关系
autowiring mode	自动装配模式
lazy-initialization mode	懒加载模式
initialization method	在创建对象时，先调用初始化方法
destruction method	在销毁对象时，在析构方法中释放资源

另外，通过 BeanDefinition 的实现类，可以看到 BeanDefinition 存储了哪些信息。

```
public abstract class AbstractBeanDefinition extends BeanMetadataAttributeAccessor
            implements BeanDefinition, Cloneable {
    private volatile Object  beanClass;
    private String  scope = SCOPE_DEFAULT;
    private boolean  abstractFlag = false;
    private boolean  lazyInit = false;
    private int  autowireMode = AUTOWIRE_NO;
    private String[]  dependsOn;
    private boolean  autowireCandidate = true;
    private boolean  primary = false;
    private final Map<String, AutowireCandidateQualifier> qualifiers =
            new LinkedHashMap<String, AutowireCandidateQualifier>(0);
    private boolean  nonPublicAccessAllowed = true;
    private boolean  lenientConstructorResolution = true;
    private String  factoryBeanName;
    private String  factoryMethodName;
    private ConstructorArgumentValues  constructorArgumentValues;
```

```
        private MutablePropertyValues  propertyValues;
        private MethodOverrides  methodOverrides = new MethodOverrides();
}
```

从 AbstractBeanDefinition 的属性信息中可以清楚地看到 Bean 的各种属性信息是如何存储的。更多 Bean 属性会在后面详细讲解。如 Object beanClass，它可以为 Class<T>或 String 类型，它不是 Bean 对象，是 Bean 的类型描述。beanClass 的赋值方式如下。

```
public void setBeanClassName(String beanClassName) {
        this.beanClass = beanClassName;
}
public void setBeanClass(Class<?> beanClass) {
        this.beanClass = beanClass;
}
```

3.4.2 id 属性和 name 属性的区别

id 和 name 都是 Bean 的属性，很容易混淆。

id 具有唯一性，用来唯一地识别一个 Bean 元素。注意，id 在同一个<beans>范围内唯一。id 识别的是 Bean 定义，不是 Bean 对象。因为 Bean 有 scope 属性，同一个配置 id 可能对应很多对象。

name 也可以称为别名，也是 Bean 的识别符，但它不一定是唯一的。可以通过 getBean(name) 在容器中基于名称查找 Bean 对象。

同一个 id 的 Bean 可以有多个别名，别名之间可以用空格、逗号或分号分隔。示例如下。

```
<bean id="helloBean" class="com.icss.biz.HelloBiz" name="h1,h2,h3"/>
```

id 和 name 可以不指定，在容器创建对象时，系统会自动生成唯一的 name。Bean 的名字必须以小写字母开头，使用驼峰命名法，如 accountManager、accountService、userDao、loginController。

使用<alias>，可以给 Bean 指定别名，对于公共组件的引用，这个方法有时很有效。

```
<alias name="subsystemA-dataSource" alias="subsystemB-dataSource"/>
<alias name="subsystemA-dataSource" alias="myApp-dataSource" />
```

3.4.3 创建 Bean 对象

BeanDefinition 本质上是创建一个或多个 Bean 对象的方法。当需要的时候，IoC 容器会通过名字从 Bean 的定义列表中找到 BeanDefinition，然后通过反射的方式创建 Bean 对象。

如果使用 XML 配置 Bean，class 属性是必需的，它会传给 BeanDefinition，这是创建 Bean 对象的关键。

典型情况下，容器通过调用 Bean Class 的构造函数直接创建对象。少数情况下，可以使用静态工厂方法创建 Bean 对象。

1. 使用默认的构造函数创建 Bean 对象

在基于 XML 的配置中，可以按照如下方式指定 Bean 的信息。

```
<bean id="exampleBean" class="examples.ExampleBean"/>
<bean name="anotherExample" class="examples.ExampleBeanTwo"/>
```

Spring 读取 class 信息，然后通过反射的方式创建指定类型的对象。这会调用 Bean 的构造函数（无参构造函数或有参构造函数），其原理如下。

```
Object obj  =  Dog.class.newInstance();
Object obj  =  Class.forName("com.icss.biz.Dog").newInstance();
Object obj  =  Dog.class. getDeclaredConstructor(String.class). .newInstance("旺财");
```

2. 通过静态工厂方法创建 Bean 对象

IoC 容器读取 BeanDefinition，使用静态工厂方法创建对象。其实使用静态工厂方法创建的 Bean 对象，没有使用反射机制。如下代码只有在特殊情况下才有用处，它与直接使用反射方式创建对象的区别是利用了已有的静态对象 clientService。使用静态工厂方法创建的对象由 IoC 容器管理，这与直接调用静态方法返回一个业务对象不同。使用静态工厂方法创建 Bean 对象的具体步骤如下。

（1）配置 Bean，传入静态工厂方法名。

```
<bean id="clientService"
      class="examples.ClientService"  factory-method="createInstance"/>
```

（2）通过静态工厂方法创建 Bean 对象。

```
public class ClientService {
    private static ClientService clientService = new ClientService();
    private ClientService() {}
    public static ClientService createInstance() {
        return clientService;
    }
    public void doSomething() {
        System.out.println("ClientService doSomething...");
    }
}
```

（3）测试。

① 测试代码如下。

```
ApplicationContext context
    = new ClassPathXmlApplicationContext("beans.xml");
ClientService client = (ClientService) context.getBean("clientService");
client.doSomething();
```

② 测试结果如下。

```
org.springframework.beans.factory.xml.XmlBeanDefinitionReader loadBeanDefinitions
信息: Loading XML bean definitions from class path resource [beans.xml]
ClientService doSomething...
```

（4）修改代码，不使用静态对象。

```
public class ClientService {
    //private static ClientService clientService = new ClientService();
    private ClientService() {
    }
    public static ClientService createInstance() {
        return new ClientService();
    }
    public void doSomething() {
        System.out.println("ClientService doSomething...");
    }
}
```

代码修改后，每次调用 createInstance( )就会创建一个新对象，但是这样静态工厂方法就失去了它的作用。

3. 通过非静态工厂方法创建 Bean 对象

与使用静态工厂方法创建 Bean 对象类似，用来进行实例化的非静态工厂方法位于另一个 Bean 中，容器将调用该 Bean 的工厂方法来创建一个新的 Bean 实例。为了使用此机制，class 属性必须为空，而 factory-bean 属性必须指定为当前（或其父）容器包含工厂方法的 Bean 的名称，而该 Bean 的工厂方法本身必须通过 factory-method 属性来设置。

示例代码如下。

```
<bean id="serviceLocator" class="examples.DefaultServiceLocator">
</bean
<bean id="clientService"
    factory-bean="serviceLocator" factory-method="createClientServiceInstance"/>
```

调用 DefaultServiceLocator 中的 createClientServiceInstance( )，创建 ClientService 的 Bean 对象。

```
public class DefaultServiceLocator {
    private static ClientService clientService
                                = new ClientServiceImpl();
    public ClientService createClientServiceInstance() {
            return clientService;
    }
}
```

上面的例子是由 Spring 官网提供的，这个写法其实有点问题。非静态工厂方法与静态工厂方法完全相反，此处建议修改成如下写法（修改原因在后面解释）。

```
public class DefaultServiceLocator {
    public ClientService createClientServiceInstance() {
        return new ClientServiceImpl();
    }
}
```

4. 配置 HelloFactoryStatic 和 HelloFactoryBean

示例 3-5：配置 HelloFactoryStatic。

首先，配置静态 factory-method 属性。

```
<bean  id="helloBean"
    class="com.icss.biz.HelloBiz" factory-method="createInstance" />
```

然后，编写静态方法 createInstance。

```
public class HelloBiz {
    private static HelloBiz helloBiz = new HelloBiz();
    public static HelloBiz createInstance() {
        System.out.println("HelloBiz createInstance....");
        return helloBiz;
    }
    public String sayHello(String name) {
         return "hello: Mr " + name;
    }
}
```

测试代码如下。

```
public static void main(String[] args) {
    ApplicationContext context
            = new ClassPathXmlApplicationContext("beans.xml");
    HelloBiz hi = (HelloBiz) context.getBean("helloBean");
    String hello = hi.sayHello("tom");
    System.out.println(hello);
}
```

测试结果如下。

```
信息: Loading XML bean definitions from class path resource [beans.xml]
HelloBiz createInstance....
hello: mr tom
```

示例 3-6：配置 HelloFactoryBean。

首先，设置配置信息。当使用 factory-bean 属性后，即使配置 class 属性，class 属性也不会有效。

```
<bean  id="helloBean"
          factory-bean="helloFactory" factory-method="createInstance" />
<bean id="helloFactory"
     class="com.icss.biz.HelloFactory"></bean>
```

然后，编写工厂类方法。

```
public class HelloFactory {
     public HelloBiz createInstance() {
          System.out.println("HelloFactory-->>createInstance....");
          return new HelloBiz();
     }
}
```

测试代码如下。

```
public static void main(String[] args) {
     ApplicationContext context
             = new ClassPathXmlApplicationContext("beans.xml");
     HelloBiz hi = (HelloBiz) context.getBean("helloBean");
     String hello = hi.sayHello("tom");
     System.out.println(hello);
}
```

测试结果如下。

```
信息: Loading XML bean definitions from class path resource [beans.xml]
HelloFactory-->>createInstance....
hello: Mr tom
```

5. 扩展 HelloFactoryBean

当业务发生变化时，为了扩展 HelloFactoryBean，可以按照如下方式操作。

（1）新增业务类 HellBizChina，它继承自 HelloBiz。

```
public class HelloBizChina extends HelloBiz{
     public String sayHello(String name) {
         return "你好: 先生  " + name;
     }
}
```

（2）新增工厂类 HelloFactoryImpl。

```
public class HelloFactoryImpl extends HelloFactory{
     public HelloBiz createInstance() {
         System.out.println("HelloFactoryImpl-->>createInstance....");
         return new HelloBizChina();
     }
}
```

（3）修改配置文件。

```
<bean  id="helloBean"  factory-bean="helloFactory"
                      factory-method="createInstance" />
<bean id="helloFactory"
              class="com.icss.biz.HelloFactoryImpl"></bean>
```

（4）测试代码与示例 3-5 相同，运行结果如下。

```
信息: Loading XML bean definitions from class path resource [beans.xml]
HelloFactoryImpl-->>createInstance....
你好: 先生  tom
```

总结：上述测试证明，Spring 中 factory-method 的配置思想与设计模式中 Factory-Method 的配置思想完全一致。

6．工厂设计模式

比较下面的配置，扩展工厂方法产生的代码与直接配置 helloBean 的实现类的区别在哪里呢？

```
<bean  id="helloBean" class="HelloBizChina">
<bean  id="helloBean"
        factory-bean="helloFactory" factory-method="createInstance" />
<bean id="helloFactory"
        class="com.icss.biz.HelloFactoryImpl"></bean>
```

要了解更多信息，可参见“四人组”编写的《设计模式》中的对象创建型模式——Factory-Method 模式。

工厂方法模式旨在定义一个用于创建对象的接口，让子类决定实例化哪一个类。Factory-Method 模式使一个类的实例化延迟到其子类。工厂方法与多态的区别是，工厂方法用于生产一系列相关产品，而多态只是一个接口实现的变体。

在如下示例中，使用工厂方法创建迷宫及相关产品。

```
public abstract class MazeFactory {
     public abstract Maze makeMaze();              //创建迷宫
     public abstract Room makeRoom(int n);         //创建屋子
     public abstract Wall makeWall();              //创建墙
     //创建门
     public abstract Door makeDoor(Room r1,Room r2);
}
```

如果只对单一产品实现多态扩展，则没有必要使用工厂方法。

3.5 HelloIoC 示例

前面介绍了 IoC 容器与 Bean 的创建，同时讲解了 IoC 的概念。IoC 的本质是通过配置信息的变化，实现程序的多态。

下面实现一个业务案例，其业务需求如下。

- 使用多种语言交互。
- 代码尽量体现灵活性、可扩展性。

3.5.1 面向接口编程

面向接口编程的基本思想是定义接口，定义接口的场景调用，在业务场景中使用的都是接口中的方法。当实际调用业务方法时，传入接口的实现类。利用方法重写的设计，实现程序的动态变化，即接口不变，实现类可以自由地扩展。程序满足开闭原则（Open Closed Principle，OCP）的设计思想。

（1）在原来的 Hello 项目的基础上新增接口。

```
public interface IHello {
     public String sayHello(String name) ;
}
```

（2）新建两个接口实现类。

```
public class HelloChina implements IHello{
     public String sayHello(String name) {
         return "你好:  " + name + " 先生";
     }
}
public class HelloEnglish implements IHello{
     public String sayHello(String name) {
          return "hello Mr.  " + name ;
     }
}
```

（3）面向接口调用。

```
public static void main(String[] args) {
     IHello hi = new HelloEnglish();    //新建对象后，无法再改变
     String hello = hi.sayHello("tom");
     System.out.println(hello);
}
```

面向接口编程增加了程序的灵活性和可扩展性，但是如何解决硬编码问题？

执行 IHello hi = new HelloEnglish( )，新建对象后，无法再改变，这称为硬编码。

为了解决硬编码的问题，早期用设计模式实现，即把 IHello 接口当成参数传入，这虽然增加了灵活性，但增加了程序的复杂度。更好的方案是采用 IoC。

在程序运行过程中，若希望动态改变对象，则需要反射技术，即通过“反射 + 接口 + XML 配置”可以实现 IoC 的变化，彻底解决硬编码问题。

3.5.2 通过 XML 和反射实现 IoC

首先自定义 XML 文件，配置 helloBean，然后读取 XML 文件，再用反射技术动态创建配置中的 class 对象，从而实现 IoC 的动态调用。操作步骤如下。

（1）定义逻辑接口和实现类。

```java
public interface IHello {
     public String sayHello(String name) ;
}
public class HelloChina implements IHello{
     public String sayHello(String name) {
        return "你好, " + name + " 先生";
     }
}
public class HelloEnglish implements IHello{
     public String sayHello(String name) {
          return "how are you " + name ;
     }
}
```

（2）使用 XML 配置 Bean。

```xml
<bean  id="helloBean"  class="com.icss.biz.HelloEnglish" />
```

（3）定义 Map，以接收所有的 Bean 对象。

```java
public class BeanFactory {
     //XML 文件，最好只读取一次，然后放在内存中随时存取（静态变量不会被回收）
     private static Map<String,Object> beans;
}
```

（4）通过 DOM 解析配置文件。

```java
//静态代码块，用于初始化静态变量
//当调用当前类中的任何信息时，会先调用静态代码块，而且静态代码块只调用一次
static {
     beans = new HashMap<>();
     //DocumentBuilderFactory 是抽象类，不能直接实例，只能从静态方法中创建
     //DocumentBuilderFactory.newInstance()创建的是 DocumentBuilderFactory 子类的对象
     DocumentBuilderFactory dbf = DocumentBuilderFactory.newInstance();
     try {
          //工厂模式
          DocumentBuilder bild = dbf.newDocumentBuilder();
          //查找 beans.xml 文件的路径（用反射技术查找）
          //注意，路径不要包含中文
          String path = BeanFactory.class.getResource("/").getPath();
          File file = new File(path+"beans.xml");
          //解析 XML 文件，得到 Document 对象
          Document  doc = bild.parse(file);
```

（5）通过反射技术创建 Bean 对象。

```java
          //根据 tag 查找所有 Bean
          NodeList nlist = doc.getElementsByTagName("bean");
          for(int i=0;i<nlist.getLength();i++) {
               Node node = nlist.item(i);
               //先判断节点类型是否是 element
               if(node instanceof Element) {
                    Element element = (Element)node;
                    //提取属性信息
                    String id = element.getAttribute("id");
                    String cname = element.getAttribute("class");
                    //用反射技术，加载类型信息
```

```
                    Class cls = Class.forName(cname);
                    //动态创建对象
                    Object obj = cls.newInstance();
                    beans.put(id, obj);
                }
            }
        } catch (Exception e) {
            e.printStackTrace();
        }
```

（6）封装 getBean( )。

```
//根据 Bean 的名字，查找对应的 Bean 对象
public static Object getBean(String name) {
    return beans.get(name);
}
```

（7）测试。

① 测试代码如下。

```
public static void main(String[] args) {
    IHello hi = (IHello) BeanFactory.getBean("helloBean");
    String hello = hi.sayHello("tom");
    System.out.println(hello);
}
```

② 测试结果如下。

```
how are you,tom
```

（8）修改配置文件。

```
<bean id="helloBean" class="com.icss.biz.HelloChina" />
```

（9）测试代码不变，输出结果如下。

```
你好， tom 先生
```

3.5.3 通过 Spring 实现 IoC

通过 XML 与反射机制，我们可以自己实现 IoC。使用 Spring 框架提供的 IoC 功能，实现起来会更加简单。具体步骤如下。

（1）逻辑代码不变。

```
public interface IHello {
    public String sayHello(String name) ;
}
public class HelloChina implements IHello{
    public String sayHello(String name) {
        return "你好， " + name + " 先生";
    }
}
```

```
public class HelloEnglish implements IHello{
    public String sayHello(String name) {
        return "how are you " + name ;
    }
}
```

（2）在项目 src 下新建 Spring 配置文件 beans.xml，配置如下。

```
<?xml version="1.0" encoding="UTF-8"?>
<beans xmlns="http://www.springframework.org/schema/beans"
xmlns:xsi="http://www.w3.org/2001/XMLSchema-instance"
xsi:schemaLocation="http://www.springframework.org/schema/beans
http://www.springframework.org/schema/beans/spring-beans.xsd">
<bean id="helloBean" class="com.icss.biz.HelloEnglish" />
</beans>
```

（3）通过实例 IoC 容器读取 Bean。

```
public static void main(String[] args) {
    //创建 Spring 容器，解析 XML 文件
    ApplicationContext context = new ClassPathXmlApplicationContext("beans.xml");
    //根据 Bean 的名字，查找 Bean 对象
    IHello  hi = (IHello)context.getBean("helloBean");
    String hello = hi.sayHello("tom");
    System.out.println(hello);
}
```

输出结果如下。

```
org.springframework.beans.factory.xml.XmlBeanDefinitionReader loadBeanDefinitions
信息: Loading XML bean definitions from class path resource [beans.xml]
hello Mr.  tom
```

（4）修改配置文件。

```
<bean  id="helloBean"  class="com.icss.biz.HelloChina" />
```

（5）测试。注意，测试代码不变。测试结果如下。

```
信息: Loading XML bean definitions from class path resource [beans.xml]
你好，tom 先生
```

3.6 依赖注入

3.6.1 依赖注入的定义

依赖注入的英文定义如下。

Dependency injection (DI) is a process whereby objects define their dependencies, that is, the other objects they work with, only through constructor arguments, arguments to a factory method, or properties that are set on the object instance after it is constructed or returned from a factory method。

一个业务系统由很多 Bean 对象组成，Bean 之间存在调用关系，那么这些 Bean 对象是如何协同工作的呢？

对象之间的调用存在依赖关系，何为依赖？下面我们以 StaffUser 员工系统为例说明依赖的含义。

3.6.2 项目案例：StaffUser 系统与 IoC

这是一个真实的项目案例，在本书中，把很多知识点会融入了这个项目中。此项目只用于功能演示，因此除必要功能外，其他功能暂不描述。

1. StaffUser 系统的需求

StaffUser 是大型企业的员工管理系统，如 SAP（System Application and Product）系统中的人资管理模块。StaffUser 也可以看成学生管理系统，如很多高校的办公系统。

员工或学生的基本信息很多，如姓名、身高、出生日期、地址、联系方式等。当把员工基本信息录入系统后，系统需要帮助该员工自动生成一个用户名。以后该员工就可以使用这个用户名登录该系统，做相关操作。

默认用户属性为员工编号、用户名、密码、角色。

用户使用员工编号登录系统，该编号唯一；用户名可以是员工的姓名或昵称；角色根据业务指定，默认为普通用户。

首次登录系统时，系统会提示用户修改密码。用户的初始别名与员工编号一致，后面可以修改。

StaffUser 系统同时支持多种数据库，如 Oracle 和 MySQL。

2. StaffUser 表的结构

当前只展现两个表，随着功能的深入，后面再添增加其他表。

员工表与用户表是一对一的关系，即一个员工只能对应一个用户。

图 3-3 为基于 MySQL 的 StaffUser 表的结构，与 Oracle 表的结构几乎完全相同，数据类型稍有变化。

3. 软件三层架构

StaffUser 系统基于 C/S 模式，采用软件三层架构搭建。软件三层架构是搭建业务系统的重要架构基础，因此要熟练掌握。

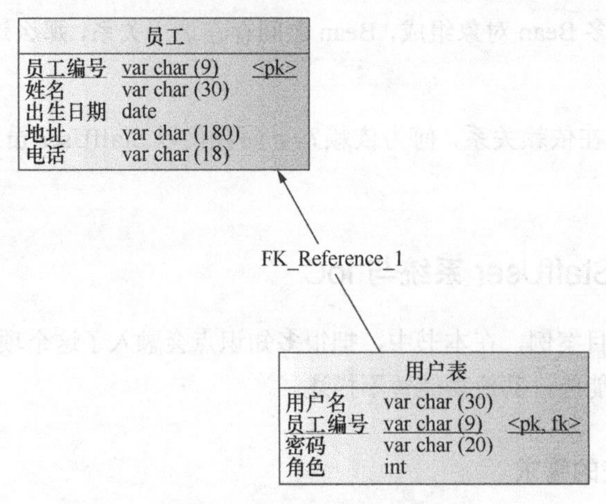

图 3-3 StaffUser 表的结构

后面的书城项目基于 B/S 模式，采用 MVC 架构搭建。MVC 架构是在软件三层架构的基础上扩展而来的。

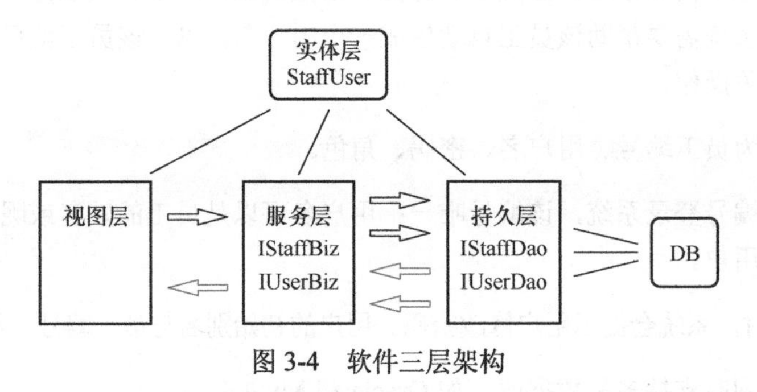

图 3-4 软件三层架构

如图 3-4 所示，三层分别为服务层、持久层、视图层。

- 服务层：负责核心业务逻辑控制，如业务流程控制、业务规则校验等。
- 持久层：需要数据访问部分的操作，简称 DAO（Data Access Object，数据访问对象）层。
- 视图层：与用户交互的部分，如输入、输出，简称 UI（User Interface，用户界面）层。
- 实体层：实体对象与表通常是映射关系，实体层可以被其他层调用，即数据的跨层传递需要使用实体（entity），实体也称为 DTO（Data Transfer Object，数据传输对象），POJO（Plain Ordinary Java Object，简单旧式 Java 对象）或 Model。

4. StaffUser 系统的接口设计

如图 3-5 所示，对于服务层的接口实现，暂时只写一个，可以扩展。持久层同时支持 MySQL

和 Oracle，还可以扩展至其他数据库。

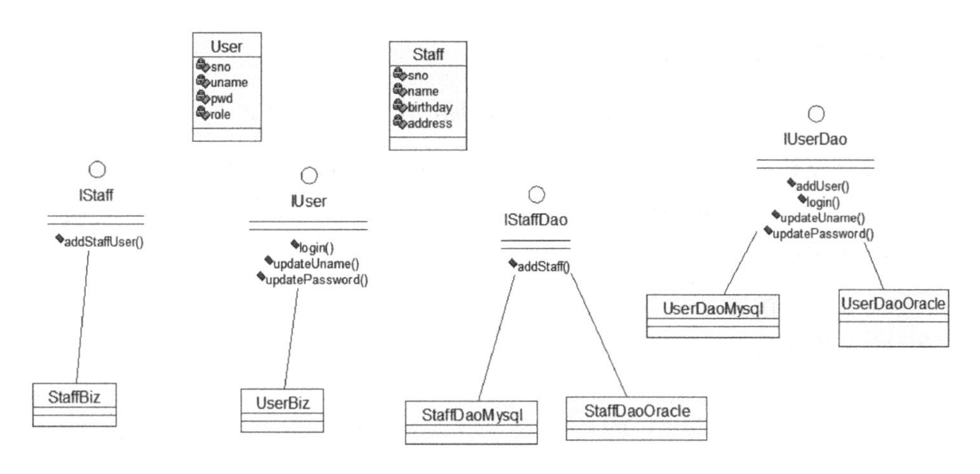

图 3-5 StaffUser 系统的接口设计

5. UML 中的依赖关系

StaffUser 系统中的依赖关系如图 3-6 所示。

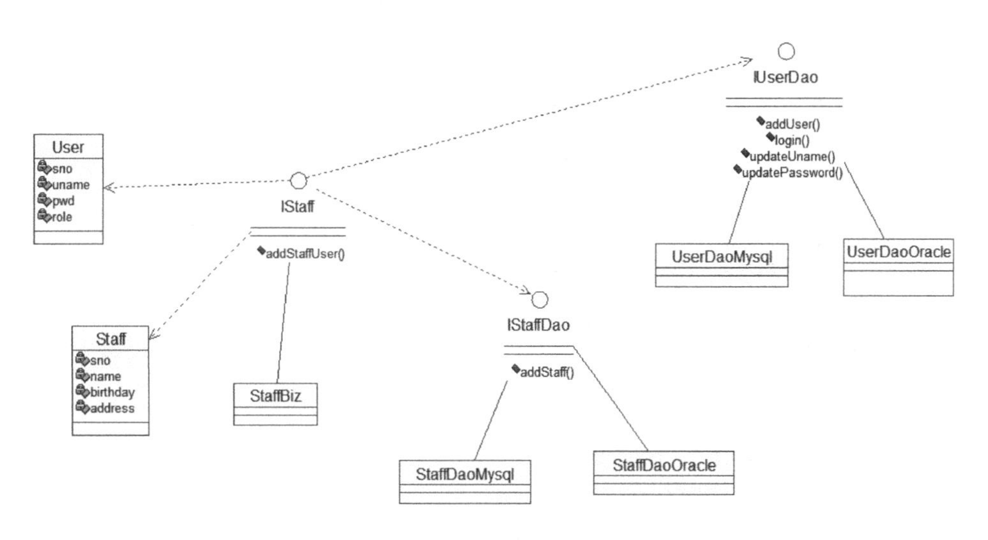

图 3-6 StaffUser 系统中的依赖关系

UML 中的依赖关系与 Spring 中的依赖关系有什么区别？

在 UML 中，对象之间有组合关系、聚合关系、依赖关系这 3 种关系，它们本质上完全相同，只是依赖性不同。其中，组合关系中的依赖性最强，依赖关系中的依赖性最弱。

Spring 中的依赖关系是 UML 中 3 种依赖关系的统称，并没有进行细分。

6. StaffUser 系统的代码实现

下面使用 Spring 实现 StaffUser 系统的接口功能，操作步骤如下。

（1）导入 Spring 核心包 Spring-core.jar、Spring-context.jar、Spring-beans.jar 和 Spring-expression.jar，导入依赖包 commons-logging.jar，导入日志包 log4j.jar，导入数据库驱动包 mysql-connector-java-8.0.11.jar、ojdbc6.jar。

（2）实现 Spring 配置文件。把服务层和持久层对象都当成 Bean，注入业务系统中。注意，UI 和实体对象无须注入。

```
<?xml version="1.0" encoding="UTF-8"?>
<beans xmlns="http://www.springframework.org/schema/beans"
     xmlns:xsi="http://www.w3.org/2001/XMLSchema-instance"
     xmlns:aop="http://www.springframework.org/schema/aop"
     xmlns:tx="http://www.springframework.org/schema/tx"
     xsi:schemaLocation="http://www.springframework.org/schema/beans
          http://www.springframework.org/schema/beans/spring-beans-4.3.xsd">
     <bean id="userDao" class="com.icss.dao.impl.UserDaoMysql"/>
     <bean id="staffDao" class="com.icss.dao.impl.StaffDaoMysql"/>
     <bean id="staffBiz" class="com.icss.biz.impl.StaffBiz"/>
    <bean id="userBiz" class="com.icss.biz.impl.UserBiz"/>
</beans>
```

（3）调用持久层（先用伪代码，后连接数据库）。

```
public User login(String sno, String pwd) throws Exception {
    User user = null;
    System.out.println("UserDaoMysql....login....");
    // 用伪代码模拟用户登录
    if (sno.equals("admin") && pwd.equals("123")) {
        user = new User();
        user.setRole(IRole.ADMIN);
        user.setUname(sno);
        user.setPwd(pwd);
    } else if (sno.equals("tom") && pwd.equals("123")) {
        user = new User();
        user.setRole(IRole.COMMON_USER);
        user.setUname(sno);
        user.setPwd(pwd);
    } else if (sno.equals("jack") && pwd.equals("123")) {
        user = new User();
        user.setRole(IRole.VIP_USER);
        user.setUname(sno);
        user.setPwd(pwd);
    } else {
    }
    return user;
}
```

(4) 调用逻辑层代码。逻辑层调用持久层，此处 IUserDao 从 IoC 容器中读取。

```
public User login(String sno, String pwd) throws Exception {
    User user;
    System.out.println("UserBiz....login....");
    // 入参校验
    if (sno == null || pwd == null || sno.trim().equals("") || pwd.trim().equals("")) {
        throw new Exception("用户名或密码为空");
    }
    try {
         ApplicationContext app = new ClassPathXmlApplicationContext("beans.xml");
         IUserDao dao = (IUserDao) app.getBean("userDao");
         user = dao.login(sno, pwd);
    } finally {
         // 释放数据库连接
    }
    return user;
}
```

(5) 测试。

① 测试代码如下。

```
public static void main(String[] args) {
    ApplicationContext app = new ClassPathXmlApplicationContext("beans.xml");
    IUser u = (IUser)app.getBean("userBiz");
    try {
        User user = u.login("admin", "123");
        if(user != null) {
             System.out.println("登录成功，身份是" + user.getRole());
         }else {
             System.out.println("登录失败");
         }
    } catch (Exception e) {
        System.out.println(e.getMessage());
     }
}
```

② 程序实现了 Spring IoC，运行正常，运行结果如下。

```
信息: Loading XML bean definitions from class path resource [beans.xml]
UserBiz....login....
org.springframework.context.support.ClassPathXmlApplicationContext
信息: Loading XML bean definitions from class path resource [beans.xml]
UserDaoMysql....login....
登录成功，身份是 1
```

思考：*若在现在的视图层和逻辑层中各创建一个 IoC 容器，则会存在什么问题？*

7. 通过单例封装 IoC 容器

前面的 StaffUser 项目使用了两个 IoC 容器，程序运行时并没有错误。但是从性能考虑，双容器模式完全是浪费，因此我们习惯上把IoC容器封装为单例模式。

```
public class BeanFactory {
      private static ApplicationContext context ;
      static {
          context = new ClassPathXmlApplicationContext("beans.xml");
      }
      public static Object getBean(String beanName) {
          return context.getBean(beanName);
      }
}
```

封装为单例模式后，UserBiz中的调用如下。

```
IUserDao dao = (IUserDao) BeanFactory.getBean("userDao");
```

封装为单例模式后，视图层的调用如下。

```
IUser biz = (IUser)BeanFactory.getBean("userBiz");
```

3.6.3 通过构造函数注入

IoC容器动态调用Bean的带参构造函数，注入依赖对象。构造函数的每个参数是一个依赖。

在Bean的构造函数中，传入要依赖的其他Bean对象引用或简单类型参数。通过构造函数注入可以明确地表明依赖者与被依赖者的关系。

1. 通过构造函数注入StaffUser持久层对象

通过构造函数注入StaffUser持久层对象的操作步骤如下。

（1）设置服务层依赖关系。

① IStaff对象需要依赖IUserDao和IStaffDao。

```
public class StaffBiz implements IStaff{
     private IUserDao userDao;
     private IStaffDao staffDao;
     public StaffBiz(IUserDao userDao,IStaffDao staffDao) {
          this.userDao = userDao;
          this.staffDao = staffDao;
     }
}
```

② IUser对象依赖IUserDao。

```
public class UserBiz implements IUser {
     private IUserDao userDao;
     public UserBiz(IUserDao userDao) {
          this.userDao = userDao;
     }
}
```

（2）配置构造函数，实现依赖注入。因为 staffBiz 对象同时依赖 IUserDao 和 IStaffDao，所以在构造函数中传入两个对象的引用。

```
<bean id="userDao" class="com.icss.dao.impl.UserDaoMysql"/>
<bean id="staffDao" class="com.icss.dao.impl.StaffDaoMysql"/>
<bean id="staffBiz" class="com.icss.biz.impl.StaffBiz">
        <constructor-arg ref="staffDao"/>
        <constructor-arg ref="userDao"/>
</bean>
```

（3）userBiz 对象需要依赖 IUserDao，因此在构造函数中传入 IUserDao 的对象引用。

```
<bean id="userDao" class="com.icss.dao.impl.UserDaoMysql"/>
   <bean id="userBiz" class="com.icss.biz.impl.UserBiz">
     <constructor-arg ref="userDao"/>
   </bean>
```

（4）调用逻辑代码。对于 UserBiz 中使用的持久层对象，不再从 BeanFactory 中查找，而直接使用注入的 userDao。

```
public User login(String sno, String pwd) throws Exception {
    User user;
    // 校验输入参数
    if (sno == null || pwd == null || sno.trim().equals("")
                                || pwd.trim().equals("")){
        throw new Exception("用户名或密码为空");
    }
    try {
        // IUserDao dao = (IUserDao) BeanFactory.getBean("userDao");
        // 此处直接使用注入的 userDao 对象
        user = userDao.login(sno, pwd);
    } finally {
        // 释放数据库
    }
    return user;
}
```

（5）调用 UI 代码。LoginTest 中的逻辑对象没有使用 DI，因此仍旧从 BeanFactory 中提取。把服务层对象注入视图层也是可以的。

```
public static void main(String[] args) {
    IUser biz = (IUser)BeanFactory.getBean("userBiz");
    try {
        User user = biz.login("admin", "123");
        if(user != null) {
            System.out.println("登录成功，身份是" + user.getRole());
        }else {
            System.out.println("登录失败");
        }
    } catch (Exception e) {
```

```
            System.out.println(e.getMessage());
        }
    }
```

测试结果如下。

```
信息: Loading XML bean definitions from class path resource [beans.xml]
UserDaoMysql....login....
登录成功，身份是 1
```

2．通过简单类型的构造函数注入

前面使用的是 Bean 对象的引用注入，IoC 容器知道 Bean 的引用类型信息。对于非 Bean 引用的简单类型，Spring 容器无法知道，因此需要在显式指定类型后注入。

示例代码如下。

```
package examples;
public class ExampleBean {
    private int years;
    private String ultimateAnswer;
    public ExampleBean(int years, String ultimateAnswer) {
        this.years = years;
        this.ultimateAnswer = ultimateAnswer;
    }
}
```

ExampleBean 类的构造函数的参数为 int 类型和 String 类型，因此需要在配置文件中显式指定类型。

```
<bean id="exampleBean" class="examples.ExampleBean">
    <constructor-arg type="int" value="7500000"/>
    <constructor-arg type="java.lang.String" value="42"/>
</bean>
```

当构造函数的多个输入参数使用的是相同类型时，按照 type 赋值的方式会出现问题，因此需要使用 index，按照参数顺序给构造函数赋值。

```
<bean id="exampleBean" class="examples.ExampleBean">
    <constructor-arg index="0" value="7500000"/>
    <constructor-arg index="1" value="42"/>
</bean>
```

直接使用 name 赋值，也可以解决按照 type 赋值出现的问题。

```
<bean id="exampleBean" class="examples.ExampleBean">
    <constructor-arg name="years" value="7500000"/>
    <constructor-arg name="ultimateAnswer" value="42"/>
</bean>
```

案例 3-1：通过构造函数注入 User 对象

User 是一个实体类，它有很多属性，如何把 User 对象注入系统中？

（1）User 类的构造函数如下。

```
public class User {。
    private String uname;
    private String sno;
    private String  pwd;
    private int   role;
    public User() {
    }
    public User(String uname,String sno,String pwd,int role) {
        this.uname = uname;
        this.sno = sno;
        this.pwd = pwd;
        this.role = role;
    }
}
```

（2）配置构造函数注入。User 依赖的属性信息都是简单类型，具体如下所示。

```
<bean id="user" class="com.icss.entity.User">
     <constructor-arg name="uname" value="tom"/>
     <constructor-arg name="pwd" value="123456"/>
     <constructor-arg name="role" value="2"/>
        <constructor-arg name="sno" value="001"/>
  </bean>
```

（3）测试。

① 测试代码如下。

```
public static void main(String[] args) {
    User user = (User)BeanFactory.getBean("user");
    System.out.println(user.getSno());
    System.out.println(user.getUname());
    System.out.println(user.getPwd());
    System.out.println(user.getRole());
}
```

② 测试结果如下。

```
信息: Loading XML bean definitions from class path resource [beans.xml]
001
tom
123456
2
```

（4）修改配置文件为索引顺序模式，测试索引配置效果。

```
<bean id="user" class="com.icss.entity.User">
     <constructor-arg index="0" value="tom"/>
     <constructor-arg index="2" value="123456"/>
```

```
        <constructor-arg index="3" value="2"/>
        <constructor-arg index="1" value="001"/>
</bean>
```

注意：如果以实体类作为 Bean 对象，则需要考虑性能问题，这在后面讨论。

3.6.4 通过 set 方法注入

通过无参构造函数实例化 Bean 对象后，IoC 容器继续调用 Bean 的 set 方法，可以把依赖对象注入运行环境。

1. 通过 set 方法注入 StaffUser 持久层对象

具体步骤如下。

（1）配置依赖关系。

```
<bean id="userDao" class="com.icss.dao.impl.UserDaoMysql"/>
<bean id="staffDao" class="com.icss.dao.impl.StaffDaoMysql"/>
<bean id="staffBiz" class="com.icss.biz.impl.StaffBiz">
    <property name="userDao" ref="userDao"></property>
    <property name="staffDao" ref="staffDao"></property>
</bean>
<bean id="userBiz" class="com.icss.biz.impl.UserBiz">
     <property name="userDao" ref="userDao"></property>
    </bean>
```

（2）通过 set 方法注入。

① 在 StaffBiz 类中注入 IUserDao 和 IStaffDao 对象。

```
public class StaffBiz implements IStaff{
    private IUserDao userDao;
    private IStaffDao staffDao;
    public void setUserDao(IUserDao userDao) {
        this.userDao = userDao;
    }
    public void setStaffDao(IStaffDao staffDao) {
        this.staffDao = staffDao;
    }
}
```

② 在 UserBiz 类中注入 IUserDao 对象。

```
public class UserBiz implements IUser {
    private IUserDao userDao;
    public void setUserDao(IUserDao userDao) {
        this.userDao = userDao;
    }
}
```

注意，set 方法的命名应该尽量规范。在上面的配置中，userBiz 对象的属性名是 userDao，因此 Set 方法的名字为 setUserDao( )。

测试结果如下。

```
信息: Loading XML bean definitions from class path resource [beans.xml]
UserDaoMysql....login....
登录成功，身份是 1
```

2. property 的配置说明

配置文件不变，UserBiz 中的属性名修改为 userDao2，参数名修改为 userDao3。

```
public class UserBiz implements IUser {
    private IUserDao userDao2;
    public void setUserDao(IUserDao userDao3) {
        this.userDao2 = userDao3;
    }
}
```

测试结果正常，没有出现错误。

注意：如果 property 元素没有 name 属性，则程序会报错。

修改配置文件中的 name="userDao2"。

```
<bean id="userBiz" class="com.icss.biz.impl.UserBiz">
     <property name="userDao2" ref="userDao"></property>
</bean>
```

测试出现异常。

```
Caused by: org.springframework.beans.NotWritablePropertyException: Invalid property
'userDao2' of bean class [com.icss.biz.impl.UserBiz]: Bean property 'userDao2' is not
writable or has an invalid setter method. Did you mean 'userDao'?
```

修改 set 方法的名称为 setUserDao2 后，正常测试。

```
public class UserBiz implements IUser {
    private IUserDao userDao;
    public void setUserDao2(IUserDao userDao) {
        this.userDao = userDao;
    }
}
```

总结：在配置文件的<property name="##">中，name 与 set 方法的名字对应，与 UserBiz 中的 userDao 属性无关，与 set 方法的参数名字也无关。

3. 通过 set 方法注入与通过构造函数注入的对比

通过 set 方法注入与通过构造函数注入各有优点，简单的对比如下。

- 通过 set 方法注入的代码更加简洁。

- 通过构造函数注入对依赖关系的表达更加清楚。
- 通过 set 方法注入可以避免循环依赖问题。

3.6.5 依赖注入的处理流程

IoC 容器按照如下步骤处理依赖注入。

（1）ApplicationContext 的创建和初始化依赖于配置信息。配置信息可以源于 XML、Java 代码或者注解。

（2）每个 Bean 的依赖以属性、构造函数参数或静态工厂方法参数的形式出现。当实际创建这些 Bean 时，这些依赖也将会提供给该 Bean。

（3）每个属性或构造函数的参数既可以是一个实际值，也可以是对该容器中另一个 Bean 的引用。

（4）每个指定的属性或构造函数参数值必须能够转换成特定的格式或构造参数所需类型。默认情况下，Spring 可以把 String 类型转换为其他各种内置类型，如 int、long、String、boolean 等。

3.6.6 依赖配置

本节将详细讲解依赖配置中的各种参数设置，如通过 property 赋值，通过 p:namespace 赋值，通过 c:namespace 赋值，以及 depends-on 属性、lazy-init 属性的用法等。

1. 直接赋值

1）用 property 直接赋值

可以使用 Bean 的<property>子元素，给 Bean 的属性直接赋值。示例代码如下。

```
<bean id="myDataSource"
    class="org.apache.commons.dbcp.BasicDataSource" destroy-method="close">
    <property name="driverClassName" value="com.mysql.jdbc.Driver"/>
    <property name="url" value="jdbc:mysql://localhost:3306/mydb"/>
    <property name="username" value="root"/>
    <property name="password" value="masterkaoli"/>
</bean>
```

注意：对比 3.6.4 节中的示例，property 的值对应的是 set 方法，不是直接给属性赋值。Property 的转换靠的是 Spring 的转换服务机制。更多信息请参见“Spring's conversion service is used to convert these values from a String to the actual type of the property or argument”文章。

2）通过 property 给 User 属性赋值

把实体类 User 配置成 Bean，并用 property 赋值。前面我们演示过通过构造函数给 User 对象赋值，这次使用的是 set 方法。示例代码如下。

（1）为 User 实体添加 set 方法。

```
public class User {
    private String uname;
    private String  sno;
    private String   pwd;
    private int    role;
    public void setUname(String uname) {
        this.uname = uname;
    }
    public void setSno(String sno) {
        this.sno = sno;
    }
    public void setPwd(String pwd) {
        this.pwd = pwd;
    }
    public void setRole(int role) {
        this.role = role;
    }
}
```

（2）用<property>给属性赋值。

```
<bean id="user" class="com.icss.entity.User">
    <property name="uname" value="tom"></property>
    <property name="sno" value="001"></property>
      <property name="pwd" value="123456"></property>
</bean>
```

（3）测试。

① 测试代码如下。

```
public static void main(String[] args) {
    ApplicationContext app = new ClassPathXmlApplicationContext("beans.xml");
    User user = (User)app.getBean("user");
    System.out.println(user.getSno());
    System.out.println(user.getUname());
}
```

② 测试结果如下。

```
信息: Loading XML bean definitions from class path resource [beans.xml]
001
tom
```

3）通过 property 给集合属性赋值

示例 3-7：邮件服务有一个邮箱列表，通过 property 与 list 给集合赋值。

（1）把 EmailService 的 blackList 属性设置为集合。

```
public class EmailService implements ApplicationEventPublisherAware{
    private ApplicationEventPublisher publisher;
    private List<String> blackList;
    public void setBlackList(List<String> blackList) {
        this.blackList = blackList;
    }
}
```

（2）在配置文件中，通过<property>给集合赋值。

```
<bean id="emailService" class="com.icss.biz.EmailService">
        <property name="blackList">
           <list>
              <value>tom1@qq.com</value>
              <value>tom2@qq.com</value>
              <value>tom3@qq.com</value>
              <value>jack1@qq.com</value>
              <value>jack2@qq.com</value>
           </list>
        </property>
</bean>
```

4）用 p:namespace 直接赋值

使用“p:属性名”，可以直接给 Bean 的属性赋值。这种赋值方式的效果等同于<property>。

注意：schema 必须要支持 p 命名空间。

下例中使用了两种属性赋值方式，首先通过标准的 XML 属性赋值，然后使用 p:namespace 进行简化的赋值。

```
<beans xmlns="http://www.springframework.org/schema/beans"
    xmlns:xsi="http://www.w3.org/2001/XMLSchema-instance"
    xmlns:p="http://www.springframework.org/schema/p"
    xsi:schemaLocation="http://www.springframework.org/schema/beans
    http://www.springframework.org/schema/beans/spring-beans.xsd">
        <bean name="classic" class="com.example.ExampleBean">
             <property name="email" value="foo@bar.com"/>
        </bean>
        <bean name="p-namespace" class="com.example.ExampleBean"
                p:email="foo@bar.com"/>
</beans>
```

5）通过“p:”给 User 属性赋值

把员工系统中的实体类 User 配置成 Bean，并用 p:namespace 赋值。代码如下。

```
<?xml version="1.0" encoding="UTF-8"?>
<beans xmlns="http://www.springframework.org/schema/beans"
    xmlns:xsi="http://www.w3.org/2001/XMLSchema-instance"
    xmlns:p="http://www.springframework.org/schema/p"
    xsi:schemaLocation="http://www.springframework.org/schema/beans
        http://www.springframework.org/schema/beans/spring-beans.xsd">
```

```
    <bean id="user"
            class="com.icss.entity.User" p:uname="tom" p:sno="001">
    </bean>
</beans>
```

2. 通过引用赋值

使用<property>的 ref 属性，可以引用其他 Bean 对象。示例代码如下。

txManager 和 sqlSessionFactory 都引用了 dataSource。

```
<bean id="dataSource"
      class="org.springframework.jdbc.datasource.DriverManagerDataSource">
    <property name="driverClassName" value="com.mysql.cj.jdbc.Driver" />
    <property name="url"    value="jdbc:mysql://localhost:3306/staff" />
    <property name="username" value="root" />
    <property name="password" value="123456" />
</bean>
<bean id="sqlSessionFactory"
     class="org.mybatis.spring.SqlSessionFactoryBean">
    <property name="configLocation" value="classpath:mybatis.xml" />
    <property name="dataSource" ref="dataSource" />
</bean>
<bean id="txManager"
    class="org.springframework.jdbc.datasource.DataSourceTransactionManager">
    <property name="dataSource" ref="dataSource" />
</bean>
```

3. 通过 c:namespace 简化配置

类似于 p:namespace，c:namespace 也是一种简化配置的方式，它可以简化构造函数的配置。

1）c:namespace 的作用

c:namespace 的作用是简化构造函数注入的操作。为了使用 c:namespace 代替构造函数，需要引入 xmlns:c 的 schema。例如，对于对象引用，需要使用 c:属性名-ref=""格式。对于简单类型，直接使用 c:属性名=""格式。

如下示例使用了两种给构造函数赋值的方式——传统赋值方式和基于 c:namespace 的简化赋值方式。

```
<beans xmlns="http://www.springframework.org/schema/beans"
    xmlns:xsi="http://www.w3.org/2001/XMLSchema-instance"
    xmlns:c="http://www.springframework.org/schema/c"
    xsi:schemaLocation="http://www.springframework.org/schema/beans
    http://www.springframework.org/schema/beans/spring-beans.xsd">
    <bean id="bar" class="x.y.Bar"/>
    <bean id="baz" class="x.y.Baz"/>
    <bean id="foo" class="x.y.Foo">
        <constructor-arg ref="bar"/>
```

```
        <constructor-arg ref="baz"/>
        <constructor-arg value="foo@bar.com"/>
    </bean>
    <bean id="foo" class="x.y.Foo" c:bar-ref="bar" c:baz-ref="baz"
        c:email="foo@bar.com"/>
</beans>
```

使用 c:_序号格式也可以按照构造函数的参数顺序赋值。示例代码如下。

```
<bean id="foo" class="x.y.Foo" c:_0-ref="bar" c:_1-ref="baz"/>
```

2）通过构造函数注入 User 对象

User 实体中有一个多参构造函数。

```
public class User {
    private String uname;
    private String  sno;
    private String  pwd;
    private int   role;
    public User(String uname,String sno,String pwd,int role) {
        this.uname = uname;
        this.sno = sno;
        this.pwd = pwd;
        this.role = role;
    }
}
```

可以使用传统的构造函数赋值方式给 User 设置初始值。

```
<bean id="user" class="com.icss.entity.User">
    <constructor-arg index="0" value="tom"></constructor-arg>
    <constructor-arg index="1" value="010111"></constructor-arg>
    <constructor-arg index="2" value="123456"></constructor-arg>
    <constructor-arg index="3" value="2"></constructor-arg>
</bean>
```

用 c:namespace 可以简化构造函数赋值，配置如下。

```
<?xml version="1.0" encoding="UTF-8"?>
<beans xmlns="http://www.springframework.org/schema/beans"
    xmlns:xsi="http://www.w3.org/2001/XMLSchema-instance"
    xmlns:c="http://www.springframework.org/schema/c"
    xsi:schemaLocation="http://www.springframework.org/schema/beans
      http://www.springframework.org/schema/beans/spring-beans.xsd">
    <bean id="user" class="com.icss.entity.User"
                c:uname="tom" c:pwd="1234" c:sno="001" c:role="2">
    </bean>
</beans>
```

4. depends-on 属性

通常情况下，一个 Bean 依赖于另一个 Bean，这可以通过属性设置来实现。但是有时候，两个 Bean 之间的依赖关系没有明确指定。使用 depends-on 属性，可以明确告诉 IoC 容器，在某个 Bean 的初始化之前，必须要先初始化它依赖的 Bean 对象。

示例 3-8：配置 jedis 连接池。

```
<!-- redis 配置信息 -->
<bean id="jedisPoolConfig" class="redis.clients.jedis.JedisPoolConfig">
    <property name="maxActive" value="20"></property>
    <property name="maxIdle" value="10"></property>
    <property name="maxWait" value="1000"></property>
</bean>
<!-- jedis 连接池信息 -->
<bean id="jedisPool"
      class="redis.clients.jedis.JedisPool"    depends-on="jedisPoolConfig">
    <constructor-arg ref="jedisPoolConfig"></constructor-arg>
    <constructor-arg value="127.0.0.1"></constructor-arg>
    <constructor-arg type="int" value="6358"></constructor-arg>
</bean>
```

上面的jedisPool定义中的depend-on="jedisPoolConfig"意味着Spring总会保证jedisPoolConfig在 jedisPool 之前实例化，总是在 jedisPool 之后销毁。

1）StaffUser 服务层对持久层的依赖

服务层对持久层有最明确的依赖关系。观察下面的代码可以发现，持久层对象一定要先于服务层对象初始化，一定要后于服务层对象关闭。

（1）配置 userBiz 依赖于 userDao。

```
<bean id="userDao" class="com.icss.dao.impl.UserDaoMysql"
                   init-method="init" destroy-method="destroy" />
<bean id="userBiz" class="com.icss.biz.impl.UserBiz"
      init-method="init"  destroy-method="destroy"  depends-on="userDao">
    <constructor-arg ref="userDao" />
</bean >
```

（2）通过代码跟踪初始化与析构过程。

```
public class UserDaoMysql implements IUserDao {
    public UserDaoMysql() {
        System.out.println("UserDaoMysql 构造....");
    }
    public void init() {
        System.out.println("UserDaoMysql 初始化....");
    }
    public void destroy() {
        System.out.println("UserDaoMysql 析构....");
    }
}
public class UserBiz implements IUser {
    private IUserDao userDao;
```

```
    public UserBiz(IUserDao userDao) {
        this.userDao = userDao;
        System.out.println("UserBiz 构造....");
    }
    public void init() {
        System.out.println("UserBiz 初始化....");
    }
    public void destroy() {
        System.out.println("UserBiz 析构....");
    }
}
```

(3) 测试。

① 测试代码如下。

```
public static void main(String[] args) {
    ApplicationContext app = new ClassPathXmlApplicationContext("beans.xml");
    IUser u = (IUser)app.getBean("userBiz");
    try {
        User user = u.login("admin", "123");
        if(user != null) {
            System.out.println("登录成功, 身份是" + user.getRole());
        }else {
            System.out.println("登录失败");
        }
    } catch (Exception e) {
        System.out.println(e.getMessage());
    }
    //关闭 IoC 容器
    ((ClassPathXmlApplicationContext) app).close();
}
```

② 测试结果如下（UserDaoMysql 先于 UserBiz 创建，后于 UserBiz 释放）。

```
信息: Loading XML bean definitions from class path resource [beans.xml]
UserDaoMysql 构造....
UserDaoMysql 初始化....
UserBiz 构造....
UserBiz 初始化....
org.springframework.context.support.AbstractApplicationContext doClose
UserDaoMysql....login....
登录成功, 身份是 1
UserBiz 析构....
UserDaoMysql 析构....
```

2）StaffUser 服务层对实体的依赖

在上面的示例中，UserBiz 和 UserDao 之间的关系在 UML 中称为聚合。这种聚合关系是强依赖关系，即使删除 depends-on 配置，执行顺序也是一样的。如图 3-7 所示，IStaff 在 addStaffUser( ) 中需要依赖实体 User 和 Staff，这是弱依赖。对于弱依赖，必须配置 depends-on 属性才能在 Bean 对象之间建立真正的依赖关系。

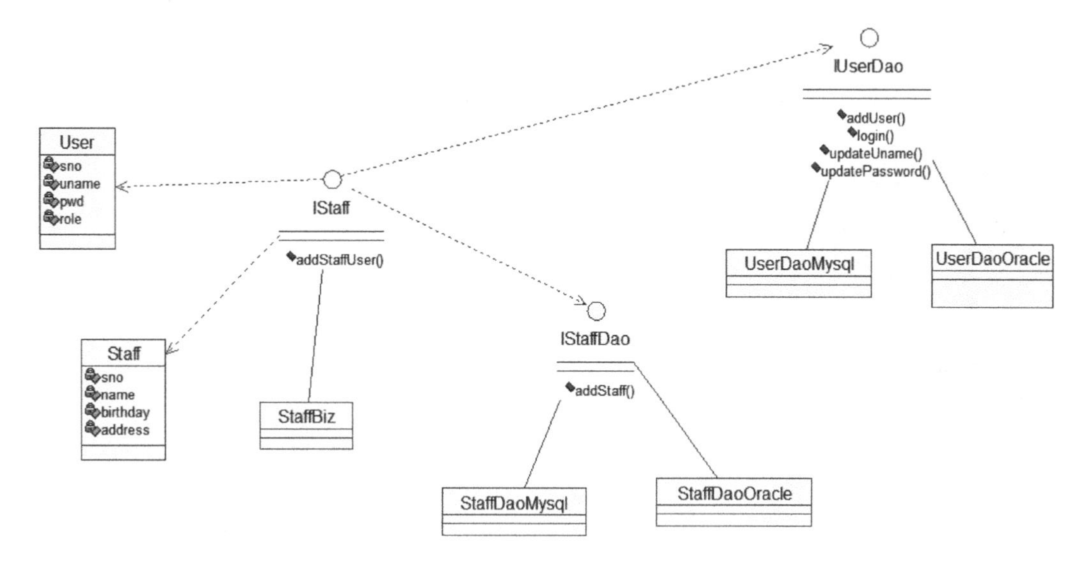

图 3-7　服务层对实体的依赖

```
public interface IStaff {
    // 添加员工信息时，系统自动创建一个默认用户
    public void addStaffUser(Staff staff, User user) throws Exception;
}
```

在 IStaff:: addStaffUser( )中，IStaff 对 User 和 Staff 实体产生了依赖，这是一种弱依赖关系。

下面测试一下 IStaff 与 User 和 Staff 的依赖关系。

（1）Staff 和 User 必须要配置成 Bean。

```
<bean id="staffBiz" class="com.icss.biz.impl.StaffBiz"
                init-method="init" destroy-method="destroy" >
</bean>
<bean id="user" class="com.icss.entity.User"
          init-method="init" destroy-method="destroy">
</bean>
<bean id="staff" class="com.icss.entity.Staff"
              init-method="init" destroy-method="destroy">
</bean>
```

（2）测试。

① 未配置依赖关系前，测试代码如下。

```
 public static void main(String[] args) {
    ApplicationContext app
                = new ClassPathXmlApplicationContext("beans.xml");
    IUser u = (IUser)app.getBean("userBiz");
    try {
        u.login("admin", "123");
```

```
        } catch (Exception e) {
            System.out.println(e.getMessage());
        }
        //关闭 IoC 容器
        ((ClassPathXmlApplicationContext) app).close();
    }
```

② 测试结果如下。

```
信息: Loading XML bean definitions from class path resource [beans.xml]
StaffBiz 构造...
StaffBiz 初始化....
User 构造...
User 初始化....
Staff 构造...
Staff 初始化....
用户登录....
org.springframework.context.support.AbstractApplicationContext doClose
Staff 析构....
User 析构....
StaffBiz 析构....
```

由测试结果可以看出，StaffBiz 先于 User 和 Staff 对象构造成功，后于 Staff 和 User 对象释放，即它们之间不存在生命周期的依赖关系。

（3）配置依赖关系。

```
<bean id="staffBiz" class="com.icss.biz.impl.StaffBiz"
    init-method="init" destroy-method="destroy" depends-on="User,Staff" >
</bean>
```

增加依赖配置后，执行结果如下。

```
信息: Loading XML bean definitions from class path resource [beans.xml]
User 构造...
User 初始化....
Staff 构造...
Staff 初始化....
StaffBiz 构造...
StaffBiz 初始化....
用户登录....
org.springframework.context.support.AbstractApplicationContext doClose
StaffBiz 析构....
Staff 析构....
User 析构....
```

总结：*在弱依赖关系中，配置 depends-on 属性后，先创建 User 与 Staff，后创建 StaffBiz；先释放 StaffBiz，后释放 Staff 和 User。*

5. 延迟初始化

ApplicationContext 实现的默认行为是在其启动过程中，将所有单例 Bean 提前进行实例化，把提前实例化 Bean 作为容器初始化过程的一部分。

ApplicationContext 实例创建并配置所有的单例 Bean，通常这是一件好事，因为这样在配置中的任何错误都会即刻被发现（否则，可能要花几小时甚至几天）。有时候这种默认处理可能并不是你想要的。如果你不想让一个单例 Bean 在容器初始化之前实例化，那么可以将 Bean 设置为延迟实例化。一个延迟初始化配置将告诉 IoC 容器是在启动时还是在第一次用到时实例化 Bean。

示例代码如下。

```
<bean id="lazy" class="com.foo.ExpensiveToCreateBean" lazy-init="true"/>
<bean name="not.lazy" class="com.foo.AnotherBean"/>
```

系统默认配置是 lazy-init="false"。如果 lazy-init="true"，则当第一次调用 Bean 对象时，才进行实例。

为了观察 StaffUser 系统中 Bean 的创建时间，操作步骤如下。

（1）系统默认配置是 lazy-init="false"，在容器启动时，就创建所有单例 Bean。

```
<bean id="userDao" class="com.icss.dao.impl.UserDaoMysql"
                 init-method="init" />
<bean id="userBiz" class="com.icss.biz.impl.UserBiz"
                 init-method="init"  lazy-init="false" >
    <property name="userDao" ref="userDao"></property>
</bean>
```

（2）测试。

① 测试代码如下。

```
public static void main(String[] args) {
    ApplicationContext app = new ClassPathXmlApplicationContext("beans.xml");
    System.out.println("IoC 容器创建完毕.........");
    System.out.println("开始 getBean 对象...");
    IUser u = (IUser)app.getBean("userBiz");
    try {
        User user = u.login("admin", "123");
        if(user != null) {
            System.out.println("登录成功，身份是" + user.getRole());
        }else {
            System.out.println("登录失败");
        }
    } catch (Exception e) {
        System.out.println(e.getMessage());
    }
}
```

② 测试结果如下。

```
信息: Loading XML bean definitions from class path resource [beans.xml]
UserDaoMysql 构造....
UserDaoMysql 初始化....
UserBiz 构造....
UserBiz 初始化....
IoC 容器创建完毕.........
开始 getBean 对象...
UserDaoMysql....login....
登录成功，身份是 1
```

测试结果显示，创建完容器之前，UserDaoMysql 和 UserBiz 的构造与初始化完成，即没有延迟初始化。

（3）增加配置 lazy-init="true"。

```
<bean id="userDao" class="com.icss.dao.impl.UserDaoMysql"
                init-method="init" lazy-init="true"/>
<bean id="userBiz" class="com.icss.biz.impl.UserBiz"
                init-method="init"  lazy-init="true" >
    <property name="userDao" ref="userDao"></property>
</bean>
```

（4）观察修改配置后的运行结果。

```
信息: Loading XML bean definitions from class path resource [beans.xml]
IoC 容器创建完毕.........
开始 getBean 对象...
UserBiz 构造....
UserDaoMysql 构造....
UserDaoMysql 初始化....
UserBiz 初始化....
UserDaoMysql....login....
登录成功，身份是 1
```

总结：如果设置了 lazy-init="true"，对于所有单例 Bean，不是在容器启动时创建 Bean 对象，而是在第一次调用 getBean( )时实例化。

3.6.7 通过 Autowire 注入

Autowire 注入模式是 Spring 企业级开发中用途最广的 DI 模式之一。相对于 XML 配置模式，使用 Autowire 注入，代码量少，灵活性高。虽然这个模式也存在一些不便之处，但这并不妨碍 Autowire 注入模式的广泛应用。

1. 通过 Autowire 注入 StaffUser 持久层对象

操作步骤如下。

（1）导入包并支持 AOP 包，见图 3-8。

Referenced Libraries
- com.springsource.org.apache.commons.logging-1.1.1.jar
- spring-beans-4.3.19.RELEASE.jar
- spring-context-4.3.19.RELEASE.jar
- spring-context-support-4.3.19.RELEASE.jar
- spring-core-4.3.19.RELEASE.jar
- spring-expression-4.3.19.RELEASE.jar
- spring-aop-4.3.19.RELEASE.jar

图 3-8　导入包并支持 AOP 包

（2）配置组件扫描位置。Spring 会扫描查找配置的包和它的所有子包是否包含组件注解。

```
<?xml version="1.0" encoding="UTF-8"?>
<beans xmlns="http://www.springframework.org/schema/beans"
    xmlns:xsi="http://www.w3.org/2001/XMLSchema-instance"
    xmlns:context="http://www.springframework.org/schema/context"
    xsi:schemaLocation="http://www.springframework.org/schema/beans
    http://www.springframework.org/schema/beans/spring-beans.xsd
    http://www.springframework.org/schema/context
    http://www.springframework.org/schema/context/spring-context.xsd">
        <context:component-scan base-package="com.icss.biz"/>
        <context:component-scan base-package="com.icss.dao"/>
 </beans>
```

（3）配置组件注解。@Component 是通用组件注解，@Service 是服务层组件注解，@Repository 是持久层组件注解。

```
@Repository("staffDao")
public class StaffDaoMysql implements IStaffDao{}
@Repository("userDao")
public class UserDaoMysql implements IUserDao {}
@Service("staffBiz")
public class StaffBiz implements IStaff{}
@Service("userBiz")
public class UserBiz implements IUser {}
```

（4）通过 Autowire 注入持久层对象。使用@Autowired 注解，在当前环境中注入依赖对象的引用。

```
@Service("userBiz")
public class UserBiz implements IUser {
    @Autowired
    private IUserDao userDao;
}
@Service("staffBiz")
public class StaffBiz implements IStaff{
    @Autowired
    private IUserDao userDao;
    @Autowired
    private IStaffDao staffDao;
}
```

（5）如果持久层的两个类使用相同名字，则程序报错。

```
@Repository("userDao")
public class UserDaoMysql implements IUserDao {}
@Repository("userDao")
public class UserDaoOracle implements IUserDao{}
```

错误消息如下。

```
Caused by:
org.springframework.context.annotation.ConflictingBeanDefinitionException: Annotation-specified bean name 'userDao' for bean class [com.icss.dao.impl.UserDaoOracle] conflicts with existing, non-compatible bean definition of same name and class [com.icss.dao.impl.UserDaoMysql]
```

为了解决问题，在上面的两个 Bean（UserDaoMysql 与 UserDaoOracle）中选一个，使它使用@Repository ("userDao")，或者使 Bean 的命名不要相同。

2. Autowire 注入模式

1）Autowire 注入模式

表 3-2 展示了 Autowire 注入模式，默认模式为 no。

表 3-2 Autowire 注入模式

Autowire 注入模式	说明
no	默认模式，没有自动装配，对于 Bean 之间的引用，必须定义 ref。对于大型业务系统，不建议改变默认配置
byName	通过属性名进行自动适配，Spring 按照属性名查找所有的 Bean 定义
byType	通过属性类型进行自动适配。如果容器中存在一个与指定属性类型相同的 Bean，那么将与该属性自动适配；否则，抛出异常
constructor	类似于 byType，但是应用于构造函数参数。如果在容器中没有找到与构造函数参数类型一致的 Bean，那么将会抛出异常

2）byType 优先

在使用@Autowired 注入依赖对象时，优先使用 byType 模式。

具体示例如下。

（1）定义持久层对象。

```
@Repository("userDao")
public class UserDaoMysql implements IUserDao {}
```

（2）在服务层注入 IUserDao，如果注入对象的名字为 userDao，与@Repository("userDao")中的命名相同，则测试正常运行。

```
@Service("userBiz")
public class UserBiz implements IUser {
    @Autowired
    private IUserDao userDao;
}
```

（3）修改注入对象的名字为 userDao2，然后测试。

```
@Service("userBiz")
public class UserBiz implements IUser {
    @Autowired
    private IUserDao userDao2;
}
```

测试结果未报错。

```
 信息: Loading XML bean definitions from class path resource [beans.xml]
UserDaoMysql....login....
登录成功，身份是 1
```

总结：通过测试发现，对在服务层注入的 IUserDao 对应的对象名无要求，可以任意选择，因为优先使用 byType 进行类型匹配。

案例 3-2：使用 byType 测试

操作步骤如下。

（1）UserDaoMysql 和 UserDaoOracle 使用不同名字的注解测试。

```
@Repository("userDao2")
public class UserDaoMysql implements IUserDao {}
@Repository("userDao3")
public class UserDaoOracle implements IUserDao{}
```

（2）在业务类中同时注入两个 IUserDao 对象。

```
@Service("userBiz")
public class UserBiz implements IUser {
    @Autowired
    private IUserDao userDao2;
    @Autowired
    private IUserDao userDao3;
}
```

运行结果报错。

```
Caused by: org.springframework.beans.factory.NoUniqueBeanDefinitionException: No qualifying
bean of type 'com.icss.dao.IUserDao' available: expected single matching bean but found 2:
userDao2,userDao3
```

总结：因为优先使用 byType，Spring 无法区分 IUser 的两个实现对象，所以系统指出期望唯一匹配的 Bean 对象，但是发现了两个 Bean 对象——userDao2 与 userDao3。

（3）删除 UserDaoOracle 的注解，只保留 userDao2。

```
@Repository("userDao2")
public class UserDaoMysql implements IUserDao {}
```

（4）服务层的注入不变。

```
@Service("userBiz")
public class UserBiz implements IUser {
    @Autowired
    private IUserDao userDao2;
    @Autowired
    private IUserDao userDao3;
}
```

测试代码如下（服务层 UserBiz 中输出 userDao2 与 userDao3 的哈希值）。

```
user = userDao2.login(sno, pwd);
System.out.println("userDao2 哈希值：" + userDao2.hashCode());
System.out.println("userDao3 哈希值：" + userDao3.hashCode());
```

测试结果如下。

```
信息：Loading XML bean definitions from class path resource [beans.xml]
UserDaoMysql....login....
userDao2 哈希值：25022727
userDao3 哈希值：25022727
登录成功，身份是 1
```

总结：通过 byType 模式注入，userDao2 和 userDao3 的哈希值相同，它们是同一对象的两个不同引用。

3）使用 byName 测试

操作步骤如下。

（1）IUserDao 的两个实现类都设置了名字。

```
@Repository("userDaoMysql")
public class UserDaoMysql implements IUserDao {}
@Repository("userDaoOracle")
public class UserDaoOracle implements IUserDao{}
```

（2）为了防止注入错误，要用@Qualifier 指明使用哪一个实现类。

```
@Service("userBiz")
public class UserBiz implements IUser {
    @Autowired
    @Qualifier("userDaoMysql")
    private IUserDao userDao;
}
@Service("staffBiz")
public class StaffBiz implements IStaff{
    @Autowired
    @Qualifier("userDaoMysql")
    private IUserDao userDao;
    @Autowired
    private IStaffDao staffDao;
}
```

总结：当存在同一接口的多个实现类注入时，byType 无法区分使用哪个实现类，因此需要用 @Qualifier 调用 byName 模式。

4）Autowire 注入模式的缺陷

Autowire 注入模式的不足如下。

（1）对于 Java 基本类型和 String 等简单类型，无法使用 Autowire 注入模式。官方解释如下。

You cannot autowire so-called simple properties such as primitives, Strings, and Classes (and arrays of such simple properties).

（2）当业务变化引发注入的配置项必须改变时，XML 配置的修改不容易。例如，原来使用的是 MySQL 数据库，现在要切换到 Oracle 库，要修改很多注解。

```
@Repository("userDaoMysql")
public class UserDaoMysql implements IUserDao {}
```

（3）同时使用同一接口的多个实现类容易引发冲突，如以下代码所示。

```
@Repository("userDaoMysql")
public class UserDaoMysql implements IUserDao {}
@Repository("userDaoOracle")
public class UserDaoOracle implements IUserDao{}
@Service("userBiz")
public class UserBiz implements IUser {
    @Autowired
    private IUserDao userDao;
}
```

（4）对于那些根据 Spring 配置文件生成文档的工具来说，Autowire 注入模式将会使这些工具无法生成依赖信息。

3.6.8 方法注入

在绝大多数业务场景下，IoC 容器中的 Bean 是单例模式。一个单例 Bean 通常依赖其他单例 Bean，或者一个非单例 Bean 依赖其他非单例 Bean。

如果依赖与被依赖的 Bean 对象的生命周期不一致怎么办？例如，如果单例 Bean A 依赖非单例 Bean B，会出什么问题？

IoC 容器只创建一次 Bean A，然后设置它的依赖对象 Bean B。以后在 Bean A 的方法调用中，Bean A 对于 Bean B 的依赖不会发生变化，即始终使用同一个 Bean B 对象。

1. userBiz 与 userDao 生命周期不一致的问题

测试 3-1：UserDaoMysql 为 prototype 模式，测试调用 getBean( )的效果。

（1）把 serDaoMysql 配置成非单例模式。

```
@Repository("userDao")
@Scope("prototype")
public class UserDaoMysql implements IUserDao {
    public UserDaoMysql() {
        System.out.println("UserDaoMysql 构造....");
    }
}
```

（2）多次调用 getBean("userDao")。

```
public static void main(String[] args) {
    for(int i=0;i<3;i++) {
        BeanFactory.getBean("userDao");
    }
}
```

运行结果如下。

```
信息: Loading XML bean definitions from class path resource [beans.xml]
UserBiz 构造...
UserDaoMysql 构造....
UserDaoMysql 构造....
UserDaoMysql 构造....
UserDaoMysql 构造....
```

结果显示，每次 getBean("userDao")都会创建新对象。调用了 getBean("userDao") 3 次，共输出了 4 次构造函数，说明创建 IoC 容器时也调用了一次 UserDaoMysql 的构造函数。

测试 3-2：UserDaoMysql 为 prototype 模式，服务层为单例，服务层依赖 userDao，测试调用 login()方法的效果。

（1）在服务层注入 userDao。服务层对象 userBiz 默认为单例模式。

```
@Service("userBiz")
public class UserBiz implements IUser {
    @Autowired
    private IUserDao userDao;
    public UserBiz() {
        System.out.println("UserBiz 构造...");
    }
    public User login(String sno, String pwd) throws Exception {
        User user;
        try {
            user = userDao.login(sno, pwd);
        }
```

```
        finally {
            // 释放数据库
        }
        return user;
    }
}
```

（2）多次调用 userBiz 的 login( )方法。

```
public static void main(String[] args) {
    for(int i=0;i<3;i++) {
        IUser u = (IUser)BeanFactory.getBean("userBiz");
        try {
            u.login("admin", "123");
        } catch (Exception e) {
        }
    }
}
```

运行结果如下（userDao 只实例化了一次）。

```
信息: Loading XML bean definitions from class path resource [beans.xml]
UserBiz 构造...
UserDaoMysql 构造....
login......
login......
login......
```

问题：若 userBiz 与 userDao 生命周期不一致，我们期望 userBiz 每一次调用 login( )时都能用新的 userDao 对象去执行，怎么办？

2. 解决方案：抛弃 DI

在 UserBiz 中注入 ApplicationContext 环境，同时需要删除 private IUserDao userDao。

```
@Service("userBiz")
public class UserBiz implements IUser ,ApplicationContextAware{
    private ApplicationContext context;
    @Override
    public void setApplicationContext(ApplicationContext arg0)
                                throws BeansException     {
        this.context = arg0;
    }
    public UserBiz() {
        System.out.println("UserBiz 构造...");
    }
}
```

不使用注入的 userDao，而采用 getBean("userDao")。

```
public User login(String sno, String pwd) throws Exception {
    User user;
    try {
        IUserDao userDao = (IUserDao)context.getBean("userDao");
        user = userDao.login(sno, pwd);
    }
```

```
        finally {
            // 释放数据库
        }
        return user;
    }
```

运行结果如下。

```
信息: Loading XML bean definitions from class path resource [beans.xml]
UserBiz 构造...
login......
UserDaoMysql 构造....
login......
UserDaoMysql 构造....
login......
UserDaoMysql 构造....
```

总结：这种方法非常简单，其实就是抛弃 Autowire 注入的对象，每次都从 IoC 容器中查找新的 userDao 对象。

3. 解决方案：重写 Lookup 方法

Lookup 方法注入的内部机制是 Spring 利用 CGLIB 库在运行时生成二进制代码，通过动态创建 Lookup 方法中 Bean 的子类而达到重写 Lookup 方法的目的。

操作步骤如下。

（1）新增如下接口。

```
@Component("lookDao")
public interface ILookDao {
    @Lookup
    public IUserDao lookUserDao();
}
```

（2）注入 lookDao，而不是 userDao。

```
@Service("userBiz")
public class UserBiz implements IUser {
    @Autowired
    private ILookDao lookDao;
    public UserBiz() {
        System.out.println("UserBiz 构造...");
    }
}
```

（3）调用业务方法。

```
IUserDao userDao = lookDao.lookUserDao();
user = userDao.login(sno, pwd);
```

（4）测试。

① 测试代码如下。

```
public static void main(String[] args) {
    for(int i=0;i<3;i++) {
        IUser u = (IUser)BeanFactory.getBean("userBiz");
        try {
            u.login("admin", "123");
        } catch (Exception e) {
        }
    }
}
```

② 测试结果如下。

```
信息: Loading XML bean definitions from class path resource [beans.xml]
UserBiz 构造...
login......
UserDaoMysql 构造....
login......
UserDaoMysql 构造....
login......
UserDaoMysql 构造....
```

总结：每次调用接口方法 lookUserDao 都会动态返回一个新的 IUserDao 对象实例。

3.6.9 依赖注入总结

1. DI 与 IoC 的关系

有人说 IoC 就是 DI，但是确切来说，IoC = 反射创建对象 + 依赖 + 注入。强依赖关系通过成员聚合建立，弱依赖关系通过 depends-on 配置。注入的概念是从 IoC 容器中查找对象，并把 Bean 对象的引用注入当前环境中。因此，依赖和注入是两个概念。

例如，服务层对象 A 依赖持久层对象 B，正常的代码创建顺序是先创建 A 对象，然后 A 在调用 B 对象的时候，创建 B 对象。IoC 的创建顺序正好相反，即先创建持久层对象 B，再创建服务层对象 A，因此说控制反转了。Bean 依赖对象的释放顺序与普通 Java 对象先统计引用计数后通过 GC 回收的顺序，也是相反的。

2. 注入方式的总结

图 3-9 总结了各种注入方式，但是不包含方法注入，因为方法注入只用于解决特殊场景下的问题。

3. 构造函数循环注入的问题

如果 Bean A 通过构造函数注入的 Bean B，同时 Bean B 需要通过构造函数注入 Bean A，会发生什么？

若 IoC 容器检测到构造函数循环注入的问题，会抛出 BeanCurrentlyInCreationException 异常。

当循环注入无法避免时，请使用 set 方法注入，不要使用构造函数注入。

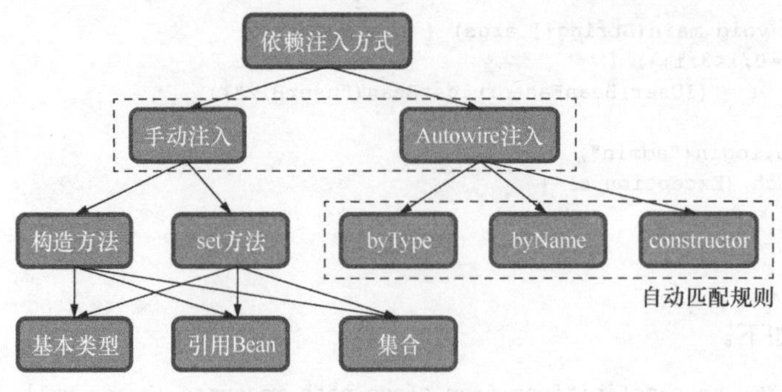

图 3-9　注入方式的总结

4. 实体类需要注入吗

实体都是有属性信息的，因此每个实体对象都有自己的特性，不能使用单例配置。

使用 EntityBean 管理 EJB 中的实体对象。每个实体 Bean 与数据库中的一条表记录形成映射关系。EntityBean 是轻易不释放的，生命周期很长。

EJB 是重量级容器，由应用服务器支持，因此可以管理有状态的实体 Bean；而 Spring 是轻量级容器，很难管理大量有状态的实体对象。

在 Spring 中可以把实体配置成 Bean，且 scope="prototype"，但是这样做的价值并不高，要根据业务场景谨慎处理。

思考：

- 如果把有属性的业务类配置成 Bean，要注意什么？
- 相对于 EJB 的会话 Bean，Spring 中 Bean 的生命周期如何管理？

3.7 Bean 对象的作用域

在 Spring 中创建一个 Bean，其实就是使用配方（recipe）创建实际的 Bean 对象。把 Bean 定义看成一个配方很有意义，这与类很相似，只要根据一张配方就可以创建多个实例。

我们不仅可以配置 Bean 的依赖关系和属性值，还可以配置 Bean 的作用域，见表 3-3。Spring Framework 支持 7 种作用域，不同的作用域适用于不同的业务层。

注意，以下 5 种作用域的适用范围。

- singleton：推荐应用于服务层和持久层对象。
- prototype：推荐应用于实体对象和有状态的业务对象。

- request：推荐应用于控制层对象。
- session：推荐应用于控制层对象。
- application：推荐应用于控制层对象。

表 3-3 Bean 的作用域

Bean 的作用域	描述
singleton	默认设置。在每个 IoC 容器中，一个 Bean 定义只能有唯一的 Bean 实例
prototype	一个 Bean 定义对应无数的 Bean 对象实例
request	一个 Bean 定义对应一个 HTTP 请求的生命周期。每个 HTTP 请求都有自己的一个 Bean 对象实例。仅在 ApplicationContext 织入（weaving）的 Web 环境下有效
session	一个 Bean 定义对应一次 HTTP 会话的生命周期。仅在 ApplicationContext 织入的 Web 环境下有效
globalSession	一个 Bean 定义对应一次全局 HTTP 会话的生命周期。通常应用于 Portlet 环境。仅在 ApplicationContext 织入的 Web 环境下有效
application	一个 Bean 定义对应一个 ServletContext。仅在 ApplicationContext 织入的 Web 环境下有效
websocket	一个 Bean 定义对应一个 WebSocket 环境。仅在 ApplicationContext 织入的 Web 环境下有效

3.7.1 配置 Bean 的作用域

使用如下两种形式均可配置 Bean 的作用域。

- 在定义 Bean 的配置文件中使用 scope。

```
<bean id="userDao" class="com.icss.dao.impl.UserDaoMysql" scope="prototype">
</bean>
```

- 在 Java 类中使用注解声明范围。

```
@Repository("userDao")
@Scope("prototype")
public class UserDaoMysql implements IUserDao {}
```

3.7.2 singleton 和 prototype 作用域

属性 scope="singleton"和 scope="prototype"可以应用于各种 ApplicationContext。它们常用于设置 Bean 的作用域。

1. singleton 作用域

在如下代码中，Bean 的默认设置是 singleton，即单例。

```
<bean id="userDao" class="com.icss.dao.impl.UserDaoMysql"></bean>
<bean id="staffDao" class="com.icss.dao.impl.StaffDaoMysql"></bean>
```

注意：在 Spring 中，Bean 的单例对应的是每个 IoC 容器里唯一的 Bean 对象，这与设计模式中的单例不同。设计模式中的单例使用静态变量存储，它是指在 JVM 范围内只有唯一的对象。

Spring 的单例 Bean 存储于缓存中，不是 JVM 中。随后对该命名 Bean 的引用都返回缓存的对象。

如图 3-10 所示，多个 Bean 引用了 accountDao，但是 accountDao 的 scope="singleton"，因此 accountDao 始终只有一个对象实例。

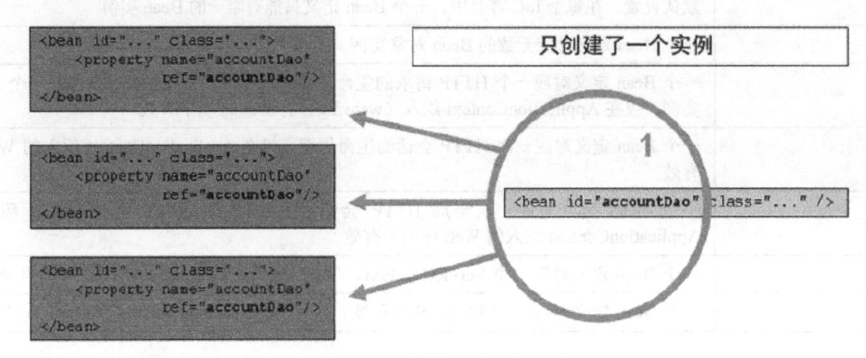

图 3-10　单例对象的引用

2．prototype 作用域

在以下代码中，当 scope="prototype"时，每次调用 getBean("userDao")都会生成一个新的对象。这与“四人组”的《设计模式》一书中 prototype 的含义基本一致。

```
<bean id="userDao"
      class="com.icss.dao.impl.UserDaoMysql" scope="prototype">
</bean>
```

如图 3-11 所示，多个 Bean 引用了 accountDao，而 accountDao 的 scope="prototype"，因此每个引用的 Bean 都可以获得不同的 accountDao 实例。

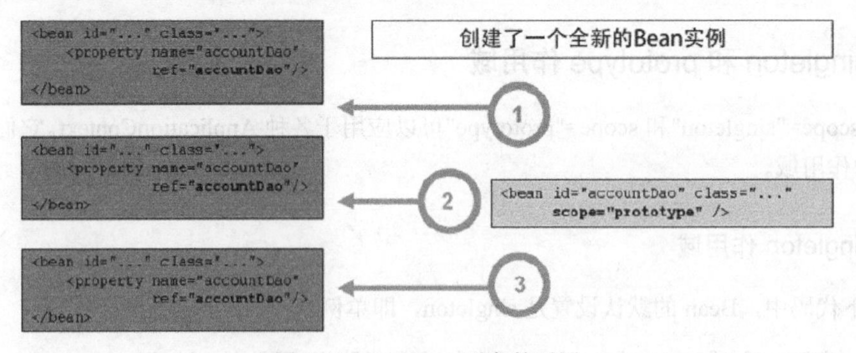

图 3-11　prototype 对象的引用

3.7.3 HelloSpringAction 示例

scope="singleton"的应用最广，也是最简单的，这里就不举例说明了。关于 scope="prototype"，我们通过实际案例来了解一下其应用场景。

1. 功能说明

在原来的 HelloSpringIoC 项目的基础上，新增一个 HelloAction 类，即允许多人同时打招呼，而且把打招呼的人和被打招呼的人都描述清楚。例如，张三说："你好，李四。"

2. 带属性的 HelloAction

编写 HelloAction 类，它有一个 pname 属性，表示打招呼的人名。

```
public class HelloAction {
    private String pname;
    private IHello helloBiz;
    public String getPname() {
        return pname;
    }
    public void setPname(String pname) {
        this.pname = pname;
    }
    public void setHelloBiz(IHello helloBiz) {
        this.helloBiz = helloBiz;
    }
    public void sayHello(String name) {
        String info = helloBiz.sayHello(name);
        System.out.println(pname + "说: " + info);
    }
}
```

配置 Bean，使用默认的 scope 属性，即 helloAction 为单例。

```
<bean id="helloBiz" class="com.icss.biz.HelloEnglish" />
<bean id="helloAction" class="com.icss.action.HelloAction" >
     <property name="helloBiz" ref="helloBiz"></property>
</bean>
```

3. 多用户并发环境测试

进行并发测试，每个线程模拟一个打招呼的人。

```
public static void main(String[] args) {
    ApplicationContext context
                    = new ClassPathXmlApplicationContext("beans.xml");
    ExecutorService pool = Executors.newCachedThreadPool();
    for(int i=0;i<50;i++) {
        pool.execute(new Runnable() {
```

```
                public void run() {
                    HelloAction  action = (HelloAction)context.getBean("helloAction");
                    action.setPname(Thread.currentThread().getName());
                    action.sayHello("tom");
                }
            });
        }
        pool.shutdown();
    }
```

因为 helloAction 是单例，在高并发环境中，后面的线程在调用 setPname 赋值时会替换前面线程的 pname 属性值，所以测试结果中有很多相同的打招呼信息。

```
信息: Loading XML bean definitions from class path resource [beans.xml]
pool-1-thread-1 说: how are you,tom
pool-1-thread-3 说: how are you,tom
pool-1-thread-2 说: how are you,tom
pool-1-thread-4 说: how are you,tom
pool-1-thread-4 说: how are you,tom
pool-1-thread-3 说: how are you,tom
pool-1-thread-1 说: how are you,tom
pool-1-thread-1 说: how are you,tom
pool-1-thread-6 说: how are you,tom
pool-1-thread-3 说: how are you,tom
```

修改配置文件，增加 scope="prototype"。

```
<bean id="helloBiz" class="com.icss.biz.HelloEnglish" />
  <bean id="helloAction" class="com.icss.action.HelloAction" scope="prototype" >
       <property name="helloBiz" ref="helloBiz"></property>
  </bean>
```

测试结果不再出现重复信息，具体如下。

```
信息: Loading XML bean definitions from class path resource [beans.xml]
pool-1-thread-34 说: how are you,tom
pool-1-thread-7 说: how are you,tom
pool-1-thread-15 说: how are you,tom
pool-1-thread-18 说: how are you,tom
pool-1-thread-31 说: how are you,tom
pool-1-thread-38 说: how are you,tom
pool-1-thread-22 说: how are you,tom
pool-1-thread-19 说: how are you,tom
pool-1-thread-10 说: how are you,tom
pool-1-thread-9 说: how are you,tom
pool-1-thread-27 说: how are you,tom
pool-1-thread-26 说: how are you,tom
```

总结：在 Struts 框架中，使用带属性的 Action 类是最常见的场景，为了使 Action 类可以正常工作，其 scope 属性必须配置成 prototype。带属性的业务对象可以称为有状态的 Bean，这些对象也应该配置成 scope="prototype"。另外，实体 Bean 也是有属性的，也要配置 scope="prototype"。

在如下示例中，userDao 为 UserBiz 的属性对象，它们的 scope 属性如何配置最合理？（注意，此处的 userDao 为 UserBiz 的属性，但是与前面不同的是它为引用。）

```
@Service("userBiz")
public class UserBiz implements IUser {
    @Autowired
    private IUserDao userDao;
}
```

3.7.4 Bean 的 Web 应用

request、session、globalSession、application 和 websocket 这几个作用域都只能应用在具有 Web 环境的 ApplicationContext（如 XmlWebApplicationContext）中。

如果在非 Web 环境（如 ClassPathXmlApplicationContext 环境）中使用上述作用域，就会抛出 IllegalStateException 异常。

request 作用域与 HTTP 的每一次请求绑定。在配置文件中，可以用 scope="request"来声明 Bean 的作用域为 request，还可以使用@RequestScope 注解声明。

```
<bean id="loginAction" class="com.foo.LoginAction" scope="request"/>
@RequestScope
@Component
public class LoginAction {
    // ...
}
```

session 作用域与一次 HTTP 会话绑定，它也有两种声明方式。

```
<bean id="userPreferences" class="com.foo.UserPreferences" scope="session"/>
@SessionScope
@Component
public class UserPreferences {
    // ...
}
```

globalSession 作用域与 session 作用域类似，但前者只能应用于 Portlet 环境。

```
<bean id="userPreferences" class="com.foo.UserPreferences" scope="globalSession"/>
```

application 作用域对应 ServletContext，这与 JSP 内置对象 application 类似。

```
<bean id="appPreferences" class="com.foo.AppPreferences" scope="application"/>
@ApplicationScope
@Component
public class AppPreferences {
    // ...
}
```

总结：结合 Bean 与 Web 环境，设置各种不同的生命周期，可以让开发人员更高效地组织项目，提高程序的灵活性。

思考：

- 对于一个带状态属性的 LoginAction，其 scope 属性是设置为 request 好还是 prototype 好呢？
- 对于一个购物车 Action，其 scope 属性是设置为 session 好还是 singleton 好呢？
- 一个全局计数器是直接存储在 ServletContext 中好还是设置一个 scope="application"的 Bean 好呢？

3.7.5 Bean 的依赖

使用注入方法的方式，可以解决 Bean 生命周期不一致的依赖问题。现在 Bean 的作用域更多了，尤其是在 Web 环境下，如何更好地解决依赖范围不一致的问题？答案是使用<aop:scoped-proxy />。

示例代码如下。

```
<bean id="userPreferences" class="com.foo.UserPreferences" scope="session">
      <aop:scoped-proxy/>
</bean>
<bean id="userService" class="com.foo.SimpleUserService">
    <property name="userPreferences" ref="userPreferences"/>
</bean>
```

userService 的 scope 属性是 singleton，如果不使用<aop:scoped-proxy/>，每次使用 userService 中注入的 userPreferences 对象，则 userPreferences 对象也是唯一的，这与一个会话有一个 userPreferences 对象的期望不符。

设置<aop:scoped-proxy/>后，userService 持有的是 userPreferences 的代理对象引用，不是实际对象的引用。每次调用，代理对象都会按照会话的识别方式，判断是否需要创建新的 userPreferences 对象。

注意：<aop:scoped-proxy/>不能和作用域为 singleton 或 prototype 的 Bean 一起使用。为作用域是 singleton 的 Bean 创建一个作用域代理将抛出 BeanCreationException 异常。

3.7.6 JavaBean 的属性范围

前面讲了 Spring 中的 Bean 有 7 种作用域，本节介绍 JavaBean 的作用域。JavaBean 可以简单划分为 Java EE Web 服务器使用的普通 Bean 和 App 服务器使用的 EJB。EJB 分为 3 种，分别是会话 Bean、实体 Bean 和消息 Bean。EJB 的生命周期各不相同，相关文档较多，这里就不叙述了。

这里简单介绍一下 Web 服务器使用的普通 JavaBean 的生命周期。

- @RequestScoped：与一次 HTTP 请求对应。

- @SessionScoped：与一次 HTTP 会话对应。
- @ApplicationScoped：与 ServletContext 对应。
- @ConversationScoped：在每次的 Servlet 请求期间激活，与传统的会话类似，可以跨越 HTTP 请求存在；它在浏览器的页签间共享状态信息。
- @Singleton：单例模式。

总结：我们的开发平台是 Java EE 7，在企业开发中，Spring 的 Bean 与 Java EE 的 Bean 可能同时存在。如何协调 IoC 容器与 Java EE 容器的关系，以及 Spring 的 Bean 与 Java EE 的 Bean 之间的关系，是一个重要的课题。

3.8 定制 Bean 的特性信息

使用初始化方法、析构方法、注册钩子函数等可以定制 Bean 的特性信息。

3.8.1 处理 Bean 的生命周期回调

实现 InitializingBean 和 DisposableBean 接口，允许 IoC 容器管理 Bean 的生命周期。容器会调用前一个接口的 afterPropertiesSet( )方法和后一个接口的 destroy( )方法，实现 Bean 的初始化和析构。

使用 JSR-250 中的注解@PostConstruct 和 @PreDestroy 是处理 Bean 的生命周期回调的最好方式，使用 Spring 的 InitializingBean 和 DisposableBean 接口也可以。使用接口的弊端是回调管理与 Spring 的代码产生了耦合，带来了不必要的麻烦。

Spring 在内部使用 BeanPostProcessor 实现来处理它能找到的任何标志接口并调用相应的方法。如果需要自定义特性或者生命周期行为，则可以实现自己的 BeanPostProcessor。

1. Bean 的初始化和析构

1）使用 init-method 属性和 destroy-method 属性

在 Spring 的 Bean 配置文件中，使用 init-method 属性与 destroy-method 属性可以快速地设置 Bean 的初始化和析构方法。

配置 UserBiz 的 init-method 属性和 destroy-method 属性。

```
<bean id="userBiz" class="com.icss.biz.impl.UserBiz"
            init-method="init" destroy-method="destroy"></bean>
```

编写初始化和析构方法。

```
public class UserBiz implements IUser{
    public UserBiz() {
```

```
        System.out.println("UserBiz 构造...");
    }
    public void destroy() {
        System.out.println("UserBiz 析构 ...");
    }
    public void init() {
        System.out.println("UserBiz 初始化...");
    }
}
```

测试代码如下。

```
public static void main(String[] args) {
    ClassPathXmlApplicationContext context
            = new ClassPathXmlApplicationContext("beans.xml");
    System.out.println("run....");
    context.close();
}
```

测试结果如下。

```
信息: Loading XML bean definitions from class path resource [beans.xml]
UserBiz 构造...
UserBiz 初始化...
run....
org.springframework.context.support.AbstractApplicationContext doClose
org.springframework.context.support.ClassPathXmlApplicationContext
UserBiz 析构 ...
```

2）使用@PostConstruct 和@PreDestroy

无须配置文件，可以直接使用@PostConstruct 和@PreDestroy 实现 Bean 的初始化与析构。示例代码如下。

```
@Service
public class UserBiz implements IUser,InitializingBean,DisposableBean {
    public UserBiz() {
        System.out.println("UserBiz 构造...");
    }
    @PreDestroy
    public void destroy() {
        System.out.println("UserBiz 析构 ...");
    }
    @PostConstruct
    public void init() {
        System.out.println("UserBiz 初始化...");
    }
}
```

3）使用 InitializingBean 和 DisposableBean

使用接口 InitializingBean 和 DisposableBean，也可以进行 Bean 的初始化和析构。这种模式对

Spring 的代码产生了耦合，不推荐使用。示例代码如下。

```
@Service
public class UserBiz implements IUser,InitializingBean,DisposableBean {
    public UserBiz() {
        System.out.println("UserBiz 构造...");
    }
    @Override
    public void destroy() {
        System.out.println("UserBiz 析构 ...");
    }
    @Override
    public void afterPropertiesSet() throws Exception {
        System.out.println("UserBiz 初始化...");
    }
}
```

2. JavaBean 的初始化和析构

前面讲了 Spring 中 Bean 的初始化和析构，JavaBean 也存在类似的处理。通过对比二者，理解会更深入。参考一下 Servlet 的设计。

```
package javax.servlet;
public interface Servlet {
    public void init(ServletConfig config) throws ServletException;
    public ServletConfig getServletConfig();
    public void destroy();
    public void service(ServletRequest req, ServletResponse res)
                        throws ServletException, IOException;
}
```

Servlet 是中量级组件，EJB 是重量级组件。这些组件都有初始化和析构方法。

简单信息（如属性的初始值）一般在构造函数中进行初始化。复杂信息、需要消耗较多时间的处理、可能出异常的处理，一般建议在初始化方法中指定。如 Servlet 可以在初始化方法中读取启动参数，但在构造函数中无法读取。EJB 作为重量级组件，其初始化和析构都比较复杂，这里不做细致介绍了。

非托管资源的释放主要在析构方法中完成。

案例 3-3：测试数据库连接的有效性

编写一个 Bean 管理数据库，在系统启动时校验数据库连接是否有效。如果配置错误，则系统及时提醒。由于测试数据连接有效性的操作耗时较多，在构造函数中处理显然不合适，因此放在 Bean 的初始化方法中。

（1）配置。配置信息如下。

```
    <context:component-scan base-package="com.icss.biz"/>
    <context:component-scan base-package="com.icss.dao"/>
    <bean id="dbFactory" class="com.icss.util.DbFactory">
      <property name="driver" value="com.mysql.cj.jdbc.Driver"></property>
      <property name="url"
              value="jdbc:mysql://localhost:3306/staff?useSSL=false">
    </property>
      <property name="username" value="root"></property>
      <property name="password" value="123456"></property>
  </bean>
```

（2）测试数据库连接。这里使用@PostConstruct 设置初始化方法。

```
public class DbFactory {
    private String driver;
    private String url;
    private String username;
    private String password;
    @PostConstruct
    public void init() {
        try {
            Class.forName(this.driver);
            DriverManager.getConnection(url, username, password);
            System.out.println("数据库配置信息正确................");
        }catch(ClassNotFoundException e) {
            System.out.println("...加载数据库驱动错误，请检查........");
        } catch (SQLException e) {
            System.out.println("...数据库连接失败，请检查..........");
        }
    }
}
```

（3）测试。

① 测试代码如下。

```
public static void main(String[] args) {
    ClassPathXmlApplicationContext context
              = new ClassPathXmlApplicationContext("beans.xml");
    System.out.println("run....");
    context.close();
}
```

② 测试结果如下。

```
信息: Loading XML bean definitions from class path resource [beans.xml]
UserBiz 构造...
UserDaoMysql 构造....
UserBiz 初始化...
数据库配置信息正确...
run....
信息: Closing org.springframework.context.support.ClassPathXmlApplicationContext
UserBiz 析构 ...
```

3. 使用钩子关闭 IoC 容器

在非 Web 环境下，使用钩子可以优雅地关闭 IoC 容器。例如，对于富客户端的桌面环境，可以向 JVM 注册一个钩子。即使程序非正常退出，钩子函数也会执行，这样在钩子函数中做环境清理工作（如关闭非托管资源）就是非常有效的方法。

为了注册一个用于关闭 IoC 容器的钩子，需要调用 ConfigurableApplicationContext 接口中的 registerShutdownHook( )方法。

HelloSpringHook 示例的代码如下。

```
public static void main(String[] args) {
    Runtime.getRuntime().addShutdownHook(new Thread() {
        public void run() {
            System.out.println("通过 hook 方法清除垃圾...");
        }
    });
    try {
        Thread.sleep(2000);
    } catch (Exception e) {
    }
    ConfigurableApplicationContext context
                = new ClassPathXmlApplicationContext("beans.xml");
    context.registerShutdownHook();
    IHello hi = (IHello) context.getBean("helloBean");
    String hello = hi.sayHello("tom");
    System.out.println(hello);
    try {
        Thread.sleep(3000);
        System.exit(1);   // 非正常退出，不影响 hook
    } catch (Exception e) {
    }
    System.out.println("系统退出....");
}
```

下面分析代码。

（1）注册钩子前，先添加处理函数。

```
Runtime.getRuntime().addShutdownHook(new Thread(){
    public void run() {
        System.out.println("通过 hook 方法清除垃圾...");
    }
  });
```

（2）注册钩子函数。

```
context.registerShutdownHook();
```

（3）程序运行中，强制退出测试。

```
System.exit(1);
```

（4）非正常退出，程序运行结果如下。

```
信息: Loading XML bean definitions from class path resource [beans.xml]
how are you,tom
通过 hook 方法清除垃圾...
```

总结：程序异常退出后，异常后面的代码不再执行，但是已经注册的 hook 方法必须要调用。

3.8.2 Aware 接口

1. Aware 接口介绍

Spring 提供了很多 Aware 接口（见表 3-4），用于向 Bean 对象提供 IoC 容器下的基础环境依赖，即在 Bean 对象内获取各种环境信息。

表 3-4 Aware 接口

Aware 接口名称	依赖注入
ApplicationContextAware	声明 ApplicationContext 环境
ApplicationEventPublisherAware	注入 ApplicationEventPublisher
BeanClassLoaderAware	注入 ClassLoader，用于装载 Bean 类
BeanFactoryAware	声明 BeanFactory
BeanNameAware	声明 Bean 的名字
BootstrapContextAware	声明资源适配器 BootstrapContext
LoadTimeWeaverAware	定义织入器，用于处理类的定义
MessageSourceAware	配置消息解析策略
NotificationPublisherAware	Spring JMX 的通知发布器
PortletConfigAware	容器运行时的当前 PortletConfig，仅在 Web 环境下有效
PortletContextAware	容器运行时的当前 PortletContext，仅在 Web 环境下有效
ResourceLoaderAware	配置访问低级资源的装载器
ServletConfigAware	容器运行时的当前 ServletConfig，仅在 Web 环境下有效
ServletContextAware	容器运行时的当前 ServletContext，仅在 Web 环境下有效

2. 在 Bean 中注入环境

Bean 实现了 ApplicationContextAware 接口，即可在 Bean 运行时获得 IoC 容器的环境信息。一个 Bean 可以实现多个 Aware 接口。这是 Bean 获取 IoC 容器环境信息的非常方便的方法。

注意，只能在 Bean 中注入 IoC 环境，在普通的业务类中不可以。

通过以下代码，可以在 UserBiz 中注入多种 IoC 环境。

```
@Service
public class UserBiz implements IUser,ApplicationContextAware,
                    MessageSourceAware,ApplicationEventPublisherAware {
```

```java
    private ApplicationContext context;
    private MessageSource messageSource;
    private ApplicationEventPublisher publisher;
    @Override
    public void setApplicationContext(ApplicationContext arg0)
                                       throws BeansException {
        this.context = arg0;
    }
    @Override
    public void setMessageSource(MessageSource arg0) {
        this.messageSource = arg0;
    }
    @Override
    public void setApplicationEventPublisher(ApplicationEventPublisher arg0) {
        this.publisher = arg0;
    }
}
```

3.9 IoC 容器扩展

通常，开发人员无须通过继承 ApplicationContext 的子类来进行功能扩展。IoC 容器可以通过插入各种集成接口的实现来进行功能扩展。

3.9.1 BeanPostProcessor 接口

IoC 容器生成 Bean 对象后，在 Bean 初始化前后，可以通过 BeanPostProcessor 接口定制业务逻辑，如日志跟踪等。

配置 BeanPostProcessor 后，Bean 的使用过程如图 3-12 所示。

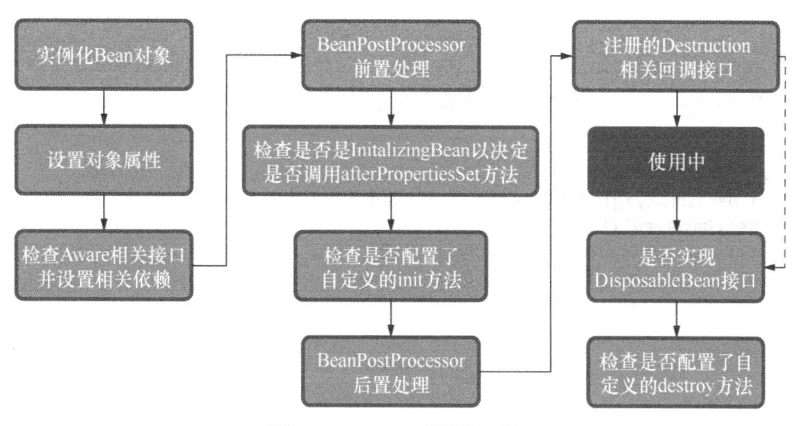

图 3-12 Bean 的使用过程

在初始化方法前，做 BeanPostProcessor 的前置处理；在初始化方法之后，做 BeanPostProcessor

的后置处理。

案例 3-4：通过日志跟踪 Bean 对象的实例

通过日志跟踪 Bean 对象的实例的操作步骤如下。

（1）通过日志跟踪所有 Bean 对象的创建过程。

```
@Component
public class LogBean implements BeanPostProcessor{
    @Override
    public Object postProcessAfterInitialization
            (Object bean, String beanName) throws BeansException {
        System.out.println("Bean: '" + beanName
                    + "' after init : " + bean.toString());
        return bean;
    }
    @Override
    public Object postProcessBeforeInitialization
            (Object bean, String beanName) throws BeansException {
        System.out.println("Bean: '" + beanName
                    + "' befor init : " + bean.toString());
        return bean;
    }
}
```

（2）配置业务 Bean。UserBiz 设置了构造方法和初始化方法，UserDao 只有构造方法。

```
@Service
public class UserBiz implements IUser {
    @Autowired
    private IUserDao userDao;
    public UserBiz() {
        System.out.println("UserBiz 构造...");
    }
    @PostConstruct
    public void init() {
        System.out.println("UserBiz 初始化...");
    }
}
@Repository("userDao")
public class UserDaoMysql implements IUserDao {
    public UserDaoMysql() {
        System.out.println("UserDaoMysql 构造...");
    }
}
```

（3）测试。

① 测试代码如下。

```
public static void main(String[] args) {
    ClassPathXmlApplicationContext context
```

```
            = new ClassPathXmlApplicationContext("beans.xml");
    System.out.println("run....");
    context.close();
}
```

② 测试结果如下。

```
信息: Loading XML bean definitions from class path resource [beans.xml]
UserBiz 构造...
UserDaoMysql 构造...
Bean: 'userBiz' befor init : com.icss.biz.impl.UserBiz@285225
UserBiz 初始化...
Bean: 'userBiz' after init : com.icss.biz.impl.UserBiz@285225
run....
org.springframework.context.support.AbstractApplicationContext doClose
```

总结：由输出日志可知，程序运行结果与流程图一致。因为 UserBiz 有初始化方法，所以在初始化前后都输出日志。

3.9.2 FactoryBean 接口

org.springframework.beans.factory.FactoryBean 接口可以作为 IoC 容器实例在逻辑上的可插入点。

FactoryBean 是对一个复杂 Bean 的包装。可以在 FactoryBean 中进行初始化，然后把初始化的值传给它包装的对象。FactoryBean 接口在 Spring Framework 框架中有大量的实现，如用于创建动态代理对象的 ProxyFactoryBean，以及 ConcurrentMapCacheFactoryBean、ConnectorServerFactoryBean、ConversionServiceFactoryBean 等。实现 FactoryBean 中的 getObject( )方法，可以返回真正需要的对象。

案例 3-5：通过 FactoryBean 包装 DbFactory

前面讲 Bean 对象的初始化时，有一个关于 DbFactory 的案例，在那个案例中测试了数据库连接的有效性。这个案例使用 Spring 的 DataSource 来测试数据库，把结果传给 DbFactory。

操作步骤如下。

（1）配置数据源。

```
<context:component-scan base-package="com.icss.biz"/>
<context:component-scan base-package="com.icss.dao"/>
<context:component-scan base-package="com.icss.util"/>
<bean id="dataSource"
      class="org.springframework.jdbc.datasource.DriverManagerDataSource">
    <property name="driverClassName" value="com.mysql.cj.jdbc.Driver" />
    <property name="url"
              value="jdbc:mysql://localhost:3306/staff?useSSL=false" />
    <property name="username" value="root" />
    <property name="password" value="123456" />
</bean>
```

（2）创建 DbFactory，在初始化时赋值。DbFactory 的数据源于 DataSource 的配置信息。

```
@Component("dbFactoryBean")
public class DbFactoryBean
        implements InitializingBean,FactoryBean<DbFactory> {
    @Autowired
    private DriverManagerDataSource dataSource;
    private DbFactory dbFactory;
    @Override
    public void afterPropertiesSet() throws Exception {
        dataSource.getConnection();          //打开数据库测试
        dbFactory = new DbFactory();
        dbFactory.setUrl(this.dataSource.getUrl());
        dbFactory.setPassword(this.dataSource.getPassword());
        dbFactory.setUsername(this.dataSource.getUsername());
    }
}
```

（3）重写 getObject( )方法。

```
public class DbFactoryBean
        implements InitializingBean,FactoryBean<DbFactory> {
    @Override
    public DbFactory getObject() throws Exception {
        return this.dbFactory;
    }
    @Override
    public Class<?> getObjectType() {
        return DbFactory.class;
    }
    @Override
    public boolean isSingleton() {
        return true;
    }
}
```

（4）测试。

① 测试代码如下。

```
public static void main(String[] args) {
    ClassPathXmlApplicationContext context
        = new ClassPathXmlApplicationContext("beans.xml");
    DbFactory db = (DbFactory)context.getBean("dbFactoryBean");
    System.out.println(db.getUsername());
    System.out.println(db.getUrl());
    context.close();
}
```

② 测试结果如下。

```
信息: Loaded JDBC driver: com.mysql.cj.jdbc.Driver
DbFactory 构造...
```

```
root
jdbc:mysql://localhost:3306/staff?useSSL=false&serverTimezone=UTC
```

分析: 下面这行代码是关键，getBean( )输入的是 dbFactoryBean，但是返回类型不是 DbFactoryBean，而是包装后的 DbFactory。

```
DbFactory db = (DbFactory)context.getBean("dbFactoryBean");
```

（5）注意，只有 DbFactoryBean 有@Component 注解，DbFactory 没有使用@Component 注解，那么，DbFactory 是不是 Spring 的 Bean 对象？为了测试，使用 LogBean 跟踪 DbFactory 的创建过程。

```
public class DbFactory {
    private String driver;
    private String url;
    private String username;
    private String password;
    public DbFactory() {
        System.out.println("DbFactory 构造...");
    }
    @PostConstruct
    public void init() {
        System.out.println("DbFactory init....");
    }
}
@Component
public class LogBean implements BeanPostProcessor{
    @Override
    public Object postProcessAfterInitialization
                (Object bean, String beanName) throws BeansException {
        System.out.println("Bean: '" + beanName
                              + "' after init : " + bean.toString());
        return bean;
    }
    @Override
    public Object postProcessBeforeInitialization
            (Object bean, String beanName) throws BeansException {
        System.out.println("Bean: '" + beanName
                              + "' befor init : " + bean.toString());
        return bean;
}
}
```

测试结果如下。

```
Bean: 'dbFactoryBean' befor init : com.icss.util.DbFactoryBean@1e4c80f
DbFactory 构造...
Bean: 'dbFactoryBean' after init : com.icss.util.DbFactoryBean@1e4c80f
```

总结：经测试发现，DbFactory 的初始化方法未执行，而且 LogBean 也没有跟踪到 DbFactory 的构造，因此可以断定，DbFactoryBean 是 Spring 的 Bean，DbFactory 不是 IoC 容器管理的 Bean。

3.10 注解配置

Spring Framework 支持很多注解，如 Spring 自定义的注解及 Spring 支持的某些 Java EE 注解。

3.10.1 与 JSR 相关的注解

JSR 属于 Java EE 规范，定义了很多注解，这些注解都需要 Java EE 容器的支持，Spring Framework 只支持其中的部分注解。

在 Spring Framework 的任何地方都支持如下注解：

- @Autowired；
- @Qualifier；
- @Resource（javax.annotation），前提是 JSR 250 存在；
- @ManagedBean（javax.annotation），前提是 JSR 250 存在；
- @Inject（javax.inject），前提是 JSR 330 存在；
- @Named（javax.inject），前提是 JSR 330 存在；
- @PersistenceContext（javax.persistence），前提是 JSR 存在；
- @PersistenceUnit（javax.persistence），前提是 JSR 存在；
- @Required；
- @Transactional。

1. JSR 330 与 Spring 对应的注解

JSR 的相关注解都定义在 javax.inject.*下面。表 3-5 罗列了功能相近的 Spring 注解与 JSR 注解，但即使是功能相近的注解，细节也是不同的。

表 3-5　功能相近的 Spring 注解与 JSR 注解

Spring 注解	JSR 注解	说明
@Autowired	@Inject	@Inject 没有 required 属性。 @Autowired 的 required 属性值默认为 true
@Component	@Named/@ManagedBean	JSR 330 没有提供组合模式，只能使用一种方法识别命名组件
@Scope("singleton")	@Singleton	JSR 330 也有@Scope，默认是 prototype，在 Spring 中默认是 singleton。因此@Singleton 无须使用，@Scope("prototype")可以用
@Qualifier	@Qualifier/@Named	javax.inject.Qualifier 仅仅是一个定制限定符的元数据。在 Spring 中可以通过 byName 方式依赖注入

续表

Spring 注解	JSR 注解	说明
@Value	—	无对应项
@Required	—	无对应项
@Lazy	—	无对应项

2. @Inject 与@Autowired

1）使用@Inject 代替@Autowired

@Inject 是 JSR 330 的注解，在使用@Autowired 的地方可以使用@Inject 代替。示例代码如下。

习惯上，我们使用 Spring 的@Autowired 注入 userDao。

```
@Service("userBiz")
public class UserBiz implements IUser {
    @Autowired
    private IUserDao userDao;
}
```

下面我们用@Inject 代替@Autowired。为此，首先需要导入 Java EE-api-7.0.jar，所有 JSR 注解都在这个包下。

```
@Service("userBiz")
public class UserBiz implements IUser {
    @Inject
    private IUserDao userDao;
}
```

测试结果如下。

```
信息：JSR-330 'javax.inject.Inject' annotation found and supported for autowiring
userBiz login......
userDao login...
tom 登录成功
```

2）required 属性

@Autowired 注解有 required 属性，@Inject 没有这个属性。required="false"表示依赖的对象可以为 null。

示例操作如下。

（1）在逻辑层注入 IUserDao 依赖。

```
@Service("userBiz")
public class UserBiz implements IUser {
    @Inject
    private IUserDao userDao;
}
```

（2）在持久层删除@Repository("userDao")注解，即没有 UserDao 这个 Bean 对象。

```
@Repository("userDao")
public class UserDaoMysql implements IUserDao {}
//删除注解
public class UserDaoMysql implements IUserDao {}
```

（3）测试后，出现如下异常。

```
Caused by:
org.springframework.beans.factory.NoSuchBeanDefinitionException:
No qualifying bean of type 'com.icss.dao.IUserDao' available: expected at least 1
bean which qualifies as autowire candidate.
```

（4）使用@Autowired 的 required 属性，把该属性设置为 false。

```
@Service("userBiz")
public class UserBiz implements IUser {
    @Autowired(required=false)
    private IUserDao userDao;
}
```

（5）required=false 表示注入的 userDao 可以为 null，因此使用 userDao 时需要判断。

```
public User login(String sno, String pwd) throws Exception {
    User user = null;
    try {
        //此处需要判断 userDao 是否为 null
        if(userDao != null) {
            user = userDao.login(sno, pwd);
        }
    } finally {
    }
    return user;
}
```

（6）测试代码如下。

```
public static void main(String[] args) {
    ApplicationContext context
            = new ClassPathXmlApplicationContext("beans.xml");
    IUser biz = (IUser)context.getBean("userBiz");
    try {
        User user = biz.login("tom", "123");
        if(user != null) {
            System.out.println("tom 登录成功");
        }else {
            System.out.println("tom 登录失败");
        }
    } catch (Exception e) {
        e.printStackTrace();
        }
}
```

（7）测试结果如下。

```
信息: Loading XML bean definitions from class path resource [beans.xml]
tom 登录失败
```

总结：@Autowired 比@Inject 增加了 required 属性，因此代码更加灵活，功能更加强大。

3）在 set 方法中使用@Autowired 注解

@Autowired 注解可以用于传统的 set 方法，示例如下。

```
public class SimpleMovieLister {
    private MovieFinder movieFinder;
    @Autowired
    public void setMovieFinder(MovieFinder movieFinder) {
        this.movieFinder = movieFinder;
    }
}
```

@Autowired 注解可以用于 set 方法，这在有些场景下是非常有用的。在如下代码中，父类为第三方已包装好的类，此处在 set 方法中用@Autowired 注解，给父类的 dataSource 属性赋值。

```
public abstract class BaseDao extends JdbcDaoSupport {
    @Autowired
    public void setDataSource(DriverManagerDataSource dataSource) {
        super.setDataSource(dataSource);
    }
}
```

3. @Name 与@ManagedBean

使用@Named 或 @ManagedBean，可以代替@Component。托管 Bean 的概念源于 Java EE 的 JSF 部分。示例代码如下。

用@Named 替代原来的@Service 和@Repository。

```
@Named("userBiz")
public class UserBiz implements IUser {
    @Autowired(required=false)
    private IUserDao userDao;
}
    @Named("userDao")
public class UserDaoMysql implements IUserDao {}
```

测试结果如下。

```
信息: JSR-330 'javax.inject.Inject' annotation found and supported for autowiring
userDao login...
tom 登录成功
```

将@Named 修改为@ManagedBean。

```
@ManagedBean("userBiz")
public class UserBiz implements IUser {
    @Autowired(required=false)
    private IUserDao userDao;
}
@ManagedBean("userDao")
public class UserDaoMysql implements IUserDao {}
```

测试结果如下。

```
信息: JSR-330 'javax.inject.Inject' annotation found and supported for autowiring
userDao login...
tom登录成功
```

总结：在 Spring 中，可以使用@Component 注解 Bean，还可以按照 Bean 所处的层进行精细注解，如@Controller 为控制层 Bean，@Service 为服务层 Bean，@Repository 为持久层 Bean。如果使用 JSR 的@Name 或@ManagedBean，所有层的 Bean 注解是相同的。很明显，Spring 的 Bean 注解更加清晰。

4. @Scope

可以直接使用 JSR 的作用域注解@RequestScoped、@SessionScoped、@Singleton、@ApplicationScoped。

如果不使用@Scope，也不使用 JSR 的作用域注解，所有 Bean 的默认作用域都是 singleton。

```
@Service("userBiz")
public class UserBiz implements IUser {
    @Inject
    private IUserDao userDao;
}
@Repository("userDao")
public class UserDaoMysql implements IUserDao {}
```

测试代码如下。

```
public static void main(String[] args) {
    ApplicationContext context
                   = new ClassPathXmlApplicationContext("beans.xml");
    for(int i=0;i<3;i++) {
        IUser biz = (IUser)context.getBean("userBiz");
        IUserDao dao = (IUserDao)context.getBean("userDao");
        System.out.println(biz.hashCode()  + "," + dao.hashCode());
    }
}
```

测试结果如下。

```
信息: Loading XML bean definitions from class path resource [beans.xml]
信息: JSR-330 'javax.inject.Inject' annotation found and supported for autowiring
7438855,20388653
7438855,20388653
7438855,20388653
```

分析：测试结果显示，多次调用 getBean( )获取的哈希对象不变，说明 userBiz 和 userDao 都是单例对象。

下面设置 userBiz 的 Scope("prototype")。

```
@Service("userBiz")
@Scope("prototype")
public class UserBiz implements IUser {
    @Autowired
    private IUserDao userDao;
}
```

测试结果如下。

```
信息: JSR-330 'javax.inject.Inject' annotation found and supported for autowiring
25694321,7381976
10296322,7381976
15658987,7381976
```

总结：测试结果显示，userBiz 使用了@Scope("prototype")注解后，每次调用 getBean( )，都获得了新的 userBiz 对象。

5. @PostConstruct 与@PreDestroy

javax.annotation.PostConstruct 和 javax.annotation.PreDestroy 这两个注解在前面已经测试过，在 Bean 的初始化和析构中，优先推荐使用这两个注解。

6. @Resource 与@Qualifier

1）@Resource

Spring 支持使用 JSR 250 @Resource 注入数据。@Resource 可以应用在属性、set 方法上，以注入数据。可以使用@Resource 代替@Inject、@Autowired。与@Autowired 相反，@Resource 默认的适配方式是 byName。第一段示例代码如下。

```
@Service("userBiz")
public class UserBiz implements IUser {
    @Resource
    private IUserDao userDao;
}
```

第二段示例代码如下。

```
@Service("userBiz")
public class UserBiz implements IUser {
    @Resource(name="userDao")
    private IUserDao userDao;
}
```

测试代码如下。

```
public static void main(String[] args) {
    ApplicationContext context
            = new ClassPathXmlApplicationContext("beans.xml");
    try {
        IUser biz = (IUser)context.getBean("userBiz");
        User user = biz.login("tom", "123");
    } catch (Exception e) {
        e.printStackTrace();
    }
}
```

测试结果如下。

```
信息: JSR-330 'javax.inject.Inject' annotation found and supported for autowiring
userBiz login...
userDao login...
```

2）通过@Resource 解决冲突

以下代码同时定义了两个 IUserDao 类型。

```
@Repository("userDaoMysql")
public class UserDaoMysql implements IUserDao {}
@Repository("userDaoOracle")
public class UserDaoOracle implements IUserDao {}
```

因此，在服务层注入持久层对象时报错。

```
@Service("userBiz")
public class UserBiz implements IUser {
    @Resource
    private IUserDao userDao;
}
Caused by: org.springframework.beans.factory.NoUniqueBeanDefinitionException: No
qualifying bean of type 'com.icss.dao.IUserDao' available: expected single matching bean
but found 2: userDaoMysql,userDaoOracle
```

为了解决上述冲突，可以使用@Resource 通过 byName 方式，指明注入哪个 userDao。

```
@Service("userBiz")
public class UserBiz implements IUser {
    @Resource(name="userDaoMysql")
    private IUserDao userDao;
}
```

3）通过@ Qualifier 解决冲突

使用@Qualifier 也可解决上面的冲突。

以下代码联合使用了@Qualifier 与@Autowired。

```
@Service("userBiz")
public class UserBiz implements IUser {
```

```
    @Autowired
    @Qualifier("userDaoMysql")
    private IUserDao userDao;
}
```

在以下代码中，在构造函数中，@Qualifier 用于指明使用哪个 Bean。注意，无须额外配置，这与通过构造函数注入不同。

```
@Service("userBiz")
public class UserBiz implements IUser {
    private IUserDao userDao;
    public UserBiz(@Qualifier("userDaoMysql") IUserDao userDao) {
        this.userDao = userDao;
    }
}
```

7. @Required

@Required 注解只能应用在 set 方法上，不能应用于属性。

对于如下设置，系统会有错误提示。

```
@Service("userBiz")
public class UserBiz implements IUser {
    @Required
    private IUserDao userDao;
}
```

如果@Required 在 set 方法中用于注入对象，那么要强制要求使用 XML 方式配置。如果使用注解方式创建 Bean 对象，则系统会报错。

注入配置属性。

```
<context:component-scan base-package="com.icss.dao"/>
<bean id="userBiz" class="com.icss.biz.impl.UserBiz">
    <property name="userDao" ref="userDao"></property>
</bean>
```

使用@Required，注入 set 方法。

```
public class UserBiz implements IUser {
    private IUserDao userDao;
    @Required
    public void setUserDao(IUserDao userDao) {
        this.userDao = userDao;
    }
}
```

总结：@Required 的使用方法过于死板，不推荐使用。

3.10.2 与 Spring 相关的注解

1．@Component

@Component 是 Bean 的通用注解。@Controller、@Service、@Repository 与@Component 等效。在不同的层，可以使用不同的注解。

- @Service：用于标注服务层 Bean。
- @Repository：用于标注持久层 Bean。
- @Controller：用于标注控制层 Bean。
- @Component：当 Bean 所属层不明确时，用这个注解。

2．@Bean 和@Configuration

@Bean 注解把方法返回的对象当成 Bean 处理。同时使用@Bean 与@Configuration 是声明 Bean 的一种常用形式。

示例操作如下。

（1）在 com.icss.util 包中定义 AppConfig。

```
@Configurable
public class AppConfig {
    @Bean("userDao")
    public IUserDao myUserDao() {
        return new UserDaoMysql();
    }
    @Bean("userBiz")
    public IUser myUserBiz() {
        return new UserBiz();
    }
}
```

（2）定义业务 Bean 的依赖。UserBiz 与 UserDaoMysql 无须使用任何注解。

```
public class UserBiz implements IUser {
    @Autowired
    private IUserDao userDao;
}
    public class UserDaoMysql implements IUserDao {}
```

（3）加载 AppConfig。

```
public static void main(String[] args) {
  ApplicationContext ctx =
        new AnnotationConfigApplicationContext(AppConfig.class);
```

```
    try {
        IUser biz = (IUser)ctx.getBean("userBiz");
        User user = biz.login("tom", "123");
    } catch (Exception e) {
        e.printStackTrace();
    }
}
```

（4）运行代码。运行结果如下。

```
org.springframework.beans.factory.annotation.AutowiredAnnotationBeanPostProcess
信息: JSR-330 'javax.inject.Inject' annotation found and supported for autowiring
userBiz login......
userDao login...
```

总之，声明 Bean 的 5 种方式如下。

- 在 XML 文件中配置<bean id="" class="" />。
- 使用@Component、@Service、@Repository、@Controller 声明 Bean。
- 同时使用@Bean 与@Configuration 声明 Bean。
- 通过@Name 声明 Bean。
- 通过@ManagedBean 声明 Bean。

3. @Primary

当使用 Autowire 方式注入对象时，可能会遇到同一个类型中存在多个对象的情况。这时使用@Primary 可以提示优先注入哪个对象。

示例代码如下。

```
@Repository("userDaoMysql")
public class UserDaoMysql implements IUserDao {}

@Repository("userDaoOracle")
public class UserDaoOracle implements IUserDao {}

@Service("userBiz")
public class UserBiz implements IUser {
    @Autowired
    private IUserDao userDao;
}
```

测试结果如下。

```
Caused by: org.springframework.beans.factory.NoUniqueBeanDefinitionException: No
qualifying bean of type 'com.icss.dao.IUserDao' available: expected single matching bean
but found 2: userDaoMysql,userDaoOracle
```

下面 3 种方法都可以解决注入对象不明确的问题。

- 使用@Primary 注解准备优先注入的对象。

```
@Repository("userDaoMysql")
@Primary
public class UserDaoMysql implements IUserDao {}
```

- 使用@Resource 标记依赖对象。

```
@Service("userBiz")
public class UserBiz implements IUser {
    @Resource(name="userDaoMysql")
    private IUserDao userDao;
}
```

- 使用@Qualifier 标记依赖对象。

```
@Service("userBiz")
public class UserBiz implements IUser {
    @Autowired
    @Qualifier("userDaoMysql")
    private IUserDao userDao;
}
```

3.11 标准事件与自定义事件

ApplicationContext 基于 Observer 模式提供了针对 Bean 的事件传播功能。通过 ApplicationContext 的 publishEvent 方法，可以将事件通知到系统内所有的 ApplicationListener。

Spring 自定义事件的注册与发布过程见图 3-13。

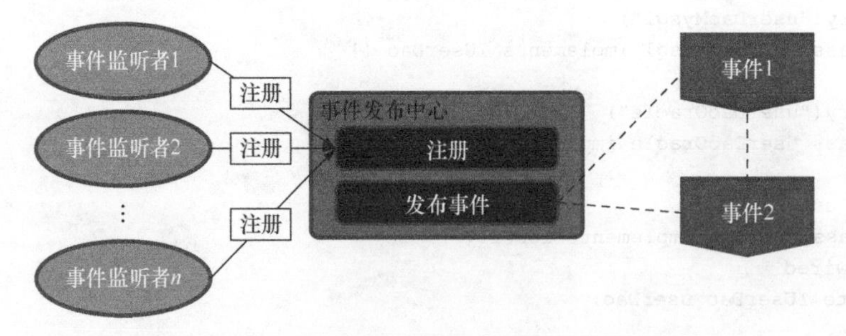

图 3-13 Spring 自定义事件的注册与发布过程

3.11.1 标准事件

Spring 标准事件见表 3-6。

表 3-6 Spring 标准事件

事件	说明
ContextRefreshedEvent	当 ApplicationContext 初始化或刷新时发送的事件。这里的初始化意味着所有的 Bean 被装载，单例被预实例化，并且 ApplicationContext 已可用
ContextStartedEvent	当容器调用 ConfigurableApplicationContext 的 Start( )方法开始/重新开启容器时触发该事件
ContextStoppedEvent	当容器调用 ConfigurableApplicationContext 的 Stop( )方法停止容器时触发该事件
ContextClosedEvent	当使用 ApplicationContext 的 close( )方法结束上下文时发送的事件。这里的结束意味着单例 Bean 被销毁
RequestHandledEvent	一个与 Web 相关的事件，告诉所有的 Bean 已经响应了一个 HTTP 请求（即在一个请求结束后会发送该事件）。注意，只有在 Spring 中使用了 DispatcherServlet 的 Web 应用才能使用这个事件

3.11.2 项目案例：打印邮件黑名单

在该项目中，当用户发送邮件时，会判断目标地址是否在黑名单中。如果目标地址在黑名单里，就发送消息通知。消息接收者把所有拉黑的消息打印出来。

操作步骤如下。

（1）定义消息。消息就是在事件中要发出的通知格式，消息类必须继承自 ApplicationEvent 下面的消息，定义邮件的目标地址和邮件的标题。

```
public class BlackListEvent extends ApplicationEvent{
    private final String address;
    private final String title;
    public String getAddress() {
        return address;
    }
    public String getTitle() {
        return title;
    }
    public BlackListEvent(Object source,String address,String title) {
        super(source);
        this.address = address;
        this.title = title;
    }
}
```

（2）通过业务对象发送消息。注入 ApplicationEventPublisher 对象，用于发送消息。此处的 EmailService 必须是 Bean。

```
public class EmailService implements ApplicationEventPublisherAware{
    private ApplicationEventPublisher publisher;
    private List<String> blackList;
    public void setBlackList(List<String> blackList) {
        this.blackList = blackList;
```

```
    }
    @Override
    public void setApplicationEventPublisher(ApplicationEventPublisher arg0) {
        this.publisher = arg0;
    }
    public void sendEmail(String address, String content,String title) {
        if (blackList.contains(address)) {
            publisher.publishEvent(new BlackListEvent(this, address,title));
            return;
        }
        System.out.println("发送邮件成功，目标：" + address
                            + ",标题：" + title + ",内容：" + content);
    }
}
```

（3）配置 Bean 和邮件列表。

```
<bean  class="com.icss.biz.BlackListNotifier"> </bean>
<bean id="emailService" class="com.icss.biz.EmailService">
   <property name="blackList">
      <list>
        <value>tom1@qq.com</value>
        <value>tom2@qq.com</value>
        <value>tom3@qq.com</value>
        <value>jack1@qq.com</value>
        <value>jack2@qq.com</value>
      </list>
   </property>
</bean>
```

（4）接收消息。实现 ApplicationListener 接口，接收消息，需要指明接收的消息类型为 BlackListEvent。此处的 BlackListNotifier 也是 Bean。

```
public class BlackListNotifier
              implements ApplicationListener<BlackListEvent>{
    public void onApplicationEvent(BlackListEvent arg0) {
              System.out.println("被拉黑的消息：标题："
                     + arg0.getTitle() + "," + arg0.getAddress()
                     + "," + new Date(arg0.getTimestamp()).toString());
    }
}
```

（5）测试。

① 测试代码如下。

```
public static void main(String[] args) {
    ApplicationContext app = new ClassPathXmlApplicationContext("beans.xml");
    EmailService e = app.getBean(EmailService.class);
    e.sendEmail("tom@sina.com","下午 3 点，全体员工在大礼堂开会", "开会通知");
    e.sendEmail("xiaohp@sina.com","下午 3 点，全体员工在大礼堂开会", "开会通知");
    e.sendEmail("tom2@qq.com","下午 3 点，全体员工在大礼堂开会", "开会通知");
}
```

② 测试结果如下。

```
信息: Loading XML bean definitions from class path resource [beans.xml]
发送邮件成功，目标：tom@sina.com,标题：开会通知,内容：下午3点，全体员工在大礼堂开会
发送邮件成功，目标：xiaohp@sina.com,标题：开会通知,内容：下午3点，全体员工在大礼堂开会
被拉黑的消息：标题：开会通知,tom2@qq.com,Mon Jan 06 09:22:43 CST 2020
```

3.11.3 项目案例：接收多类型消息

在该项目中，同一个消息接收者可以同时接收多种类型的消息。

操作步骤如下。

（1）定义消息接收者。为了接收多种消息，此处使用的消息类型为消息的抽象父类 ApplicationEvent。

```
public class BlackListNotifier
            implements ApplicationListener<ApplicationEvent>{
    @Override
    public void onApplicationEvent(ApplicationEvent event) {
        if(event instanceof BlackListEvent) {
            BlackListEvent arg0 = (BlackListEvent)event;
            System.out.println("拉黑的消息：" + arg0.getSource().toString() + ","
                    + new Date(arg0.getTimestamp()).toString()
                  + "," + arg0.getTitle() + "," + arg0.getAddress());
        }else if(event instanceof ContextRefreshedEvent) {
            ContextRefreshedEvent arg0 = (ContextRefreshedEvent)event;
            System.out.println(" 服务器刷新...." + new
                                    Date(arg0.getTimestamp()).toString() );
        }else if(event instanceof ContextClosedEvent) {
            ContextClosedEvent arg0 = (ContextClosedEvent)event;
            System.out.println(" 服务器关闭...." +
                                   new Date(arg0.getTimestamp()).toString() );
        }
    }
}
```

（2）发送消息（有自定义消息和系统消息）。

```
public static void main(String[] args) {
    ConfigurableApplicationContext app =
                    new ClassPathXmlApplicationContext("beans.xml");
    EmailService e = app.getBean(EmailService.class);
    e.sendEmail("aa@sina.com","下午3点，全体开会", "开会通知");
    e.sendEmail("tom2@qq.com","下午3点，全体开会", "开会通知");
    app.close();
}
```

（3）运行代码。运行结果如下。

```
信息: Loading XML bean definitions from class path resource [beans.xml]
服务器刷新....Mon Jan 06 09:32:42 CST 2020
```

```
发送邮件，目标：aa@sina.com,内容：下午 3 点，全体开会
拉黑的消息：com.icss.biz.EmailService@1caeb3e,Mon Jan 06 09:32:42 CST 2020,开会通知,tom2@qq.com
org.springframework.context.support.AbstractApplicationContext doClose
org.springframework.context.support.ClassPathXmlApplicationContext: startup
服务器关闭....Mon Jan 06 09:32:42 CST 2020
```

3.12 Bean 工厂

3.12.1 BeanFactory 接口

BeanFactory 接口是 Spring 组件的注册中心和配置中心。

BeanFactory 接口的实现类需要管理大量 Bean 的定义，每个 Bean 定义了唯一的 ID。

在配置 Spring Bean 对象时，最好使用 DI 方式，而不是从 BeanFactory 查找配置。

BeanFactory 的 API 描述如下。

```
org.springframework.beans.factory  Interface BeanFactory
```

已知子接口包括 ApplicationContext、AutowireCapableBeanFactory、Configurable ApplicationContext、ConfigurableBeanFactory、ConfigurableListableBeanFactory、ConfigurablePortlet ApplicationContext、ConfigurableWebApplicationContext、HierarchicalBeanFactory、ListableBean Factory 和 WebApplication Context。

BeanFactory 是访问 Spring 容器的根接口。这个接口的实例拥有很多具有唯一识别名的 Bean 对象。BeanFactory 的实现应该尽可能地支持标准 Bean 生命周期管理的接口。

```
public interface BeanFactory {
    Object getBean(String name) throws BeansException;
    <T> T getBean(Class<T> requiredType) throws BeansException;
}
```

BeanFactory 可以通过名字查找 Bean 对象，或通过 Bean 的类型查找 Bean 对象。

注意：如果 Bean 是单例模式，则直接返回 Bean 对象的引用。对于其他类型的作用域，查找其实是 Bean 对象的实例化过程。

3.12.2 HierarchicalBeanFactory 接口

HierarchicalBeanFactory 接口是 BeanFactory 的子接口，它扩展了 BeanFactory 接口，使其成为树状结构。

```
public interface HierarchicalBeanFactory extends BeanFactory {
    BeanFactory getParentBeanFactory();
}
```

HierarchicalBeanFactory 是所有 IoC 容器 ApplicationContext 的父接口，因此 IoC 容器也是树状结构，即当前容器可能存在父容器。

当使用 getBean( )查找 Bean 对象时，如果在当前容器中未找到，则马上到父工厂中查找。

3.12.3 ListableBeanFactory 接口

可以用 getBean( )的方式查找 Bean 对象。通过 ListableBeanFactory 接口，也可以使用迭代方式，找到 Bean 工厂中的所有 Bean 实例。

```
public interface ListableBeanFactory extends BeanFactory {
    int getBeanDefinitionCount();
    String[] getBeanDefinitionNames();
    String[] getBeanNamesForType(Class<?> type);
}
```

ListableBeanFactory 接口的主要方法如下。

- int getBeanDefinitionCount( )：获得 Bean 定义的数量。
- String[] getBeanDefinitionNames( )：获得所有 Bean 定义的名字。

根据类型，可能找到多个 Bean 名字。

```
String[] getBeanNamesForType(Class<?> type)
```

3.12.4 DefaultListableBeanFactory 类

DefaultListableBeanFactory 是 BeanFactory 接口的重要实现类，前一个类的定义如下。

```
public class DefaultListableBeanFactory
               extends AbstractAutowireCapableBeanFactory
               implements ConfigurableListableBeanFactory,
                          BeanDefinitionRegistry, Serializable {}
public interface ConfigurableBeanFactory
               extends HierarchicalBeanFactory, SingletonBeanRegistry {}
```

DefaultListableBeanFactory 的成员信息如下。

```
//键为依赖注入类型，值是 autowired 对象的引用
Map<Class<?>, Object> resolvableDependencies
                = new ConcurrentHashMap<Class<?>, Object>(16);
//键为 Bean 的名字
Map<String, BeanDefinition> beanDefinitionMap
                = new ConcurrentHashMap<String, BeanDefinition>(256);
//键是依赖类型，值是单例和非单例 Bean 的名字
Map<Class<?>, String[]> allBeanNamesByType
                = new ConcurrentHashMap<Class<?>, String[]>(64);
// 键是依赖类型，值是单例 Bean 的名字
Map<Class<?>, String[]> singletonBeanNamesByType
```

```
                    = new ConcurrentHashMap<Class<?>, String[]>(64);
//所有 Bean 定义的名字
List<String> beanDefinitionNames = new ArrayList<String>(256);
```

在 DefaultListableBeanFactory 中，beanDefinitionMap 存储 IoC 容器中的所有 Bean 定义。resolvableDependencies 存储了当前容器的 Bean 之间的依赖关系。allBeanNamesByType 表明多个名字可能指向同一个 Bean 类型。beanDefinitionNames 是 IoC 容器中所有 Bean 定义的名字。

3.12.5 Bean 与 BeanFactory

BeanDefinition 接口用于描述 Bean，该接口的定义如下。

```
public interface BeanDefinition
        extends AttributeAccessor, BeanMetadataElement {}
```

AbstractBeanDefinition 是 BeanDefinition 的唯一直接实现类。

```
public abstract class AbstractBeanDefinition
                extends BeanMetadataAttributeAccessor
                implements BeanDefinition, Cloneable {}
```

BeanFactory 存储的是 BeanDefinition 实例，不是 Bean 对象（参见 DefaultListableBeanFactory 的成员信息）。单例的 Bean 对象在 IoC 容器创建时就默认创建。而非单例的 Bean 对象的创建时间点不一致，可能是在调用 getBean( )时创建，也可能是在接收 HTTP 请求时创建。

3.12.6 IoC 容器与 BeanFactory

在以下代码中，ApplicationContext 接口代表 IoC 容器，BeanFactory 是 ApplicationContext 接口最重要的父接口。

```
public interface ApplicationContext
        extends EnvironmentCapable, ListableBeanFactory, HierarchicalBeanFactory,
        MessageSource, ApplicationEventPublisher, ResourcePatternResolver {}
```

通过 ApplicationContext 接口可以看到，IoC 容器在 BeanFactory 的基础上又增加了环境管理、消息管理、资源管理、事件发布等功能。BeanFactory 与 ApplicationContext 的功能对比见表 3-7。

表 3-7　BeanFactory 与 ApplicationContext 的功能对比

功能	BeanFactory	ApplicationContext
Bean 实例/织入（weave）	支持	支持
集成生命周期管理	不支持	支持
BeanPostProcessor 自动注册	不支持	支持
BeanFactoryPostProcessor 自动注册	不支持	支持
便利的消息源访问（用于国际化）	不支持	支持
ApplicationEvent 消息发布机制	不支持	支持

思考：

（1）Spring 支持 JSR 注解，如@Name 被 Spring 接管后，它成了 Spring 的 Bean，那么 Web 服务器是否还要处理呢？若没有 Spring 参与，Java EE 容器应该把它作为 JavaBean, Spring 容器与 Java EE 容器是如何协同管理 JSR 标记对象的？

（2）在高并发环境下，Spring 的有状态 Bean 是否会影响性能？

（3）IoC 的多容器管理模式会带来哪些影响？

第 4 章
SpEL

4.1 SpEL 的基本概念

在 Spring 产品中，Spring 表达式语言（Spring Expression Language，SpEL）是表达式计算的基础。SpEL 是强大的表达式语言，支持运行时查询、操纵一个对象图等功能。SpEL 的语法类似于表达式语言（Expression Language，EL），但提供了更多的功能，主要有显式方法调用和基本字符串模板函数。

同很多可用的 Java 表达式语言（如 OGNL、MVEL 和 JBoss EL）相比，SpEL 的诞生是为了给 Spring 目录中所有产品提供单种良好支持的表达式语言。其语言特性由 Spring 目录中的项目需求（包括基于 Eclipse 的 SpringSource 套件中的代码补全工具需求）驱动。也就是说，SpEL 是一个基于技术中立的 API，允许根据需要与其他表达式语言集成。

SpEL 与 Spring 不是直接绑定关系，SpEL 可以独立存在，并应用到其他平台。

SpEL 支持如下表达式。

- 基本表达式：包括字面量表达式、关系表达式、逻辑与算术运算表达式、字符串连接及截取表达式、三目运算及 Elivis 表达式、正则表达式、带括号的表达式。
- 类的相关表达式：包括类型表达式、instanceof 表达式、赋值表达式、对象属性存取与安全导航表达式，以及类实例化、变量定义与引用、对象方法调用、Bean 引用过程中的表达式。

- 集合的相关表达式：支持内联列表、内联数组、集合、列表，字典，字典访问、数组修改、集合投影、集合选择。不支持多维内联数组初始化，不支持内联字典定义。
- 其他表达式：如模板表达式。

4.2 SpEL 的基本语法

SpEL 的基本语法如下。

- XML 中使用 #{ 表达式 }。
- Bean 中使用 @Value("#{ 表达式 }")。

注意，SpEL 可以应用于 Java 代码中，还可以应用于 XML 配置中。从项目开发的角度考虑，本书中的案例以通过 XML 调用 SpEL 为主。SpEL 支持的运算符类型见表 4-1。

表 4-1　　SpEL 支持的运算符类型

运算符类型	说明
算术运算符	包括+、−、*、/、%、^
比较运算符	符号形式包括<、>、==、<=、>=，文本形式包括 lt、gt、eq、le、ge
逻辑运算符	包括 and、or、not
条件运算符	包括?:（ternary）、?:（Elvis）
正则表达式	matches
计算集合	包括[]、.?[]、.^[]、.$[]、.![]

4.2.1 算术运算符

SpEL 中的算术运算符主要有+、−、*、/、%、^。下面给出一个关于算术运算符的示例。

（1）定义两个业务 Bean。

```
public class MyCount {
    private int counter;
    private double radius;
}
public class MyTest {
    private int adjustAmount;
    private double circum;
    private double area;
}
```

（2）在 XML 配置中，使用 SpEL 进行算术运算。在 SpEL 中，使用 T( )运算符会调用类作用域中的静态方法或常量。例如，如果在 SpEL 中使用 Java 的 Math 类，就可以像下面的示例一样使用 T()运算符——T(java.lang.Math)。注意，在 XML 配置中，#{}之间不能出现空格，否则会报错。

```
<bean id="myCount" class="com.icss.s1.MyCount">
    <property name="counter" value="100"></property>
    <property name="radius" value="10"></property>
</bean>
<bean id="myTest" class="com.icss.s1.MyTest">
        <property name="adjustAmount"
                value="#{myCount.counter + 20}"></property>
        <property name="circum"
                value="#{2*T(java.lang.Math).PI * myCount.radius }">
        </property>
        <property name="area"
                value="#{myCount.counter/3.5}"></property>
</bean>
```

（3）测试。

① 测试代码如下。

```
public static void main(String[] args) {
        ApplicationContext context
                = new ClassPathXmlApplicationContext("beans.xml");
        MyTest t1 = (MyTest)context.getBean("myTest");
        System.out.println(t1.getAdjustAmount());
        System.out.println(t1.getCircum());
        System.out.println(t1.getArea());
}
```

② 测试结果如下。

```
信息: Loading XML bean definitions from class path resource [beans.xml]
120
62.83185307179586
28.571428571428573
```

4.2.2 比较运算符

SpEL 中的比较运算符包括 <、>、==、<=、>=、lt、gt、eq、le 和 ge。

下面给出一个关于比较运算符的示例。

（1）在 MyTest 中新增两个布尔值。

```
public class MyTest {
    private boolean isOK;
    private boolean isRight;
    private int adjustAmount;
    private double circum;
    private double area;
}
```

（2）使用 SpEL 比较布尔值。注意，在 XML 中，限制使用很多比较运算符，因此要使用转义符。

```
<bean id="myTest" class="com.icss.s1.MyTest">
    <property name="isOK" value="#{myCount.counter==100}"></property>
    <property name="isRight" value="#{myCount.counter lt 100}"></property>
</bean>
```

（3）测试。

① 测试代码如下。

```
public static void main(String[] args) {
    ApplicationContext context
            = new ClassPathXmlApplicationContext("beans.xml");
    MyTest t1 = (MyTest)context.getBean("myTest");
    System.out.println(t1.isOK());
    System.out.println(t1.isRight());
}
```

② 测试结果如下。

```
信息: Loading XML bean definitions from class path resource [beans.xml]
true
false
```

4.2.3 逻辑运算符

SpEL 中的逻辑运算符包括 and、or、not。下面给出一个关于逻辑运算符的示例。

（1）在 MyCount 中新增一个返回布尔值的方法 conterTest。

```
public class MyCount {
    private int counter;
    private double radius;
    public boolean counterTest() {
        if(this.counter > 100)
            return true;
        else
            return false;
    }
}
```

（2）使用 SpEL 进行逻辑运算。在 SpEL 中既可以调用对象的属性，也可以直接调用 Bean 对象的方法。

```
<bean id="myTest" class="com.icss.s1.MyTest">
    <property name="isOK"
              value="#{myCount.counter==100 and myCount.counter lt 100}">
    </property>
     <property name="isRight" value="#{not myCount.counterTest()}"></property>
</bean>
```

（3）测试代码与 4.2.2 节相同。测试结果如下。

```
信息: Loading XML bean definitions from class path resource [beans.xml]
false
true
```

4.2.4 其他运算符

其他运算符包括加号、条件运算符、正则表达式等。加号还可以用于字符串连接。

```
<constructor-arg value="perform.firstName + '' + perform.LastName">
</constructor-arg>
```

条件运算符的用法示例如下。

```
<constructor-arg value="songSelector.selectSong()=='winter'?'s1':'s2'">
</constructor-arg>
```

在 JavaScript、SpEL 等很多语言中，可以使用正则表达式。表 4-2 展示了正则表达式中使用的符号。

表 4-2　　正则表达式中使用的符号

符号	说明
/.../	代表一个模式的开始和结束
^	匹配字符串的开始
$	匹配字符串的结束
\s	代表任何空白字符
\S	代表任何非空白字符
\d	匹配一个数字字符，等价于[0-9]
\D	匹配除了数字之外的任何字符，等价于[^0-9]
\w	匹配数字、下划线或字母，等价于[A-Za-z0-9]
\W	代表任何非单字字符，等价于[^a-zA-z0-9]
.	代表除了换行符之外的任意字符
{*n*}	匹配前一项 *n* 次
{*n*,}	匹配前一项 *n* 次，或者多次
{*n*,*m*}	匹配前一项至少 *n* 次，但是不能超过 *m* 次
*	匹配前一项 0 次或多次，等价于{0,}
+	匹配前一项 1 次或多次，等价于{1,}
?	匹配前一项 0 次或 1 次，也就是说，前一项是可选的，等价于{0,1}

JavaScript 中的正则表达式示例如下。

- 要表示 6 位数字的邮政编码，可以使用/^\d{6}$/。
- 要表示 11 位电话号码，并且首位必须为 1，可以使用/^1\d{10}$/。
- 要表示 E-mail 格式，可以使用/^\w+@\w+.[a-zA-Z]{2,3}$/。

在 SpEL 中，使用 matches 调用正则表达式。

```
#{admin.email matches '[a-zA-Z0-9._%+-]+@[a-zA-Z0-9._%+-]+.com'}  <construtor-arg
 value="#{admin.email matches '[a-zA-Z0-9._%+-]+@[a-zA-Z0-9._%+-]+.com'} ">
```

下面用正则表达式校验输入值是否合格。如果正确，则返回真；否则，返回假。

```
boolean falseValue = parser.parseExpression(
        "'5.0067' matches '^-?\\d+(\\.\\d{2})?$'").getValue(Boolean.class);
```

4.3 ExpressionParser

在 Java 代码中，不仅可以使用 ExpressionParser 调用 SpEL，还可以动态调用 Bean 对象的属性和方法。

4.3.1 在代码中调用 SpEL

示例 4-1：调用字面量表达式。

具体代码如下。

```
public static void main(String[] args) {
    ExpressionParser parser = new SpelExpressionParser();
    Expression exp = parser.parseExpression("'Hello World'.concat('!')");
    String message = (String) exp.getValue();
    System.out.println(message);
}
```

其中使用反射机制，动态调用了 String 的 concat 方法，运行结果如下。

```
Hello World!
```

示例 4-2：动态调用 JDK 的 String 对象方法。

具体代码如下。

```
public static void main(String[] args) {
    ExpressionParser parser = new SpelExpressionParser();
    Expression exp =
        parser.parseExpression("new String('hello world').toUpperCase()");
    String message = exp.getValue(String.class);
    System.out.println(message);
}
```

运行结果如下。

```
HELLO WORLD
```

示例 4-3：动态调用 JDK 的 Date 对象方法。

具体代码如下。

```
public static void main(String[] args) {
    ExpressionParser parser = new SpelExpressionParser();
    Expression exp = parser.parseExpression("new java.util.Date()");
    String message = exp.getValue(String.class);
    System.out.println(message);
}
```

运行结果如下。

```
Thu Jan 16 11:44:06 CST 2020
```

4.3.2 在代码中调用 Bean 对象的属性

在 Java 代码中，使用 ExpressionParser 动态调用 Bean 对象的属性和方法。示例代码如下。

```
public class User {
    private String uname;
    private String pwd;
    private int role;
}
<bean id="user" class="com.icss.entity.User">
    <property name="uname" value="tom"></property>
    <property name="pwd" value="123456"></property>
    <property name="role" value="2"></property>
</bean>
```

使用 SpelExpressionParser 调用 Bean 对象比较麻烦。

```
ApplicationContext app = new ClassPathXmlApplicationContext("beans.xml");
    User user = (User)app.getBean(User.class);
    EvaluationContext ctx = new StandardEvaluationContext();
    ctx.setVariable("user", user);
    ExpressionParser parser = new SpelExpressionParser();
    Expression exp = parser.parseExpression("#user.getUname()");
                 //或者 parser.parseExpression("#user.uname");
    String uname = (String)exp.getValue(ctx);
    System.out.println(uname);
```

运行结果如下。

```
信息: Loading XML bean definitions from class path resource [beans.xml]
tom
```

4.4 基于 XML 的 SpEL 应用

在 XML 配置文件中，通常通过#{}调用 SpEL。操作步骤如下。

(1) 配置 Bean，给属性赋值。

```
<bean id="numberGuess" class="org.spring.samples.NumberGuess">
    <property name="randomNumber"
              value="#{ T(java.lang.Math).random() * 100.0 }" />
</bean>
<bean id="shapeGuess" class="org.spring.samples.ShapeGuess">
    <property name="initialShapeSeed"
              value="#{ numberGuess.randomNumber }" />
</bean>
<bean id="taxCalculator" class="org.spring.samples.TaxCalculator">
    <property name="defaultLocale"
             value="#{ systemProperties['user.country'] }" />
</bean>
```

(2) 编写 Bean 的类。

```
public class NumberGuess {
    private int randomNumber;
    public int getRandomNumber() {
        return randomNumber;
    }
    public void setRandomNumber(int randomNumber) {
        this.randomNumber = randomNumber;
    }
    public void doSomething() {
        System.out.println("randomNumber=" + randomNumber);
    }
}
public class ShapeGuess {
    private int initialShapeSeed;
    public int getInitialShapeSeed() {
        return initialShapeSeed;
    }
    public void setInitialShapeSeed(int initialShapeSeed) {
        this.initialShapeSeed = initialShapeSeed;
    }
    public void doSomething() {
        System.out.println("initialShapeSeed=" + initialShapeSeed);
    }
}
public class TaxCalculator {
    private String defaultLocale;
    public String getDefaultLocale() {
        return defaultLocale;
    }
    public void setDefaultLocale(String defaultLocale) {
        this.defaultLocale = defaultLocale;
```

```
    }
    public void doSomething() {
        System.out.println("defaultLocale=" + defaultLocale);
    }
}
```

（3）测试。

① 测试代码如下。

```
public static void main(String[] args) {
    ApplicationContext context = new ClassPathXmlApplicationContext("beans.xml");
    NumberGuess  ng = (NumberGuess)context.getBean("numberGuess");
    ng.doSomething();
    ShapeGuess  sg = (ShapeGuess)context.getBean("shapeGuess");
    sg.doSomething();
    TaxCalculator  tc = (TaxCalculator)context.getBean("taxCalculator");
    tc.doSomething();
    System.out.println(System.getProperty("user.country"));
}
```

② 测试结果如下。

```
信息: Loading XML bean definitions from class path resource [beans.xml]
randomNumber=14
initialShapeSeed=14
defaultLocale=CN
CN
```

4.5 通过正则表达式校验邮箱

可以使用正则表达式校验邮箱、密码、用户名、电话等的格式。下面展示一个用 SpEL 校验邮箱的示例。

校验 E-mail 的格式如下。

```
/^\w+@\w+.[a-zA-Z]{2,3}$/
#{admin.email matches '[a-zA-Z0-9._%+-]+@[a-zA-Z0-9._%+-]+.com'}
```

假设 MyCounter 中有一个邮箱，OtherBean 希望使用 MyCounter 的邮箱，但是前提是邮箱格式符合 OtherBean 的要求。

```
public class MyCounter {
    private int total;
    private String email;
}
public class OtherBean {
    private int adjustedCount;
    private String email;
}
```

使用正则表达式和三元运算符校验邮箱格式，代码如下。

```
<bean id="myCounter" class="com.icss.biz.MyCounter">
    <property name="total" value="100"></property>
    <property name="email" value="xiaohp@qq.com"></property>
</bean>
<bean id="otherBean" class="com.icss.biz.OtherBean">
    <property name="adjustedCount"
        value="#{2* T(java.lang.Math).PI * myCounter.total}"></property>
    <property name="email"
        value="#{myCounter.email matches
            '[a-zA-Z0-9._%+-]+@[a-zA-Z0-9._%+-]+.com'?myCounter.email:false}">
    </property>
</bean>
```

myCounter.email matches 使用 counter 的邮箱匹配正则表达式。如果成功，则返回 true，否则，返回 false。

配合三元运算符，如果 myCounter.email matches 返回 true，则 otherBean 的 email 引用 myCounter.email；否则，返回 false。

测试代码如下。

```
public static void main(String[] args) {
    ApplicationContext app
            = new ClassPathXmlApplicationContext("beans.xml");
    OtherBean otherBean = (OtherBean)app.getBean("otherBean");
    System.out.println(otherBean.getAdjustedCount());
    System.out.println(otherBean.getEmail());
}
```

测试结果如下。

```
信息: Loading XML bean definitions from class path resource [beans.xml]
628
xiaohp@qq.com
```

4.6 项目案例：基于@Value 注解的应用

使用@Value 在注解中声明 SpEL 的效果等同于在 XML 中配置 SpEL。操作步骤如下。

（1）配置组件扫描。

```
<?xml version="1.0" encoding="UTF-8"?>
<beans xmlns="http://www.springframework.org/schema/beans"
    xmlns:xsi="http://www.w3.org/2001/XMLSchema-instance"
    xmlns:context="http://www.springframework.org/schema/context"
    xsi:schemaLocation="http://www.springframework.org/schema/beans
    http://www.springframework.org/schema/beans/spring-beans-4.3.xsd
    http://www.springframework.org/schema/context
    http://www.springframework.org/schema/context/spring-context-4.3.xsd">
<context:component-scan base-package="org.spring.samples"/>
</beans>
```

（2）使用@Value 和 SpEL 赋值。

```
@Component
public class NumberGuess {
    @Value("#{ T(java.lang.Math).random() * 100.0 }")
    private int randomNumber;
    public int getRandomNumber() {
        return randomNumber;
    }
    public void setRandomNumber(int randomNumber) {
        this.randomNumber = randomNumber;
    }
    public void doSomething() {
        System.out.println("randomNumber=" + randomNumber);
    }
}
@Component
public class ShapeGuess {
    @Value("#{ numberGuess.randomNumber }")
    private int initialShapeSeed;
    public int getInitialShapeSeed() {
        return initialShapeSeed;
    }
    public void setInitialShapeSeed(int initialShapeSeed) {
        this.initialShapeSeed = initialShapeSeed;
    }
    public void doSomething() {
        System.out.println("initialShapeSeed=" + initialShapeSeed);
    }
}
@Component
public class TaxCalculator {
    @Value("#{ systemProperties['user.country'] }")
    private String defaultLocale;
    public String getDefaultLocale() {
        return defaultLocale;
    }
    public void setDefaultLocale(String defaultLocale) {
        this.defaultLocale = defaultLocale;
    }
    public void doSomething() {
        System.out.println("defaultLocale=" + defaultLocale);
    }
}
```

（3）测试。

① 测试代码如下。

```
public static void main(String[] args) {
    ApplicationContext context
            = new ClassPathXmlApplicationContext("beans.xml");
    NumberGuess ng = (NumberGuess)context.getBean("numberGuess");
```

```
    ng.doSomething();
    ShapeGuess  sg = (ShapeGuess)context.getBean("shapeGuess");
    sg.doSomething();
    TaxCalculator  tc = (TaxCalculator)context.getBean("taxCalculator");
    tc.doSomething();
    System.out.println(System.getProperty("user.country"));
}
```

② 测试结果如下。

```
信息: Loading XML bean definitions from class path resource [beans.xml]
randomNumber=38
initialShapeSeed=38
defaultLocale=CN
CN
```

第 5 章 AOP

5.1 AOP 概述

面向切面编程（Aspect-Oriented Programming，AOP）是继面向对象编程（Object-Oriented Programming，OOP）之后的一种重要的编程。与 OOP 不同，AOP 提供了一种完全不同的编程思想。当然，OOP 的应用场景更广，AOP 只能应用于特殊场景。

AOP 不属于必需功能，Spring 的 IoC 容器可以不依赖 AOP，如果不需要，可以不导入 AOP 的相关包。

Spring AOP 提供了两种模式，分别是基于 XML 的模式、基于@AspectJ 的注解模式。

基于 Spring AOP 的重要应用如下。

- 用 AOP 声明性事务代替 EJB 的企业服务。
- 用 AOP 进行日志处理。
- 用 AOP 进行权限控制，如 Spring Security。

5.1.1 AOP 中的专业术语

下面介绍 AOP 中的专业术语。

- 切面（aspect）：横切关注点的模块化，关注多个类的共性处理。事务管理是 J2EE 应用中关于横切关注点的很好的例子。
- 连接点（join point）：在程序执行过程中某个特定的点，如某方法调用的时候或者处理异常的时候。在 Spring AOP 中，一个连接点总是表示一个方法的执行。
- 通知（advice）：在切面的某个特定的连接点上执行的动作。其中包括了 around、before 和 after 等不同类型的通知。许多 AOP 框架（包括 Spring）以拦截器作为通知模型，并维护一个以连接点为中心的拦截器链。
- 切入点（pointcut）：由切入点表达式和签名组成。切入点如何与连接点匹配是 AOP 的核心，Spring 默认使用 AspectJ 切入点语法。
- 引入（introduction）：用来给一个类型声明添加额外的方法或属性。Spring 允许给任何织入的对象引入新的接口。例如，通过引入，可以使一个 Bean 实现 IsModified 接口，以便简化缓存机制。
- 目标对象（target object）：被一个或者多个切面通知的对象，也称作被通知对象。既然 Spring AOP 是通过动态代理实现的，这个对象就永远是一个被代理对象。
- AOP 代理（AOP proxy）：动态代理，在 Spring 中，AOP 代理可以是 JDK 代理或者 CGLIB。
- 织入（weaving）：创建一个通知者，把切面连接到其他应用程序类型或者对象上。这些操作可以在编译时、加载类时和运行时完成。Spring 和其他纯 Java AOP 框架一样，在运行时完成织入。

5.1.2 通知的类型

通知的类型如下。

- 前置通知（before advice）：在某连接点之前执行的通知，但这个通知不能阻止连接点之前的执行流程（除非它抛出一个异常）。
- 后置返回通知（after returning advice）：在某连接点正常完成后执行的通知。例如，一个方法没有抛出任何异常，正常返回。
- 异常通知（after throwing advice）：在方法抛出异常退出时执行的通知。
- 最终通知［after (finally) advice］：当某连接点退出的时候执行的通知（不论是正常返回还是异常退出）。
- 环绕通知（around advice）：包围一个连接点的通知，如方法调用。这是一种最强大的通知类型。环绕通知可以在方法调用前后完成自定义的行为。它也会选择是否继续执行连接点或直接返回它自己的返回值或抛出异常来结束执行。

通过切入点匹配连接点的概念是AOP的关键，这使得AOP不同于其他仅仅提供拦截功能的旧技术。

环绕通知是最常用的通知类型。和 AspectJ 一样，Spring 提供所有类型的通知，我们推荐使用尽可能简单的通知类型来实现需要的功能。例如，如果只需要一个方法的返回值来更新缓存，则最好使用最终通知，而不是环绕通知（尽管环绕通知也能完成同样的事情）。用最合适的通知类型可以使得编程模型变得简单，并且能够避免很多潜在的错误。例如，不在连接点上调用用于环绕通知的 proceed( )方法，从而避免调用的问题。

5.1.3 AOP 动态代理的选择

AOP 的基础是动态代理。Spring 默认使用标准 JDK 动态代理（dynamic proxy）来作为 AOP 的代理，这种模式的代理对象的返回类型只能是接口。

Spring 也可以使用 CGLIB 代理。在需要代理类但是没有接口的情况下，CGLIB 代理是很有必要的。如果一个业务对象并没有实现一个接口，则默认使用 CGLIB 代理。

JDK 代理或 CGLIB 代理对象都通过 ProxyFactoryBean 创建。

```
public class ProxyFactoryBean extends ProxyCreatorSupport
    implements FactoryBean<Object>, BeanClassLoaderAware, BeanFactoryAware {
    public Object getObject() throws BeansException {
        initializeAdvisorChain();
        if (isSingleton()) {
            return getSingletonInstance();
        }else {
            return newPrototypeInstance();
        }
    }
}
```

ProxyFactoryBean 用于动态创建 ProxyFactory 对象。前面讲过 FactoryBean 模式，调用 ProxyFactoryBean::getObject( )，返回的对象是 ProxyFactory。

```
public class ProxyFactory extends ProxyCreatorSupport {
    public Object getProxy() {
        return createAopProxy().getProxy();
    }
    public Object getProxy(ClassLoader classLoader) {
        return createAopProxy().getProxy(classLoader);
    }
}
```

调用 ProxyFactory::getProxy( )即可返回动态代理。

注意：Spring 4.3 已经打包了 CGLIB 3.2.4 库，无须额外导入其他版本的 CGLIB 库。在与 Hibernate 或 MyBatis 集成时，要解决 CGLIB 的版本冲突问题。

5.2 支持@AspectJ

5.2.1 @AspectJ

@AspectJ 是一种风格样式，可以把 Java 的普通类声明为一个切面。为了在 Spring 4.3 中使用@AspectJ，需要导入 spring-aspects-4.3.jar，还需要导入依赖包 aspectjweaver.jar（1.6.8 及以上版本）。

5.2.2 autoproxying 配置

@AspectJ 需要 autoproxying 配置，这样 IoC 容器启动时，就可以查找 AspectJ 注解了。如下两种配置模式均可。

- Java 类配置模式。

```
@Configuration
@EnableAspectJAutoProxy
public class AppConfig {
}
```

- XML 配置模式。

```
<beans xmlns="http://www.springframework.org/schema/beans"
   xmlns:xsi="http://www.w3.org/2001/XMLSchema-instance"
   xmlns:aop="http://www.springframework.org/schema/aop"
   xsi:schemaLocation="http://www.springframework.org/schema/beans
       http://www.springframework.org/schema/beans/spring-beans-4.3.xsd
       http://www.springframework.org/schema/aop
       http://www.springframework.org/schema/aop/spring-aop-4.3.xsd">
   <aop:aspectj-autoproxy />
</beans>
```

5.2.3 声明切面

切面中包含切入点、通知、引入等信息，因此@AspectJ 开发从定义切面开始。一个系统中可以定义很多切面，每个切面中的内容相互独立。切面的定义方法如下。

（1）新建一个 Bean。

```
@Component
public class DbProxy {}
```

（2）使用@Aspect 声明一个切面。

```
@Component
@Aspect
public class DbProxy {}
```

和普通类一样，在@Aspect 声明的类中可以添加属性和方法，还可以包含切入点、通知等内容。

5.2.4 声明切入点

切入点的声明包含两个部分——签名和切入点表达式。

- 签名：由一个名字和多个参数组成。
- 切入点表达式：表达式的配置。

下面给出一段关于切入点的代码。

```
@Component
@Aspect
public class DbProxy {
    @Pointcut("execution(public * com.icss.biz.*.*(..))")
    private void businessOperate() {}
}
```

其中，@Pointcut("execution(public * com.icss.biz.*.*(..))")为切入点表达式，businessOperate 是签名。

注意事项如下。

- 切入点对应的签名必须返回 void。
- 签名对应的方法不会被调用，我们需要的仅仅是方法的名字，因此方法中无须写任何代码。
- 签名方法可以使用任何可见性修饰符，因为签名可以在其他切面中引用。
- 可以把所有公用的切入点定义在一个切面中，以供其他切面调用。

5.2.5 切入点表达式

1. 切入点中的指示符

Spring AOP 支持在切入点表达式中使用如下 AspectJ 切入点指示符。

- execution：匹配方法执行的连接点，这是 Spring 中主要的切入点指示符。
- within：限定匹配特定类型的连接点（在使用 Spring AOP 的时候方法的执行）。
- this：限定匹配特定的连接点（使用 Spring AOP 的时候方法的执行），其中 Bean reference（Spring AOP 代理）是指定类型的实例。
- target：限定匹配特定的连接点（使用 Spring AOP 的时候方法的执行），其中目标对象（被代理的应用对象）是指定类型的实例。
- args：限定匹配特定的连接点（使用 Spring AOP 的时候方法的执行），其中参数是指定类型的实例。

- @target：限定匹配特定的连接点（使用 Spring AOP 的时候方法的执行），其中正执行对象的类持有指定类型的注解。
- @args：限定匹配特定的连接点（使用 Spring AOP 的时候方法的执行），其中实际传入参数的运行时类持有指定类型的注解。
- @within：限定匹配特定的连接点，其中连接点所在类型已指定注解（在使用 Spring AOP 的时候，所执行的方法所在类型已指定注解）。
- @annotation：限定匹配特定的连接点（使用 Spring AOP 的时候方法的执行），其中连接点的主题持有指定的注解。

注意：对于 JDK 动态代理，只有 public 接口的方法调用能被拦截；对于 CGLIB 动态代理，public、protected 方法的调用可以被拦截。对于切入点定义，可以应用于任何非公有方法。

2．联合使用切入点表达式

可以通过&&、||和感叹号（!）联合使用多个切入点表达式。示例如下。

```
@Pointcut("execution(public * *(..))")
private void anyPublicOperation() {}
@Pointcut("within(com.xyz.someapp.trading..*)")
private void inTrading() {}
```

通过如下代码可以联合使用上面的两个切入点表达式。

```
@Pointcut("anyPublicOperation() && inTrading()")
private void tradingOperation() {}
```

3．共享通用的切入点表达式

通过以下代码，可以共享通用的切入点表达式。

```
@Aspect
public class SystemArchitecture {
    /* 用于业务层连接点*/
    @Pointcut("execution(* com.xyz.someapp..service.*.*(..))")
    public void businessService() {}
    /* 用于数据访问层连接点*/
    @Pointcut("execution(* com.xyz.someapp.dao.*.*(..))")
    public void dataAccessOperation() {}
    }
}
```

切入点定义后，一般在通知中调用，公用的切入点在其他切面中的通知上调用。有关通知的内容稍后讲解。

4．execution 指示符

在 Spring AOP 中，使用最多的指示符就是 execution 指示符，格式如下。

```
execution(修饰符 返回类型  方法名(方法参数) 异常)
```

- 返回类型：使用最频繁的返回类型模式是*，它代表匹配任意的返回类型。
- 方法名：使用*，可以通配全部或部分名字。
- 方法参数：()表示无参；(..)表示零个或任意个参数；(*)表示一个参数，类型任意；(*,String)，表示两个参数，第一个参数的类型可以随意指定。

下面给出一些通用切入点表达式的例子。

- execution(public * *(..))：任意公共方法的执行。
- execution(* set*(..))：任何一个名字以“set”开始的方法的执行。
- execution(* com.xyz.service.AccountService.*(..))：AccountService 接口定义的任意方法的执行。
- execution(* com.xyz.service.*.*(..))：在 service 包中定义的任意方法的执行。
- execution(* com.xyz.service..*.*(..))：在 service 包或其子包中定义的任意方法的执行。
- within(com.xyz.service.*)：在 service 包中的任意连接点（在 Spring AOP 中只是方法的执行）。
- within(com.xyz.service..*)：在 service 包或其子包中的任意连接点（在 Spring AOP 中只是方法的执行）。
- this(com.xyz.service. AccountService)：实现了 AccountService 接口的代理对象的任意连接点。this 在绑定表单中更加常用，请参见 5.2.6 节以了解如何使代理对象在通知体内可用。
- target(com.xyz.service.AccountService)：实现 AccountService 接口的目标对象的任意连接点。target 在绑定表单中更加常用，请参见 5.2.6 节以了解如何使目标对象在通知体内可用。
- args(java.io.Serializable)：任何一个只接受一个参数并且运行时传入的参数是 Serializable 接口的连接点。args 在绑定表单中更加常用（5.2.6 节会介绍如何使方法参数在通知体内可用）。

注意，例子中给出的切入点不同于 execution(* *(java.io.Serializable))：args 版本只有在动态运行过程中传入的参数可序列化时才匹配，而 execution 版本在方法签名中声明了一个 Serializable

类型的参数时匹配。

- @target(org.springframework. transaction.annotation.Transactional)：目标对象中有一个@Transactional 注解的任意连接点。@target 在绑定表单中更加常用，5.2.6 节会介绍如何使得注解对象在通知体内可用。
- @within(org. springframework.transaction.annotation.Transactional)：任何一个目标对象声明的类型有一个@Transactional 注解的连接点。@within 在绑定表单中更加常用，参见 5.2.6 节以了解如何使注解对象在通知体内可用。
- @annotation(org.springframework. transaction.annotation.Transactional)：任何一个执行的方法有一个@Transactional 注解的连接点。@annotation 在绑定表单中更加常用，参见 5.2.6 节以了解如何使注解对象在通知体内可用。
- @args(com.xyz.security.Classified)：任何一个只接受一个参数并且运行时所传入的参数类型具有@Classified 注解的连接点（在 Spring AOP 中只是方法的执行）。@args 在绑定表单中更加常用，参见 5.2.6 节以了解如何使得注解对象在通知体内可用。
- bean(tradeService)：任何一个在名为 tradeService 的 Spring Bean 之上的连接点。
- bean(*Service)：任何一个在名字匹配通配符表达式*Service 的 Spring Bean 之上的连接点。

5.2.6 声明基于注解的通知

通知与切入点表达式关联，在切入点表达式匹配的方法执行之前、之后执行。可以直接使用切入点表达式，也可以通过名字引用已定义的切入点表达式。

1. 声明通知的方法

1）声明前置通知

当切入点匹配连接点方法时，在连接点之前执行的通知即声明前置通知，这个通知不能阻止连接点之前的执行流程（除非它抛出一个异常）。@Before 表示前置通知。

@Before 通过名字引用公用的切入点表达式。切入点的引用格式为“包名.类名.切入点签名”。注意，签名的可见性最好设置为 public。

声明前置通知的示例如下。

（1）在通知中调用其他切面中的切入点。

```
@Aspect
public class BeforeExample {
```

```
    @Before("com.xyz.myapp.SystemArchitecture.dataAccessOperation()")
    public void doAccessCheck() {
        // ...
    }
}
```

（2）直接定义切入点表达式。

```
@Aspect
public class BeforeExample {
    @Before("execution(* com.xyz.myapp.dao.*.*(..))")
    public void doAccessCheck() {
        // ...
    }
}
```

2）声明后置返回通知

当切入点匹配连接点方法时，方法正常执行后执行的通知即声明后置返回通知。注意，方法正常执行后，不能抛出任何异常。@AfterReturning 表示最终通知。当切入点匹配代理方法时，正常执行方法后，触发通知。最终通知声明为@AfterReturning。

```
@Aspect
public class AfterReturningExample {
    @AfterReturning("com.xyz.myapp.SystemArchitecture.businessService()")
    public void doAccessCheck() {
        // ...
    }
}
```

3）声明异常通知

当切入点匹配连接点方法时，异常结束，触发声明异常通知。异常通知声明为@After Throwing。

```
@Aspect
public class AfterThrowingExample {
    @AfterThrowing("com.xyz.myapp.SystemArchitecture.dataAccessOperation()")
    public void doRecoveryActions() {
        // ...
}
```

4）声明最终通知

不论一个连接点方法如何结束，是否存在异常，最终通知都会运行。使用@After 注解来声明最终通知。最终通知必须准备处理正常返回和异常返回两种情况。

通常用@After 来释放资源，如数据库连接。

```
@Aspect
public class AfterFinallyExample {
    @After("com.xyz.myapp.SystemArchitecture.dataAccessOperation()")
```

```
    public void doReleaseLock() {
        // ...此处释放非托管资源
    }
}
```

5）环绕通知

当切入点匹配连接点方法时，方法执行前和执行后都触发环绕通知。环绕通知使用@Around注解来声明。通知的第一个参数必须是ProceedingJoinPoint类型。调用ProceedingJoinPoint的proceed( )方法，触发连接点方法的执行。

```
@Aspect
public class AroundExample {
@Around("com.xyz.myapp.SystemArchitecture.businessService()")
public Object doBasicProfiling(ProceedingJoinPoint pjp) throws Throwable {
    // 开始观察
    Object retVal = pjp.proceed();
    // 停止观察
    return retVal;
}
}
```

2. 向通知传递参数

在向通知传递参数时，org.aspectj.lang.JoinPoint 常作为第一个参数使用。作为 JoinPoint 的子类，org.aspectj.lang. ProceedingJoinPoint 只能用于环绕通知。

JoinPoint 中的方法如下。

- getArgs( )：返回代理对象的方法参数。
- getTarget( )：返回目标对象。
- getSignature( )：返回被通知方法的信息。
- toString( )：输出被通知方法的有用信息。

下面给出一个示例。

（1）定义切面和通知。

```
import org.aspectj.lang.JoinPoint;
import org.aspectj.lang.annotation.Aspect;
import org.aspectj.lang.annotation.Before;
import org.springframework.stereotype.Component;
@Component
@Aspect
public class BeforeExample {
    @Before("execution (* com.icss.biz.*.*(..))")
    public void doAccessCheck(JoinPoint jp) {
```

```
        System.out.println("before:" + jp.getTarget().toString());
        System.out.println("before:" + jp.toString());
        System.out.println("before:" + jp.getArgs()[0].toString());
        System.out.println("before:" + jp.getSignature().toString());
    }
}
```

（2）测试。

① 测试代码如下。

```
public static void main(String[] args) {
    UserBiz iuser = (UserBiz)BeanFactory.getBean("userBiz");
    try {
        User user = iuser.login("admin", "123");
        if(user != null) {
            System.out.println("登录成功，身份是" + user.getRole());
        }else {
            System.out.println("登录失败");
        }
    } catch (Exception e) {
        System.out.println(e.getMessage());
    }
}
```

② 测试结果如下。

```
INFO - Loading XML bean definitions from class path resource [beans.xml]
before:com.icss.biz.UserBiz@6647c2
before:execution(User com.icss.biz.UserBiz.login(String,String))
before:admin
before:User com.icss.biz.UserBiz.login(String,String)
```

注意：在导入包时很容易出错，应该使用 org.aspectj 包，下面的两个连接点区分大小写。

```
import org.aopalliance.intercept.Joinpoint;
import org.aspectj.lang.JoinPoint;
```

5.2.7 管理 StaffUser 日志

基于 AOP 的日志管理是 AOP 的典型应用场景。下面给出一个示例。

（1）使用环绕通知配置日志代理。

```
@Component
@Aspect
public class LogProxy {
    @Around("execution(public * com.icss.biz.*.*(..))")
    public Object logging(ProceedingJoinPoint pjp) throws Throwable{
        Object obj = null;
        Log.logger.info(pjp.getSignature() + new Date().toString());
        obj = pjp.proceed();                        //必须有返回值
```

```
        Log.logger.info(pjp.toString() + new Date().toString());
        return obj;
    }
}
```

（2）测试登录功能。

① 测试代码如下。

```
public static void main(String[] args) throws Exception{
    UserBiz iuser = (UserBiz)BeanFactory.getBean("userBiz");
    User user = iuser.login("admin", "123");
    if(user != null) {
        System.out.println("登录成功，身份是" + user.getRole());
    }else {
        System.out.println("登录失败");
    }
}
```

② 测试结果如下。

```
INFO - User com.icss.biz.UserBiz.login(String,String)
UserBizMysql....login....
UserDaoMysql....login....
INFO - execution(User com.icss.biz.UserBiz.login(String,String))
execution(User com.icss.biz.UserBiz.login(String,String))
登录成功，身份是 1
```

5.2.8 管理 StaffUser 数据库的连接

使用 AOP，可以实现所有业务层方法，自动打开数据库，自动关闭数据库。具体步骤如下。

（1）使用@Before 打开数据库，使用@After 关闭数据库。

```
@Component
@Aspect
public class DbProxy {
    @Pointcut("execution(public * com.icss.biz.*.*(..))")
    private void businessOperate() {
    }
    @Before("businessOperate()")
    public void openDataBase(JoinPoint jp) {
        System.out.println("-----------打开数据库--------------");
    }
    @After("businessOperate()")
    public void closeDataBase(JoinPoint jp) {
        System.out.println(jp.toString());
        System.out.println("-------------关闭数据库----------");
    }
}
```

（2）测试。即使出现异常，也要确保关闭数据库。

① 测试代码如下。

```
public static void main(String[] args) throws Exception{
    UserBiz iuser = (UserBiz)BeanFactory.getBean("userBiz");
    User user = iuser.login("admin", "123");
    if(user != null) {
        System.out.println("登录成功，身份是" + user.getRole());
    }else {
        System.out.println("登录失败");
    }
}
```

② 测试异常，结果如下。

```
------------打开数据库--------------
INFO - User com.icss.biz.UserBiz.login(String,String)
UserBizMysql....login....
UserDaoMysql....login....
execution(User com.icss.biz.UserBiz.login(String,String))
--------------关闭数据库----------
Exception in thread "main" java.lang.Exception: 异常测试...
    at com.icss.dao.UserDaoMysql.login(UserDaoMysql.java:40)
    at com.icss.biz.UserBiz.login(UserBiz.java:28)
```

5.3 基于 XML 的 AOP 配置

对于 XML 格式，Spring 通过“aop”命名标签，支持面向切面编程。XML 同样支持与@AspectJ 风格一致的切入点表达式和通知。在 Spring 的配置文件中，所有的切面和通知都必须定义在<aop:config>元素内部（context 可以包含多个 <aop:config>）。

一个<aop:config>可以包含切入点、通知器和 aspect 元素 （注意，这 3 个元素必须按照这个顺序进行声明）。

警告：<aop:config>风格的配置使得 Spring 的 auto-proxying 机制变得很笨重。如果已经通过 BeanNameAutoProxyCreator 或类似的东西显式使用 auto-proxying，则可能会出现问题（如通知没有被织入）。

推荐的做法是仅仅使用<aop:config>风格，或者仅仅使用 AutoProxyCreator 风格。

5.3.1 声明切面

使用<aop:aspect>来声明切面，通过 ref 可以引用支撑 Bean。

```
<aop:config>
    <aop:aspect id="myAspect" ref="aBean">
        ...
    </aop:aspect>
```

```
</aop:config>
<bean id="aBean" class="...">
    ...
</bean>
```

注意：切面中要配置通知，通知需要通过代码实现，因此必须要有一个支撑 Bean。

5.3.2 声明切入点

<aop:pointcut>应该声明在<aop:config>内，这样其他切面和通知器就可以共享这个切入点。

```
<aop:config>
    <aop:pointcut id="businessService"
        expression="execution(* com.xyz.myapp.service.*.*(..))"/>
</aop:config>
```

在 XML 中切入点表达式的语法与在@AspectJ 中使用切入点表达式的语法完全一致。XML 配置中的 id 就是切入点的签名（它和方法无关，就是一个名称识别），expression 为切入点表达式。

也可以通过名字引用其他切入点。下面给出一个示例。

（1）引用@AspectJ 定义的切入点。

```
<aop:config>
    <aop:pointcut id="businessService"
       expression="com.xyz.myapp.SystemArchitecture.businessService()"/>
</aop:config>
```

（2）引用 XML 中定义的切入点。

```
<aop:config>
    <aop:aspect id="myAspect" ref="aBean">
    <aop:pointcut id="businessService"
       expression="execution(* com.xyz.myapp.service..(..)) and this(service)"/>
    <aop:before pointcut-ref="businessService" method="monitor"/>
    ...
    </aop:aspect>
</aop:config>
```

5.3.3 声明基于 XML 的通知

XML 支持在@AspectJ 中定义的 5 个通知类型，而且这些通知的语义相同。

下面给出一个示例。

（1）声明前置通知。

```
<aop:aspect id="beforeExample" ref="aBean">
    <aop:before
        pointcut-ref="dataAccessOperation" method="doAccessCheck"/>
```

```
    <aop:before
        pointcut="execution(* com.xyz.myapp.dao.*.*(..))"
        method="doAccessCheck"/>
</aop:aspect>
```

（2）声明后置返回通知。

```
<aop:aspect id="afterReturningExample" ref="aBean">
    <aop:after-returning
            pointcut-ref="dataAccessOperation"
            method="doAccessCheck"/>
</aop:aspect>
```

（3）声明异常通知。

```
<aop:aspect id="afterThrowingExample" ref="aBean">
    <aop:after-throwing
        pointcut-ref="dataAccessOperation"
        method="doRecoveryActions"/>
</aop:aspect>
```

（4）声明最终通知。

```
<aop:aspect id="afterFinallyExample" ref="aBean">
    <aop:after
            pointcut-ref="dataAccessOperation"
            method="doReleaseLock"/>
    ...
</aop:aspect>
```

（5）声明环绕通知。

```
<aop:aspect id="aroundExample" ref="aBean">
    <aop:around
                pointcut-ref="businessService"
                method="doBasicProfiling"/>
    ...
</aop:aspect>
```

（6）在支撑 Bean 中定义通知要响应的方法。

```
public class ABean {
    public void doAccessCheck(JoinPoint jp) {
    }
    public void doRecoveryActions(JoinPoint jp ) {
    }
    public void doReleaseLock() {
    }
    public Object doBasicProfiling(ProceedingJoinPoint pjp) throws Throwable{
        return pjp.proceed();
    }
}
```

5.3.4 使用通知器

通知器（advisor）的概念源于Spring对AOP的支持，在@AspectJ中没有等价内容。一个通知器是一个自包含的切面，通知器只能有一个通知。Aspect需要一个Bean的支持，通知器则不需要，这样配置更加简单。示例代码如下。

```
<aop:config>
    <aop:pointcut id="businessService"
            expression="execution(* com.xyz.myapp.service.*.*(..))"/>
    <aop:advisor
            pointcut-ref="businessService"
            advice-ref="tx-advice"/>
</aop:config>
<tx:advice id="tx-advice">
    <tx:attributes>
        <tx:method name="*" propagation="REQUIRED"/>
    </tx:attributes>
</tx:advice>
```

在以下代码中，可以协同使用通知器与事务通知策略。

```
<bean id="txManager" class="org.springframework
                        .jdbc.datasource.DataSourceTransactionManager">
    <property name="dataSource" ref="dataSource" />
</bean>
<aop:config>
    <aop:pointcut id="serviceOperation"
            expression="execution(* com.icss.biz.*.*(..)))" />
    <aop:advisor advice-ref="txAdvice" pointcut-ref="serviceOperation" />
</aop:config>
<tx:advice id="txAdvice" transaction-manager="txManager">
    <tx:attributes>
        <tx:method name="add*"  rollback-for="Throwable" />
        <tx:method name="*"  read-only="true" />
    </tx:attributes>
</tx:advice>
```

5.3.5 管理StaffUser系统的日志

前面介绍了StaffUser系统，下面使用XML配置切面、切入点、通知，实现日志管理，对员工系统的服务层方法和持久层方法调用都输出日志。操作步骤如下。

（1）编写日志处理代码。

```
@Around("execution(public * com.icss.biz.*.*(..))")
public Object logging(ProceedingJoinPoint pjp) throws Throwable{
    Object obj = null;
    Log.logger.info(pjp.getSignature() + new Date().toString());
    obj = pjp.proceed();              //必须要有返回值
    Log.logger.info(pjp.toString() + new Date().toString());
```

```
        return obj;
    }
```

（2）配置切面和切入点。

```
<aop:config>
        <aop:aspect id="logAspect" ref="logBean">
            <aop:around method="logging" pointcut-ref="businessOperate"/>
    </aop:aspect>
    </aop:config>
    <bean id="logBean" class="com.icss.biz.LogProxy"></bean>
```

（3）测试服务层方法的输出日志。

① 测试代码如下。

```
public static void main(String[] args) throws Exception{
    UserBiz iuser = (UserBiz)BeanFactory.getBean("userBiz");
    User user = iuser.login("admin", "123");
    if(user != null) {
        System.out.println("登录成功，身份是" + user.getRole());
    }else {
        System.out.println("登录失败");
    }
}
```

② 测试结果如下。

```
INFO - User com.icss.biz.UserBiz.login(String,String)
UserBizMysql....login....
UserDaoMysql....login....
INFO - execution(User com.icss.biz.UserBiz.login(String,String))
execution(User com.icss.biz.UserBiz.login(String,String))
登录成功，身份是1
```

（4）测试持久层方法的输出日志。在 XML 中可以使用&（&表示&&）、and、or 连接多个条件。

① 测试代码如下。

```
<aop:config>
    <aop:pointcut id="allLayer"
            expression="execution(public * com.icss.biz.*.*(..)) or
                        execution(public * com.icss.dao.*.*(..)) "/>
    <aop:aspect id="logAspect" ref="logBean">
        <aop:around method="logging" pointcut-ref="allLayer"/>
    </aop:aspect>
  </aop:config>
  <bean id="logBean" class="com.icss.biz.LogProxy"></bean>
```

② 测试结果如下。

```
INFO - User com.icss.biz.UserBiz.login(String,String)
UserBizMysql....login....
INFO - User com.icss.dao.IUserDao.login(String,String)
```

```
UserDaoMysql....login....
INFO - execution(User com.icss.dao.IUserDao.login(String,String))
INFO - execution(User com.icss.biz.UserBiz.login(String,String))
execution(User com.icss.biz.UserBiz.login(String,String))
登录成功，身份是 1
```

5.3.6 管理 StaffUser 系统中的数据库连接

前面介绍了 StaffUser 系统，下面使用 XML 配置切面、切入点、通知，管理数据库连接，自动打开数据库连接，自动关闭数据库连接。

操作步骤如下。

（1）编写打开、关闭数据库的代码。

```
public class DbProxy {
    public void openDataBase(JoinPoint jp) {
        System.out.println("-----------打开数据库--------------");
    }
    public void closeDataBase(JoinPoint jp) {
        System.out.println(jp.toString());
        System.out.println("-------------关闭数据库----------");
    }
}
```

（2）编写 XML 配置信息。

```
<aop:config>
         <aop:pointcut expression="execution(public * com.icss.biz.*.*(..))"
                        id="businessOperate"/>
         <aop:aspect id="dbAspect" ref="dbBean">
      <aop:before method="openDataBase" pointcut-ref="businessOperate"/>
      <aop:after method="closeDataBase" pointcut-ref="businessOperate"/>
    </aop:aspect>
    </aop:config>
    <bean id="dbBean" class="com.icss.biz.DbProxy"></bean>
```

（3）运行代码。运行结果如下。

```
 -----------打开数据库--------------
UserBizMysql....login....
UserDaoMysql....login....
-------------关闭数据库----------
登录成功，身份是 1
```

5.4 代理机制

Spring 的 AOP 部分使用 JDK 动态代理，部分使用 CGLIB 来为目标对象创建代理。若被代理的目标对象实现了至少一个接口，则会使用 JDK 动态代理；若该目标对象没有实现任何接口，则创建一个 CGLIB 代理。

CGLIB 是第三方包，从 Spring 4.3 开始，已经在 Spring-core.jar 中加入了 CGLIB 3.2.4.jar，因此无须再次导入。

5.4.1 静态代理

动态代理是在静态代理模式的基础上发展起来的，因此学习动态代理之前，需要先明白静态代理模式。

1. 设计模式之静态代理

GOF 静态代理模式表示为其他对象提供一种代理，以控制对这个对象的访问。图 5-1 为代理模式的类图。

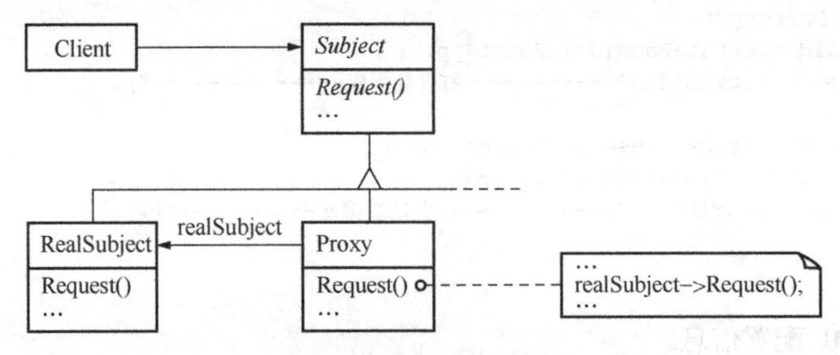

图 5-1 代理模式的类图

客户向代理发出请求，真正做事的是被代理者（见图 5-2）。代理模式应用广泛，如 Hibernate 的懒加载机制。

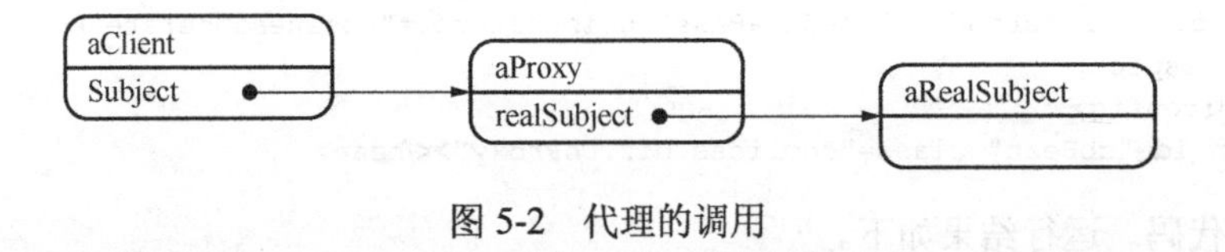

图 5-2 代理的调用

2. 关于演员与经纪人的示例

要邀请演员拍电影、参加娱乐活动等，需要与其经纪人联系，但实际参加活动的是演员。这里需要注意的是，经纪人不是简单的委托人，他会做一些额外的事情。

下面给出一个关于演员和经纪人的示例。

（1）设计接口，定义演员的 3 个行为——拍电影、拍广告、参加娱乐活动。

```
public interface IStar {
    public void playMovie();                    //拍电影
    public void playAd();                       //拍广告
```

```
    public void playActive();        //参加娱乐活动
}
```

（2）定义 MovieStar 业务类，实现 IStar 接口。

```
public class MovieStar implements IStar{
    private String name;
    public String getName() {
        return name;
    }
    public MovieStar(String name) {
        this.name = name;
    }
    @Override
    public void playMovie() {
        System.out.println(this.name + " 正在拍电影...");
    }
    @Override
    public void playAd() {
        System.out.println(this.name + "正在拍广告...");
    }
    @Override
    public void playActive() {
        System.out.println(this.name + "正在参加活动...");
    }
}
```

（3）定义静态代理类。代理类与被代理类必须要实现相同的接口。代理人在被代理人实际参与的活动中会做一些额外的事情。

```
public class StarProxy  implements IStar{
    private String name;
    private MovieStar star;
    public StarProxy(String name,MovieStar star) {
        this.name = name;
        this.star = star;
        System.out.println(this.name + "的代理" + star.getName() + "在工作");
    }
    @Override
    public void playMovie() {
        star.playMovie();
        System.out.println(this.name + "主动照顾"
                          + star.getName() + "家属...");
    }
    @Override
    public void playAd() {
        star.playAd();
        System.out.println(this.name + "获得 3 万元....");
    }
    @Override
    public void playActive() {
        star.playActive();
        System.out.println(this.name + "获得 8 万元....");
    }
}
```

（4）测试。

① 测试代码如下（发起活动的都是代理者，实际执行操作的是被代理者）。

```
public static void main(String[] args) {
    MovieStar star = new MovieStar("张三");
    IStar proxy = new StarProxy("李四",star);
    proxy.playActive();
    proxy.playAd();
    proxy.playMovie();
}
```

② 测试结果如下。

```
李四的代理张三在工作
张三正在参加活动...
李四获得 8 万元....
张三正在拍广告...
李四获得 3 万元....
张三正在拍电影...
李四获得 15 万元，主动照顾张三家属...
```

5.4.2 动态代理

1. 静态代理的日志实现

在如下示例中，使用代理模式为所有的业务操作输出日志。

（1）定义业务接口 IBook 和业务行为。

```
public interface IBook {
    public void getBookInfo();
    public void buyBook();
    public void updateBook();
}
```

（2）定义业务实现类 BookBiz，实现接口 IBook。

```
public class BookBiz implements IBook{
    @Override
    public void getBookInfo() {
        System.out.println("getBookInfo...");
    }
    @Override
    public void buyBook() {
        System.out.println("buyBook...");
    }
    @Override
    public void updateBook() {
```

```
        System.out.println("updateBook...");
    }
}
```

（3）定义代理类 BookLog，它也实现 IBook 接口，同时在构造函数中传入被代理的对象 BookBiz。

```
public class BookLog implements IBook{
    private BookBiz bookBiz;
    public BookLog(BookBiz bookBiz) {
        this.bookBiz = bookBiz;
    }
    @Override
    public void getBookInfo() {
        Log.logger.info("getBookInfo...begin...");
        bookBiz.getBookInfo();
        Log.logger.info("getBookInfo...end...");
    }
    @Override
    public void buyBook() {
        Log.logger.info("buyBook...begin...");
        bookBiz.buyBook();
        Log.logger.info("buyBook...end...");
    }
    @Override
    public void updateBook() {
        Log.logger.info("updateBook...begin...");
        bookBiz.updateBook();
        Log.logger.info("updateBook...end...");
    }
}
```

（4）测试。

① 测试代码如下。

```
public static void main(String[] args) {
    BookBiz bookBiz = new BookBiz();
    IBook ibook = new BookLog(bookBiz);
    ibook.buyBook();
}
```

② 测试结果如下。

```
 INFO - buyBook...begin...
buyBook...
INFO - buyBook...end...
```

对于接口 IUser 中的方法，如果也要实现日志，该怎么办？如果还使用静态代理，又需要给 IUser 增加 UserLog 实现类，是不是太麻烦了？如何解决？

```
public interface IUser {
    public void regist(User user) throws Exception;
    public void login(String uname,String pwd) throws Exception;
}
```

2. Proxy 类和 InvocationHandler 接口

java.lang.reflect .Proxy 是 JDK 提供的反射工具类。Proxy 提供了创建动态代理类和实例的静态方法，它也是创建所有动态代理类的超类。

```
public class Proxy implements java.io.Serializable {
     public static Object newProxyInstance(ClassLoader loader,
                                    Class<?>[] interfaces, InvocationHandler h){}
}
```

Proxy.newProxyInstance( )返回的对象就是动态代理对象。

```
InvocationHandler handler = new MyInvocationHandler(...);
Foo f = (Foo) Proxy.newProxyInstance(Foo.class.getClassLoader(),
                                          new Class<?>[] { Foo.class },
                                          handler);
```

案例 5-1：实现日志的动态代理

为了使用 java.lang.reflect.Proxy 实现日志的动态代理，给所有业务类添加日志功能。我们创建的每个动态代理类都是对 Proxy 的包装，有专门的用途。

（1）编写日志动态代理类，实现 InvocationHandler 接口。

```
public class LogDynamic implements InvocationHandler{
    private Object target;
    /**
    * 传入被代理对象，返回代理对象
    */
    public Object bind(Object target) {
        this.target = target;
            return Proxy.newProxyInstance(target.getClass().getClassLoader(),
                target.getClass().getInterfaces(),this);
    }
    /**
     * 当调用被代理对象的方法时，这个 invoke 方法会自动激活
    */
    public Object invoke(Object proxy, Method method, Object[] args)
                                        throws Throwable {
        Log.logger.info(method.getName() + " beging....");
        //动态调用被代理对象的方法
        Object result = method.invoke(target, args);
        Log.logger.info(method.getName() + " ending....");
        return result;
    }
}
```

（2）创建 LogDynamic 动态代理对象，分别绑定 UserBiz 对象和 BookBiz 对象，进行测试。

① 测试代码如下。

```
public static void main(String[] args) {
    LogDynamic logd = new LogDynamic();
    IUser proxy = (IUser)logd.bind(new UserBiz());
    try {
        proxy.login("tom", "123445");
    } catch (Exception e) {
    }
    IBook bookProxy = (IBook)logd.bind(new BookBiz());
    bookProxy.buyBook();
}
```

② 测试结果如下。

```
INFO - login beging....
tom is logging....
INFO - login ending....
INFO - buyBook beging....
buyBook...
INFO - buyBook ending....
```

测试结果显示，IUser 和 IBook 对象绑定后，都可以动态输出日志。

5.4.3 项目案例：自动管理 StaffUser 系统中的数据库连接

前面讲过的员工管理项目使用的都是伪代码。从本节开始，连接真正的数据库。首先考虑如何管理数据库连接，如何通过业务类动态打开、关闭数据库。注意，在实际项目中，打开数据库的位置一般在持久层，但是关闭数据库的位置一般在服务层。后面解释为何这样使用数据库连接。

StaffUser 系统同时支持 Oracle 和 MySQL，演示的时候以 MySQL 为主。

1. 配置 MySQL 数据库环境

配置 MySQL 数据库环境的步骤如下。

（1）安装 MySQL 8 数据库。

（2）通过 create database staff 语句，创建库；通过 use staff 语句，打开库。

（3）创建表。

```
create table TStaff (
     sno                    varchar(9) not null,
     name                   varchar(30),
     birthday               date,
```

```
        address                 varchar(180),
        tel                     varchar(18),
        primary key (sno)
);
create table TUser (
        uname                   varchar(30) not null,
        sno                     varchar(9) not null,
        pwd                     varchar(20),
        role                    int,
        primary key (sno),
        key AK_Key_2 (uname)
);
alter table TUser add constraint FK_Reference_1 foreign key (sno)
                                 references TStaff (sno);
```

（4）在 StaffUser 系统中导入 MySQL 驱动包 mysql-connector-java-8.0.11.jar。

（5）配置数据库的连接信息，新建 db.properties 文件。

```
driver=com.mysql.cj.jdbc.Driver
url=jdbc:mysql://localhost:3306/staff?useSSL=false
          &serverTimezone=UTC&allowPublicKeyRetrieval=true
username=root
password=123456
```

（6）封装单例类 DbInfo，读取数据库连接信息。

```
public class DbInfo {
private static DbInfo info;         //单例
private String driver;
private String url;
private String uname;
private String pwd;
public String getDriver() {
    return driver;
}
public String getUrl() {
    return url;
}
public String getUname() {
    return uname;
}
public String getPwd() {
    return pwd;
}
private DbInfo() {
    //构造函数私有化
}
static {
    info = new DbInfo();
    InputStream fis = null;
    try {
        String fname = DbInfo.class.getResource("/").getPath() +"db.properties";
```

```java
            Properties prop = new Properties();
            fis = new FileInputStream(new File(fname));
            prop.load(fis);
            info.driver = prop.getProperty("driver");
            info.url = prop.getProperty("url");
            info.uname = prop.getProperty("username");
            info.pwd = prop.getProperty("password");
        } catch (Exception e) {
            Log.logger.error(e.getMessage());
        }finally{
            if (fis != null) {
                try {
                    fis.close();
                } catch (IOException e) {
                    Log.logger.error(e.getMessage());
                }
            }
        }
    }
    public static DbInfo instance(){
        return info;
    }
}
```

2. 封装数据库连接

使用 ThreadLocal 封装数据库的连接，保证每个线程最多持有一个数据库连接。这个封装非常重要，可以解决在持久层打开数据库、在服务层关闭数据库的问题。在 Hibernate 模式下，由于懒加载特性的存在，需要在视图层关闭数据库连接——使用 ThreadLocal 可以关闭。

```java
    public class DbFactory {
        private static ThreadLocal<Connection> tlocal = new ThreadLocal<>();
        public static Connection openConnection() throws
                                ClassNotFoundException,SQLException{
            Connection conn = tlocal.get();
            try {
                if(conn == null || conn.isClosed()) {
                    DbInfo db = DbInfo.instance();
                    Class.forName(db.getDriver());
                    conn =  DriverManager.getConnection(
                                        db.getUrl(),db.getUname(),db.getPwd());
                    tlocal.set(conn);
                    Log.logger.info(Thread.currentThread().getId()
                                    + "打开数据库,生成一个新连接......");
                }else {
                    Log.logger.info(Thread.currentThread().getId()
                                    + "打开数据库,使用原有连接......");
                }
            } catch (ClassNotFoundException e) {
```

```
                Log.logger.error(e.getMessage(),e);
                throw e;
            }catch( SQLException e) {
                Log.logger.error(e.getMessage(),e);
                throw e;
            }
            return conn;
        }
        public static void closeConnection(){
            Log.logger.info(Thread.currentThread().getId() + "关闭数据库......");
            Connection conn = tlocal.get();
            tlocal.set(null);
            if(conn != null) {
                try {
                    conn.close();
                } catch (Exception e) {
                    Log.logger.error(e.getMessage(),e);
                }
            }
        }
    }
```

3. 实现数据库的自动打开和关闭

基于前面讲解的 DbFactory 和 Proxy，编写动态代理类，实现数据库连接的自动打开和关闭。操作步骤如下。

（1）编写代理类。

```
public class DbProxy implements InvocationHandler {
    private Object target;
    public Object bind(Object target) {
        this.target = target;
        return Proxy.newProxyInstance(target.getClass().getClassLoader(),
                            target.getClass().getInterfaces(), this);
    }
    @Override
    public Object invoke(Object proxy, Method method, Object[] args)
                                                throws Throwable {
        Object result = null;
        DbFactory.openConnection();
        try {
            result = method.invoke(target, args);
        } finally {
            DbFactory.closeConnection();
        }
        return result;
    }
}
```

（2）定义服务层。

```
public class UserBiz implements IUser{
    @Override
```

```
    public User login(String uname, String pwd) throws Exception {
        Log.logger.info(Thread.currentThread().getId() + ":" + uname + " login ..");
        IUserDao dao = new UserDaoMysql();
        return dao.login(uname, pwd);
    }
}
```

（3）定义持久层。

```
public class UserDaoMysql implements IUserDao {
    public User login(String name, String pwd) throws Exception {
        User user = null;
        DbFactory.openConnection();    //获取动态代理已打开的数据库连接
        // 用伪代码模拟用户登录
        if (name.equals("admin") && pwd.equals("123")) {
            user = new User();
            user.setRole(IRole.ADMIN);
            user.setUname(name);
            user.setPwd(pwd);
        } else if (name.equals("tom") && pwd.equals("123")) {
            user = new User();
            user.setRole(IRole.COMMON_USER);
            user.setUname(name);
            user.setPwd(pwd);
        } else if (name.equals("jack") && pwd.equals("123")) {
            user = new User();
            user.setRole(IRole.VIP_USER);
            user.setUname(name);
            user.setPwd(pwd);
        } else {
        }
        return user;
    }
}
```

（4）调用 UI 代码，使用多线程模拟多用户并发访问。

```
public static void main(String[] args) {
        ExecutorService pool = Executors.newCachedThreadPool();
        for(int i=0;i<3;i++) {
            pool.execute(new Runnable() {
                @Override
                public void run() {
                    DbProxy proxy = new DbProxy();
                    IUser userProxy = (IUser)proxy.bind(new UserBiz());
                    try {
                         userProxy.login("tom", "123456");
                    } catch (Exception e) {
                    }
                }
            });
        }
        pool.shutdown();
}
```

测试结果如下。

```
INFO - 10 打开数据库,生成一个新连接......
INFO - 8 打开数据库,生成一个新连接......
INFO - 9 打开数据库,生成一个新连接......
INFO - 8:tom login ....
INFO - 10:tom login ....
INFO - 9:tom login ....
INFO - 8 打开数据库,使用原有连接......
INFO - 8 关闭数据库......
INFO - 9 打开数据库,使用原有连接......
INFO - 9 关闭数据库......
INFO - 10 打开数据库,使用原有连接......
INFO - 10 关闭数据库..
```

通过测试发现，即使业务操作出现异常，也不影响数据库连接的获得与释放。在高并发的环境下，每个线程使用自己的数据库连接，线程间不产生冲突。

5.4.4 项目案例：基于动态代理实现 StaffUser 系统的事务处理

事务管理是所有业务系统的核心操作，也是 Spring Framework 要解决的核心问题。

1. 事务的 ACID 特性与本地事务的管理

事务具有原子性（Atomicity）、一致性（Consistency）、隔离性（Isolation）、持久性（Durability），简称 ACID 特性。本地事务（local transaction）使用单个资源管理器管理本地资源。本地事务的管理见图 5-3。

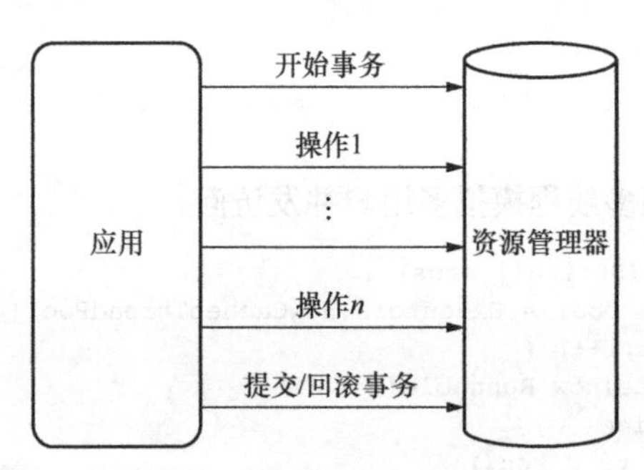

图 5-3 本地事务的管理

2. 通过本地事务新增员工和用户

在添加新员工时，自动添加系统的一个用户。用户名默认为员工编号，密码默认为身份证号码的后 6 位。用户第一次登录时可提示修改密码。用户名为用户的姓名或昵称，用于显示，不做唯一性校验。用事务的 ACID 特性保证员工和默认用户同时添加成功。通过本地事务新增员工和用户的具体操作如下。

（1）在 DbFactory 中封装 JDBC 的事务操作。

```
public class DbFactory {
    private static ThreadLocal<Connection> tlocal = new ThreadLocal<>();
    /**
     * 开启事务
     * @throws Exception
     */
    public static void beginTransaction() throws Exception{
       Connection conn = openConnection();
       conn.setAutoCommit(false);
    }
    /**
     * 提交事务
     * @throws Exception
     */
    public static void commit() throws Exception {
         Connection conn = openConnection();
         conn.commit();
    }
    /**
     * 回滚事务
     * @throws Exception
     */
    public static void rollback() throws Exception{
         Connection conn = openConnection();
         conn.rollback();
    }
```

（2）在服务层控制事务，保证同时添加员工和用户。

```
public class StaffBiz implements IStaff{
      /**
       * 使用本地事务，保证员工和用户信息同时写入成功
       */
      public void addStaffUser(TStaff staff) throws Exception {
           TUser user = new TUser();
           user.setUname(staff.getSno());
           user.setPwd("123");
           user.setRole(2);
           user.setSno(staff.getSno());
           //打开数据库
           DbFactory.openConnection();
           //开启事务
           DbFactory.beginTransaction();
           StaffDao staffDao = new StaffDao();
           UserDao userDao = new UserDao();
           try {
                staffDao.addStaff(staff);
                userDao.addUser(user);
                //若无异常，提交事务
                DbFactory.commit();
           } catch (Exception e) {
                //若出现异常，回滚事务
                DbFactory.rollback();
```

```
            throw e;    //异常信息要再次抛出
        }finally{
            DbFactory.closeConnection(); //关闭数据库
        }
    }
}
```

（3）在持久层分别添加员工和用户。添加员工的操作在 StaffDao 中完成，添加用户的操作在 UserDao 中完成。多数情况下，一个 DAO 类对应一个表的操作。

```
public class StaffDao {
    //添加员工
    public void addStaff(TStaff staff) throws Exception{
        String sql = "insert into tstaff values(?,?,?,?,?)";
        Connection conn = DbFactory.openConnection();
        PreparedStatement ps = conn.prepareStatement(sql);
        ps.setString(1, staff.getSno());
        ps.setString(2, staff.getName());
        ps.setDate(3, new java.sql.Date(staff.getBirthday().getTime()));
        ps.setString(4, staff.getAddress());
        ps.setString(5,staff.getTel());
        ps.executeUpdate();
        ps.close();
    }
}
public class UserDao {
    //添加用户
    public void addUser(TUser user) throws Exception {
        String sql = "insert into tuser values(?,?,?,?)";
        Connection conn = DbFactory.openConnection();
        PreparedStatement ps = conn.prepareStatement(sql);
        ps.setString(1, user.getUname());
        ps.setString(2, user.getSno());
        ps.setString(3, user.getPwd());
        ps.setInt(4,user.getRole());
        ps.executeUpdate();
        ps.close();
    }
}
```

（4）测试视图层。

```
public static void main(String[] args) {
    StaffBiz biz = new StaffBiz();
    TStaff staff = new TStaff();
    staff.setSno("121000127");
    staff.setName("tom");
    SimpleDateFormat sdf = new SimpleDateFormat("yyyy-MM-dd");
    try {
        staff.setBirthday(sdf.parse("1995-10-1"));
    } catch (Exception e) {
        e.printStackTrace();
    }
```

```
        staff.setAddress("北京朝阳区建国门");
        staff.setTel("13522454666");
        try {
            biz.addStaffUser(staff);
            System.out.println(staff.getSno() + "创建成功....");
        } catch (Exception e) {
            e.printStackTrace();
        }
    }
```

测试结果显示，员工和用户同时添加成功。

（5）在 addUser 中主动抛出异常，测试事务回滚。

① 测试代码如下。

```
public class UserDao {
    //添加用户
    public void addUser(TUser user) throws Exception {
        String sql = "insert into tuser values(?,?,?,?)";
        Connection conn = DbFactory.openConnection();
        PreparedStatement ps = conn.prepareStatement(sql);
        ps.setString(1, user.getUname());
        ps.setString(2, user.getSno());
        ps.setString(3, user.getPwd());
        ps.setInt(4,user.getRole());
        ps.executeUpdate();
        ps.close();
        throw new RuntimeException("事务回滚测试...");
    }
}
```

② 测试结果如下。

```
log4j:WARN Please initialize the log4j system properly.
java.lang.RuntimeException: 事务回滚测试...
    at com.icss.dao.UserDao.addUser(UserDao.java:21)
    at com.icss.biz.StaffBiz.addStaffUser(StaffBiz.java:27)
    at com.icss.ui.Test.main(Test.java:23)
```

检查数据库中的数据，若发现没有出现错误数据，表明事务正确回滚。

3. 通过动态代理管理事务

前面采用编程式事务管理模式完成了添加员工的操作，下面演示如何通过动态代理管理事务。这两种事务管理模式都非常重要。具体操作如下。

（1）实现动态代理。

```
public class TransacationProxy implements InvocationHandler {
    private Object target;
    public Object bind(Object target) {
```

```
        this.target = target;
        return Proxy.newProxyInstance(target.getClass().getClassLoader(),
                                        target.getClass().getInterfaces(), this);
    }
    @Override
    public Object invoke(Object proxy, Method method, Object[] args)
                                                throws Throwable {
        Object result = null;
        // 打开数据库
        DbFactory.openConnection();
        // 开启事务
        DbFactory.beginTransaction();
        try {
            result = method.invoke(target, args);
            // 事务提交
            DbFactory.commit();
        } catch (Exception e) {
            // 事务回滚
            DbFactory.rollback();
            throw e; // 第二次抛出异常
        } finally {
            // 关闭数据库
            DbFactory.closeConnection();
        }
        return result;
    }
}
```

（2）把新增员工和用户的业务逻辑代码简化为自动控制事务。通过动态代理管理事务大幅减少了业务部分的代码，使开发人员的精力可以集中在业务逻辑本身上。

```
public class StaffBiz implements IStaff{
    /**
     * 通过动态代理管理本地事务，保证员工和用户信息同时写入成功
     */
    public void addStaffUser(Staff staff) throws Exception {
        Log.logger.info(Thread.currentThread().getId() + ":addStaffUser()");
        User user = new User();
        user.setSno(staff.getSno());
        user.setUname(staff.getSno());
        user.setRole(2);
        user.setPwd("1234");
        StaffDao staffDao = new StaffDao();
        UserDao userDao = new UserDao();
        staffDao.addStaff(staff);
        userDao.addUser(user);
    }
}
```

（3）持久层的代码不变，视图层使用动态代理类进行测试。

① 测试代码如下。

```
public class TestAddStaffUser {
    public static void main(String[] args) {
        Staff staff = new Staff();
        staff.setSno("101000129");
        staff.setName("jack");
        SimpleDateFormat sdf = new SimpleDateFormat("yyyy-MM-dd");
        try {
            staff.setBirthday(sdf.parse("1995-10-1"));
        } catch (Exception e) {
        }
        staff.setAddress("北京朝阳区建国门");
        staff.setTel("13522454666");
        //绑定动态代理
        TransacationProxy proxy = new TransacationProxy();
        IStaff staffProxy = (IStaff)proxy.bind(new StaffBiz() );
        try {
            staffProxy.addStaffUser(staff);
            System.out.println(staff.getSno() + "创建成功....");
        } catch (Exception e) {
            e.printStackTrace();
        }
    }
}
```

② 测试结果如下。

```
INFO - 1 打开数据库,生成一个新连接......
INFO - 1 打开数据库,使用原有连接......
INFO - 1 开启事务
INFO - 1:addStaffUser()
INFO - 1:addStaff()
INFO - 1 打开数据库,使用原有连接......
INFO - 1:addUser()
INFO - 1 打开数据库,使用原有连接......
INFO - 1 打开数据库,使用原有连接......
INFO - 1 提交事务
INFO - 1 关闭数据库......
101000129 创建成功....
```

（4）异常测试。如果正常添加员工、添加用户，数据提交前主动抛出异常，观察已添加的员工和用户数据是否能够回滚。

① 测试代码如下。

```
public class UserDao {
    public void addUser(User user) throws Exception {
        Log.logger.info(Thread.currentThread().getId() + ":addUser()");
        String sql = "insert into tuser values(?,?,?,?)";
        Connection conn = DbFactory.openConnection();
        PreparedStatement ps = conn.prepareStatement(sql);
        ps.setString(1, user.getUname());
        ps.setString(2, user.getSno());
        ps.setString(3, user.getPwd());
        ps.setInt(4,user.getRole());
```

```
            ps.executeUpdate();
            ps.close();
            throw new RuntimeException("事务回滚测试...");
        }
    }
```

② 测试结果如下。

```
INFO - 1打开数据库,生成一个新连接......
INFO - 1打开数据库,使用原有连接......
INFO - 1开启事务
INFO - 1:addStaffUser()
INFO - 1:addStaff()
INFO - 1打开数据库,使用原有连接......
INFO - 1:addUser()
INFO - 1打开数据库,使用原有连接......
INFO - 1打开数据库,使用原有连接......
INFO - 1回滚事务
INFO - 1关闭数据库......
Caused by: java.lang.RuntimeException: 事务回滚测试...
    at com.icss.dao.UserDao.addUser(UserDao.java:37)
    at com.icss.biz.impl.StaffBiz.addStaffUser(StaffBiz.java:24)
```

检查数据库中的数据，如果没有出现错误数据，表明事务正确回滚。

4. 自定义注解@Transaction

在 StaffUser 系统的业务操作中，addStaffUser( )需要事务处理，而 login( )只需要数据库连接，不需要事务处理。如果所有的方法都需要事务处理，则系统性能会受到很大影响。同理，有些业务方法是不需要数据库连接的，我们可以采用相同的识别方式。

如何识别一个业务方法是否需要事务处理呢？可以采用非侵入式 XML 配置和自定义注解。下面演示基于自定义注解的解决方案。

（1）新增注解@Transaction。

```
@Target(ElementType.METHOD)
@Retention(RetentionPolicy.RUNTIME)
public @interface Transaction {
}
```

（2）在接口上使用注解。

```
public interface IStaff {
    @Transaction
    public void addStaffUser(Staff staff) throws Exception;
}
```

（3）在事务的动态代理类 TransacationProxy 中增加注解判断。

```
public Object invoke(Object proxy, Method method, Object[] args)
                                        throws Throwable {
```

```
        Object result = null;
        //打开数据库
        DbFactory.openConnection();
        if (method.isAnnotationPresent(Transaction.class)) {
            //开启事务
            DbFactory.beginTransaction();
        }
        try {
            result = method.invoke(target, args);
            if (method.isAnnotationPresent(Transaction.class)) {
                //提交事务
                DbFactory.commit();
            }
        } catch (Exception e) {
            if (method.isAnnotationPresent(Transaction.class)) {
                //回滚事务
                DbFactory.rollback();
            }
            throw e; //再次抛出异常
        } finally {
            //关闭数据库
            DbFactory.closeConnection();
        }
        return result;
    }
```

（4）添加新员工测试（测试代码参见本节中“通过动态代理管理事务”小节的测试代码）。测试结果如下。

```
INFO - 1 打开数据库,生成一个新连接......
INFO - 1 打开数据库,使用原有连接......
INFO - 1 开启事务
INFO - 1:addStaffUser()
INFO - 1:addStaff()
INFO - 1 打开数据库,使用原有连接......
INFO - 1:addUser()
INFO - 1 打开数据库,使用原有连接......
INFO - 1 打开数据库,使用原有连接......
INFO - 1 提交事务
INFO - 1 关闭数据库......
101000130 创建成功....
```

（5）进行用户登录测试。

① 测试代码如下。

```
public static void main(String[] args) {
    TransacationProxy proxy = new TransacationProxy();
    IUser userProxy = (IUser) proxy.bind(new UserBiz());
    try {
        userProxy.login("tom", "123456");
    } catch (Exception e) {
    }
}
```

② 测试结果如下。

```
INFO - 1打开数据库,生成一个新连接......
INFO - 1:UserBiz-->>login()
INFO - 1:UserDao-->>login()
INFO - 1关闭数据库......
```

总结：使用自定义注解@Transaction，可以有效识别业务方法是否需要事务管理，这对业务系统的性能提升有很大好处。

5.4.5 项目案例：基于 AspectJ 实现动态的事务管理

前面介绍了如何使用动态代理实现 StaffUser 系统的事务管理，现在我们基于 Spring 的 AspectJ 来实现动态的事务管理。

如果使用动态代理实现员工系统的事务管理，每次调用业务方法前，都需要动态绑定并创建动态代理，这样操作很麻烦。在基于 AspectJ 的方案中，没有动态绑定，只需要指明切入点和通知即可，代码会更加简洁。

1．配置扫描信息

配置 AspectJ 支持，以及业务 Bean 扫描。

为了管理业务层的事务，所有业务类必须是 Bean 对象，而持久层对象是否是 Bean 对象不影响操作。因此，此处配置只扫描业务层。

```
<beans xmlns="http://www.springframework.org/schema/beans...">
    <aop:aspectj-autoproxy />
    <context:component-scan      base-package="com.icss.biz" />
</beans>
```

2．编写切面和通知

为了在环绕通知中做事务处理，所有资源的释放必须放在最终通知中。对于不需要事务操作的方法，在前置通知中打开数据库。

```
@Component
@Aspect
public class DbProxy {
    @Pointcut("execution(public * com.icss.biz.*.*(..))")
    private void businessOperate() {
    }
    @Before("businessOperate()")
    public void openDataBase(JoinPoint jp) throws Exception{
        Log.logger.info(Thread.currentThread().getId()
                    + ":openDataBase..." + jp.toString());
        DbFactory.openConnection();
```

```
    }
    @After("businessOperate()")
    public void closeDataBase(JoinPoint jp) {
        Log.logger.info(Thread.currentThread().getId()
                        + ":closeDataBase..." + jp.toString());
        DbFactory.closeConnection();   //在最终通知中关闭数据库
    }
    @Around("businessOperate()")
    public Object transaction(ProceedingJoinPoint pjp) throws Throwable{
        Object obj = null;
        Log.logger.info(pjp.toString());
        MethodSignature ms = (MethodSignature)pjp.getSignature();
       Method m = ms.getMethod();
        try {
            if(m.isAnnotationPresent(Transaction.class)) {
                DbFactory.beginTransaction();     //开启事务
            }
             obj = pjp.proceed();            //必须要有返回值
            if(m.isAnnotationPresent(Transaction.class)) {
              DbFactory.commit();                 //提交事务
            }
        } catch (Throwable e) {
              if(m.isAnnotationPresent(Transaction.class)) {
                  DbFactory.rollback();           //回滚事务
              }
             throw e;                            //必须要抛出异常
        }
        return obj;
    }
}
```

3. 编写逻辑代码

为了使用 JDK 的 Proxy 作为动态代理，所有业务类必须实现接口。用 AspectJ 控制事务，业务类可以没有接口，它会选择使用 CGLIB 作为动态代理。

IStaff 接口与类 StaffBiz 的实现与 5.4.5 节的最后一个示例相同。

```
public interface IStaff {
    @Transaction
    public void addStaffUser(Staff staff) throws Exception;
}
@Service
public class StaffBiz implements IStaff{}
```

4. 编写测试代码

测试代码如下。

```
public static void main(String[] args) {
    Staff staff = new Staff();
```

```
    staff.setSno("121000128");
    staff.setName("jack");
    SimpleDateFormat sdf = new SimpleDateFormat("yyyy-MM-dd");
    try {
        staff.setBirthday(sdf.parse("1995-10-1"));
    } catch (Exception e) {
    }
    staff.setAddress("北京朝阳区建国门");
    staff.setTel("13522454666");
     //获得Spring的Bean对象
    IStaff staffProxy = (IStaff)BeanFactory.getBean(IStaff.class);
    try {
        staffProxy.addStaffUser(staff);
        System.out.println(staff.getSno() + "创建成功....");
    } catch (Exception e) {
        e.printStackTrace();
    }
}
```

测试结果如下。

```
INFO - 1打开数据库,生成一个新连接......
INFO - 1开启事务
INFO - 1:openDataBase..execution(void com.icss.biz.IStaff.addStaffUser(Staff))
INFO - 1打开数据库,使用原有连接......
INFO - 1:addStaffUser()
INFO - 1:addStaff()
INFO - 1打开数据库,使用原有连接......
INFO - 1:addUser()
INFO - 1打开数据库,使用原有连接......
INFO - 1打开数据库,使用原有连接......
INFO - 1提交事务
INFO - 1:closeDataBase..execution(void com.icss.biz.IStaff.addStaffUser(Staff))
INFO - 1关闭数据库......
121000128创建成功....
```

5. 事务异常测试

成功添加员工后，如果在添加用户时主动抛出异常，观察新增的员工数据是否回滚了。

```
public void addUser(User user) throws Exception {
    Log.logger.info(Thread.currentThread().getId() + ":addUser()");
    String sql = "insert into tuser values(?,?,?,?)";
    Connection conn = DbFactory.openConnection();
    PreparedStatement ps = conn.prepareStatement(sql);
    ps.setString(1, user.getUname());
    ps.setString(2, user.getSno());
    ps.setString(3, user.getPwd());
    ps.setInt(4,user.getRole());
    ps.executeUpdate();
    ps.close();
    throw new RuntimeException("事务回滚测试...");
}
```

测试结果如下。

```
INFO - 1 打开数据库,生成一个新连接......
INFO - 1 开启事务
INFO - 1:openDataBase..execution(void com.icss.biz.IStaff.addStaffUser(Staff))
INFO - 1 打开数据库,使用原有连接......
INFO - 1:addStaffUser()
INFO - 1:addStaff()
INFO - 1 打开数据库,使用原有连接......
INFO - 1:addUser()
INFO - 1 打开数据库,使用原有连接......
INFO - 1 打开数据库,使用原有连接......
INFO - 1 回滚事务
INFO - 1:closeDataBase..execution(void com.icss.biz.IStaff.addStaffUser(Staff))
INFO - 1 关闭数据库......
    java.lang.RuntimeException: 事务回滚测试...
   at com.icss.dao.UserDao.addUser(UserDao.java:37)
   at com.icss.biz.impl.StaffBiz.addStaffUser(StaffBiz.java:28)
```

检查数据库数据，如果没有错误数据，表明事务正确回滚。

6. 用户登录测试

对于没有事务操作的用户登录，只需要数据库连接，如果增加事务操作，会带来性能压力。

测试代码如下。

```
public static void main(String[] args) {
    //获得 Spring 的 Bean 对象
    IUser userProxy = (IUser)BeanFactory.getBean(IUser.class);
    try {
        userProxy.login("tom", "123456");
    } catch (Exception e) {
    }
}
```

测试结果如下。

```
INFO - 1 打开数据库,生成一个新连接......
INFO - 1:UserBiz-->>login()
INFO - 1:UserDao-->>login()
INFO - 1:closeDataBase..execution(User com.icss.biz.IUser.login(String,String))
INFO - 1 关闭数据库......
```

第 6 章 整合数据层

Spring 整合了数据层访问技术，如整合 JDBC、Hibernate、MyBatis 等。整合数据层的核心是使用 Spring 的统一事务管理模型，管理数据访问层中各种框架的事务。

6.1 事务分类

关系型数据库的核心是事务。事务可以分为本地事务和全局事务。

- 本地事务（local transaction）：使用单一资源管理器，管理本地资源。
- 全局事务（global transaction）：通过事务管理器和多种资源管理器，管理多种不同类型的资源，如同时管理 Java 数据库连接（Java Database Connection，JDBC）资源和 Java 消息服务（Java Message Service，JMS）资源。

事务还可以细分为编程式事务、声明性事务、JTA 事务、容器管理事务等。

- 编程式事务：通过编码方式开启事务、提交事务、回滚事务。
- 声明性事务：通过 XML 配置或注解，实现事务管理。Spring AOP 和 EJB 都是声明性事务。
- JTA 事务：Java 事务 API 使用 javax.transaction.UserTransaction 接口，访问多种资源管理器。JTA 事务可以被容器或应用组件调用。

- 容器管理事务（Container Management Transaction，CMT）：通过容器自动控制事务的开启、提交和回滚。开发人员不需要手工编写代码，配置注解，由容器来控制事务的边界。

关于 CMT 的示例代码如下。

```
<enterprise-beans>
    <session>
        <ejb-name>BookStore-BookBean</ejb-name>
        <ejb-class>com.icss.test.biz.BookBiz</ejb-class>
        <remote>com.icss.test.biz.BookRemote</remote>
        <local>com.icss.test.biz.BookLocal</local>
        <sesion-type>Stateless</sesion-type>
    </session>
</enterprise-beans>
@Stateless
public class BookBiz implements BookLocal,BookRemote{    }
```

关于 JTA 事务的示例代码如下。

```
@Resource UserTransacton tx;
public  void  updateData(...){
    tx.begin();         //开启 JTA 事务
      ......            //实现业务操作
      tx.commit();      //提交事务
}
```

Java EE 事务模型（见图 6-1）使用 EJB 解决事务。注意，EJB 之间存在调用关系，在 Dubbo 和 Spring Cloud 架构中，服务之间的调用仍然很关键。

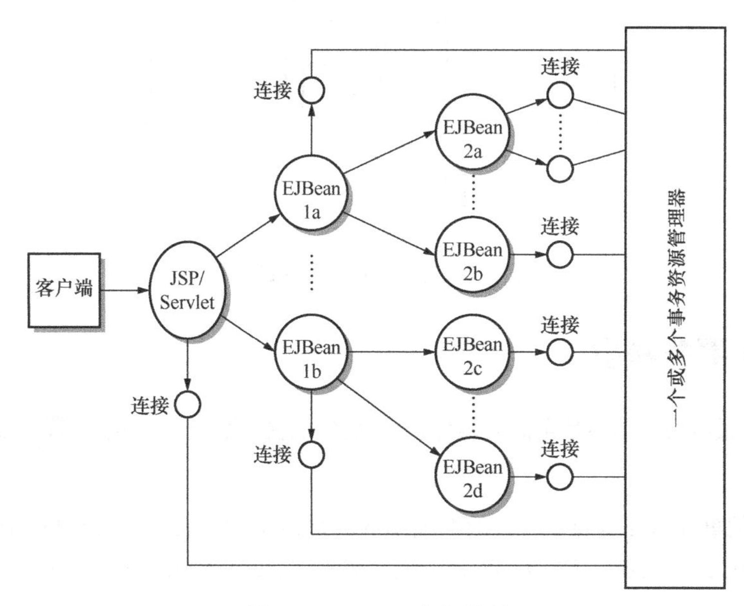

图 6-1　Java EE 事务模型

XA 事务模型（见图 6-2（a）与（b））使用 XA 的两段提交协议来管理全局事务。提交 XA 事务的过程如图 6-3 所示，事务管理器和资源管理器的分离是关键，只有多个资源管理器都提交成功，事务管理器才最终提交。

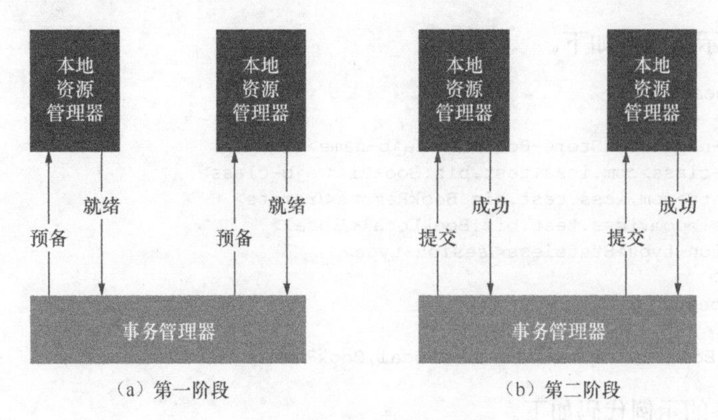

图 6-2　XA 事务模型

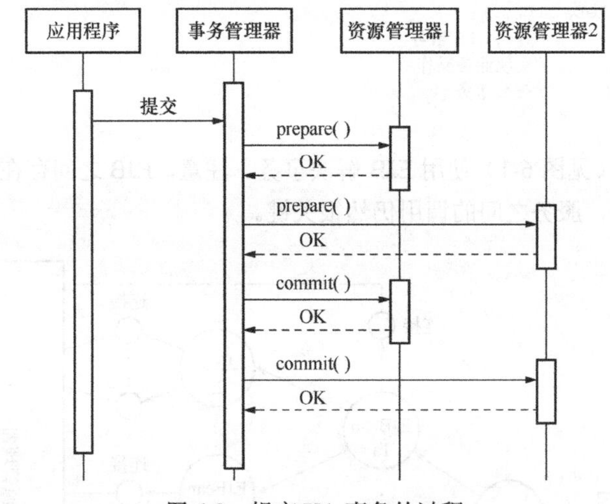

图 6-3　提交 XA 事务的过程

6.2 Spring 事务模型

Spring 框架最核心的功能是全面的事务管理功能。它提供了一致的事务管理抽象，这带来了以下好处。

- 为复杂的事务 API 提供了一致的编程模型，如 JTA、JDBC、Hibernate、Java 持久化 API（Java Persistence API，JPA）和 Java 数据对象（Java Data Object，JDO）。
- 支持声明式事务管理。
- 提供比大多数复杂的事务 API（如 JTA）更简单的、更易于使用的编程式事务管理 API。

- 支持整合各种数据访问层抽象。

全局事务有一个缺陷，即代码需要使用 JTA，这是一套笨重的 API（因为它的异常模型）。此外，JTA 的 UserTransaction 通常需要从 JNDI 获得，这意味着为了 JTA，需要同时使用 Java 命名和目录接口（Java Naming and Directory Interface，JNDI）和 JTA。JTA 需要应用服务器环境的支持。基于 EJB 的 CMT 管理是声明性事务模式，EJB 3 之前的配置过于复杂，常被人诟病。

关于 JTA 与 JNDI 的示例代码如下。

```
public void updateData(...) {
    Context  initCtx = new InitialContext();
    UserTransaction tx = (UserTransaction)initCtx.lookup("java:comp/UserTransaction");
    tx.begin();
        ......
    tx.commit();
}
```

本地事务容易使用，但也有明显的缺点，即它们不能用于多个事务性资源。例如，使用 JDBC 事务管理的代码不能用于全局 JTA 事务。另一个缺点是本地事务为侵入式编程模型，编程麻烦。

Spring 解决了这些问题。它使应用开发者能够在不同环境下使用一致的编程模型。用户只需要写一次代码，就可以在不同事务环境下的不同事务策略切换时享受好处。

Spring 框架同时提供声明式事务管理和编程式事务管理，声明式事务管理是多数使用者的首选。

有了 Spring 事务管理，还需要使用应用服务器进行事务管理吗？

Spring 框架对事务的管理，改变了传统上认为企业级 Java 应用必须使用应用服务器的认识。尤其对于中小型企业，使用笨重的 EJB 太过麻烦。大型企业到底选择 Spring，还是 EJB，就是仁者见仁、智者见智的事情了。

典型情况下，只有需要管理多个资源的事务时，才需要用到应用服务器的 JTA 功能。Spring Framework 允许用户在需要时，把代码很容易地迁移到应用服务器。用 EJB 的 CMT 或 JTA 管理本地事务（如 JDBC），不再是唯一的选择。

总之，Spring 提供了轻量级的声明性事务方案，解除了事务对重量级应用服务器的依赖；解决了编程式事务的代码耦合；通过一致的编程模型，解决了不同事务环境的迁移问题。

6.3 Spring 事务抽象模型

Spring 事务策略统一由 PlatformTransactionManager 接口描述。

```
package org.springframework.transaction;
public interface PlatformTransactionManager {
    TransactionStatus getTransaction(TransactionDefinition definition)
                                        throws TransactionException;
```

```
    void commit(TransactionStatus status) throws TransactionException;
    void rollback(TransactionStatus status) throws TransactionException;
}
```

对于不同的数据层访问技术，Spring 提供了具体的事务管理器，如 HibernateTransactionManager、JpaTransactionManager、DataSourceTransactionManager 和 JtaTransactionManager 等。Spring 事务的统一接口见图 6-4。

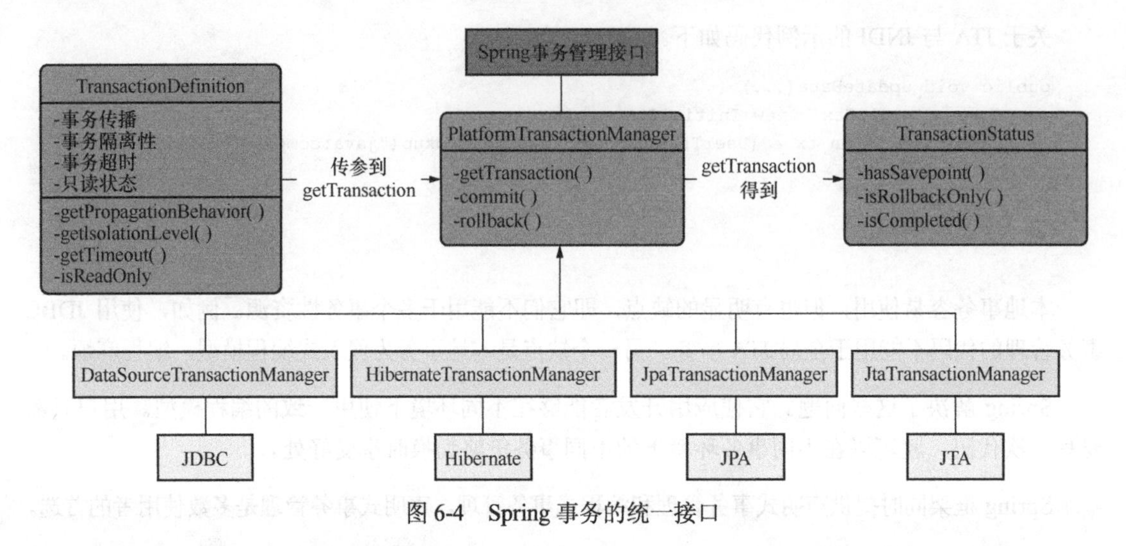

图 6-4　Spring 事务的统一接口

TransactionStatus 用于描述事务状态。

```
public interface TransactionStatus extends SavepointManager, Flushable {
    boolean isNewTransaction();
    boolean hasSavepoint();
    void setRollbackOnly();
    boolean isRollbackOnly();
    void flush();
    boolean isCompleted();
}
```

TransactionDefinition 接口定义了事务的属性信息。

- 事务隔离性：当前事务和其他事务的隔离程度。例如，这个事务能否看到其他事务未提交的写数据。
- 事务传播：当一个事务方法被另一个事务方法调用时，应该如何执行这个事务。如果一个事务上下文已经存在，有几个选项可以指定一个事务方法的执行方式。例如，简单地在现有的事务中继续运行，或者挂起现有事务，创建一个新的事务。Spring 提供 EJB CMT 中常见的事务传播选项。
- 事务超时：事务在超时前能运行多久。

- 只读状态：只读事务不修改任何数据。只读事务在某些情况下（如当使用 Hibernate 时）是一种非常有用的优化。

TransactionDefinition 定义了以下 7 个事务传播属性和 4 个事务隔离级别。

```
    int PROPAGATION_REQUIRED = 0;
    int PROPAGATION_SUPPORTS = 1;
    int PROPAGATION_MANDATORY = 2;
    int PROPAGATION_REQUIRES_NEW = 3;
    int PROPAGATION_NOT_SUPPORTED = 4;
    int PROPAGATION_NEVER = 5;
    int PROPAGATION_NESTED = 6;
    int ISOLATION_READ_UNCOMMITTED
                 = Connection.TRANSACTION_READ_UNCOMMITTED;
    int ISOLATION_READ_COMMITTED
                 = Connection.TRANSACTION_READ_COMMITTED;
    int ISOLATION_REPEATABLE_READ
                 = Connection.TRANSACTION_REPEATABLE_READ;
    int ISOLATION_SERIALIZABLE
                 = Connection.TRANSACTION_SERIALIZABLE;
}
```

DataSourceTransactionManager 是 PlatformTransactionManager 的实现类，可作为 JDBC 和 MyBatis 访问数据库的事务管理器。

```
<bean id="dataSource"
    class="org.springframework.jdbc.datasource.DriverManagerDataSource">
    <property name="driverClassName"
             value="com.mysql.cj.jdbc.Driver" />
    <property name="url"
            value="jdbc:mysql://localhost:3306/bk?useSSL=false" />
    <property name="username" value="root" />
    <property name="password" value="123456" />
</bean>
<bean id="txManager"
    class="org.springframework.jdbc.datasource.DataSourceTransactionManager">
    <property name="dataSource" ref="dataSource" />
</bean>
```

HibernateTransactionManager 是 PlatformTransactionManager 的实现类，可作为 Spring 管理 Hibernate 的事务管理器。

```
<bean id="sessionFactory"
    class="org.springframework.orm.hibernate3.LocalSessionFactoryBean">
    <property name="configLocation"
        value="classpath:hibernate.cfg.xml">
    </property>
   </bean>
   <bean id="txManager"
    class="org.springframework.orm.hibernate3.HibernateTransactionManager">
    <property name="sessionFactory" ref="sessionFactory" />
</bean>
```

JtaTransactionManager 是 PlatformTransactionManager 的实现类，可作为 Spring 集成数据源被应用服务器管理的 JTA 事务时使用的事务管理器。

```
<?xml version="1.0" encoding="UTF-8"?>
    <beans xmlns="http://www.springframework.org/schema/beans"
    xmlns:xsi="http://www.w3.org/2001/XMLSchema-instance"
    xmlns:jee="http://www.springframework.org/schema/jee"
    xsi:schemaLocation="
    http://www.springframework.org/schema/beans
    http://www.springframework.org/schema/beans/spring-beans.xsd
    http://www.springframework.org/schema/jee
    http://www.springframework.org/schema/jee/spring-jee.xsd">
        <jee:jndi-lookup id="dataSource" jndi-name="jdbc/jpetstore"/>
        <bean id="txManager"
              class="org.springframework.transaction.jta.JtaTransactionManager" />
    ......
</beans>
```

关于 JpaTransactionManager，本节不展开讨论。

6.4 事务与资源管理

使用了不同的事务管理器后，具体的资源如何控制呢？如果使用 DataSourceTransactionManager 封装了 JDBC 访问数据库的事务操作，那么封装之后，是使用传统的 JDBC 接口操作数据库，还是有其他更好的解决方案？

方案 1：使用高层级抽象，对底层资源的本地 API 进行封装，提供模板方法，如 JdbcTemplate、HibernateTemplate、JdoTemplate、SqlSessionTemplate 等。

方案 2：直接使用资源的本地持久化原生 API。Spring 使用 AOP 接管了数据库的打开、关闭，以及事务的提交、回滚等操作。要使用原生 API，需要从 Spring 的上下文中提取原始的 Connection、SqlSession、Hibernate Session 等原生对象。

Spring 使用包装类对这些原生对象进行管理，如 DataSourceUtils（用于 JDBC）、EntityManagerFactoryUtils（用于 JPA）、SessionFactoryUtils（用于 Hibernate）、PersistenceManager FactoryUtils（用于 JDO）。

在以下示例代码中，从 DataSourceUtils 中提取原生的 Connection 对象。

```
public abstract class DataSourceUtils {
    public static Connection getConnection(DataSource dataSource){
        return doGetConnection(dataSource);
    }
}
```

总之，JdbcTemplate 封装 JDBC 操作后，代码并不简洁，开发难度增加，好处不明显。HibernateTemplate 封装 Hibernate 操作后，对于使用复杂的原生 SQL，如复杂的多表多条件嵌套查询，使用 HibernateTemplate 的局限性很大，开发好处更不明显。推荐使用原生 API 操作资源数据。

6.5 Spring 声明性事务

Spring 的声明式事务管理是通过 Spring AOP 实现的。

Spring 声明式事务管理可以在任何环境下使用。只需要更改配置文件，该事务管理机制就可以和 JDBC、JDO、Hibernate 或其他的事务机制一起使用。

Spring 的事务代理步骤见图 6-5。

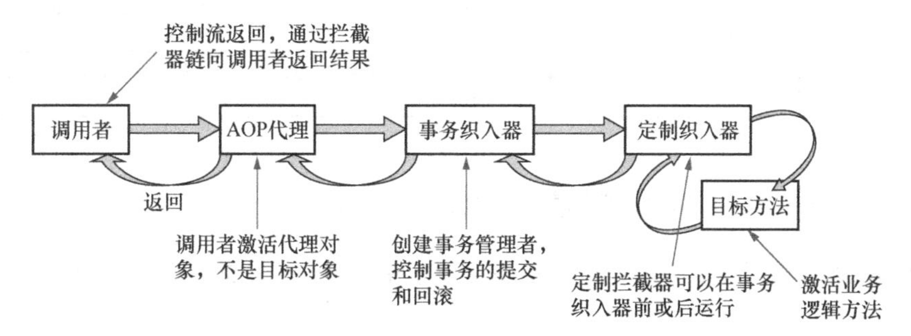

图 6-5 Spring 的事务处理步骤

6.5.1 使用 XML 管理声明性事务

使用 XML 管理声明性事务、配置事务策略，是非常有用的事务管理模式。首先，根据管理的资源，选择不同的事务管理器，如 DataSourceTransactionManager、HibernateTransactionManager、JtaTransactionManager 等。然后，配置切面和事务管理策略。前面讲述了如何在 XML 中配置切面、切入点、通知等。这里需要使用事务织入器<tx:advice>，配置事务策略。

从以下几方面设置<tx:advice>的属性。

- 设置事务传播行为，默认为 REQUIRED。
- 设置事务隔离级别。不同数据库的隔离级别不同，如 Oracle 的默认隔离级别是 read-committed，MySQL 的默认隔离级别是 repeatable-read。
- 设置是只读事务还是写事务。
- 设置事务超时时间，通常使用默认值，超时时间的长短依赖于事务系统。
- 设置异常回滚策略。Spring 的默认设置是任何 RuntimeException 将触发事务回滚，但是任何已检查的异常将不触发事务回滚。

最终，<tx:advice>的属性如表 6-1 所示。

表 6-1 <tx:advice>的属性

属性	是否需要	默认值	描述
name	是		与事务属性关联的方法名。通配符（*）可以用来指定一批关联到相同事务属性的方法。 如 get*、handle*、on*Event 等
propagation	不	REQUIRED	事务传播行为
isolation	不	DEFAULT	事务隔离级别
timeout	不	−1	事务超时时间（以秒为单位）
read-only	不	false	指定事务是否只读
rollback-for	不		将触发并回滚的异常，以逗号分开，如 com.foo.MyBusinessException，ServletException
no-rollback-for	不		将触发但不进行回滚的异常，以逗号分开，如 com.foo.MyBusinessException，ServletException

6.5.2 项目案例：使用 XML 配置 StaffUser 事务

使用 XML 配置声明性事务，管理 StaffUser 系统中员工与用户的同时插入操作。

1．导入模块和包

在图 6-6 所示的 Spring 模块中，JDBC、事务、Bean、Core、上下文、AOP 模块为 Spring 访问数据需要导入的模块。

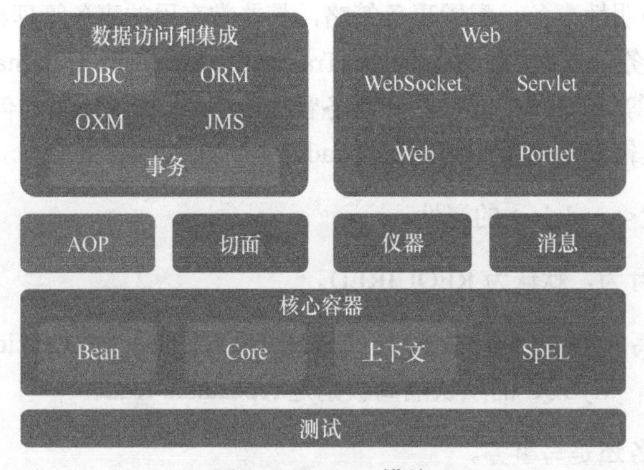

图 6-6 Spring 模块

导入核心包——spring-beans-4.3.19.jar、spring-core-4.3.19.jar、spring-context-4.3.19.jar、spring-context-support-4.3.19.jar，导入 AOP 包——spring-aop-4.3.19.jar，导入数据整合包——spring-jdbc-4.3.19.jar、spring-tx-4.3.19.jar，导入依赖包——commons.logging-1.1.jar、org.aspectj.weaver-1.6.8.jar。

2. 配置 Spring 的核心配置文件

配置 Spring 的核心配置文件 beans.xml，并配置相关内容。

（1）配置 schema，以支持 tx 命名空间。

```
<beans xmlns="http://www.springframework.org/schema/beans"
    xmlns:xsi="http://www.w3.org/2001/XMLSchema-instance"
    xmlns:aop="http://www.springframework.org/schema/aop"
    xmlns:tx="http://www.springframework.org/schema/tx"
    xsi:schemaLocation="http://www.springframework.org/schema/beans
        http://www.springframework.org/schema/beans/spring-beans-4.3.xsd
        http://www.springframework.org/schema/tx
        http://www.springframework.org/schema/tx/spring-tx-4.3.xsd
        http://www.springframework.org/schema/aop
        http://www.springframework.org/schema/aop/spring-aop-4.3.xsd">
</beans>
```

（2）配置数据源。

```
<bean id="dataSource"
     class="org.springframework.jdbc.datasource.DriverManagerDataSource">
    <property name="driverClassName" value="com.mysql.cj.jdbc.Driver" />
    <property name="url"
          value="jdbc:mysql://localhost:3306/staff?useSSL=false" />
    <property name="username" value="root" />
    <property name="password" value="123456" />
</bean>
```

（3）配置事务管理器，使用 DataSourceTransactionManager 管理 JDBC 事务。

```
<bean id="txManager"
    class="org.springframework.jdbc.datasource.DataSourceTransactionManager">
    <property name="dataSource" ref="dataSource" />
</bean>
```

（4）配置切面和事务策略。

注意：对于 Spring Framework 的事务，默认只有运行时异常和未检查的异常才会进行事务回滚。如果从事务方法中抛出已检查的异常，将不会进行事务回滚。因此，配置 rollback-for=Throwable 非常有必要。

read-only 表示只读事务，只读事务并不是一个强制选项，只是一个暗示，提示数据库系统，这个事务并不包含更改数据的操作，因此 JDBC 驱动程序和数据库就有可能根据这种情况对该事务进行一些特定的优化，例如，不安排相应的数据库锁，以减轻事务对数据库的压力。并不是所有数据库都支持只读事务，不同的数据库下会有不同的结果。

```
<aop:config>
    <aop:pointcut id="serviceOperation"
```

```
            expression="execution(* com.icss.biz.*.*(..)))" />
    <aop:advisor advice-ref="txAdvice" pointcut-ref="serviceOperation" />
</aop:config>
   <tx:advice id="txAdvice" transaction-manager="txManager">
    <tx:attributes>
        <tx:method name="delete*"  rollback-for="Throwable" />
        <tx:method name="update*"  rollback-for="Throwable" />
         <tx:method name="add*"  rollback-for="Throwable" />
         <tx:method name="*"  read-only="true" />
    </tx:attributes>
</tx:advice>
```

在上述配置中，除了写操作的事务外，还会使用只读事务处理，如果有些方法不需要数据库操作怎么办？这无疑会浪费数据库的连接。所以，要用 XML 方式精确管理声明性事务，配置会是比较复杂的。

（5）配置 Bean。

```
<bean id="userDao" class="com.icss.dao.UserDao">
    <property name="dataSource" ref="dataSource"></property>
</bean>
<bean id="staffDao" class="com.icss.dao.StaffDao">
    <property name="dataSource" ref="dataSource"></property>
</bean>
<bean id="staffBiz" class="com.icss.biz.StaffBiz">
    <property name="userDao" ref="userDao"></property>
    <property name="staffDao" ref="staffDao"></property>
</bean>
```

3. 逻辑层事务控制

逻辑层的操作（打开数据库、关闭数据库、开启事务、提交事务、回滚事务等操作）都被 XML 中配置的事务管理器所控制。因此，这里只需要重点关注业务本身即可。

```
public class StaffBiz implements IStaff{
    private StaffDao staffDao;
    private UserDao userDao;
    public void setStaffDao(StaffDao staffDao) {
        this.staffDao = staffDao;
    }
    public void setUserDao(UserDao userDao) {
        this.userDao = userDao;
    }

    public void addStaffUser(TStaff staff) throws Exception {
        Log.logger.info("StaffBiz-->>addStaffUser()");
        TUser user = new TUser();
        user.setSno(staff.getSno());
        user.setUname(staff.getSno());
        user.setRole(2);
        user.setPwd("1234");
```

```
            staffDao.addStaff(staff);
            userDao.addUser(user);
        }
    }
```

4. 在持久层中使用原生 API

首先，在 BaseDao 中继承 JDBCDaoSupport。此处采用 set 方法注入模式，需要注入 dataSource，供 JDBCDaoSupport 使用。

```
<bean id="userDao" class="com.icss.dao.UserDao">
  <property name="dataSource" ref="dataSource"></property>
</bean>
<bean id="staffDao" class="com.icss.dao.StaffDao">
   <property name="dataSource" ref="dataSource"></property>
</bean>
public abstract class BaseDao extends JdbcDaoSupport {

    public void setDataSource(org.springframework.jdbc
                        .datasource.DriverManagerDataSource dataSource) {
            super.setDataSource(dataSource);
    }
    public Connection openConnection() {
         return this.getConnection();
    }
}
```

然后，从 JDBCDaoSupport 中获取 Spring 上下文中存储的 Connection 对象，进行原生 API 操作。

```
public class StaffDao extends BaseDao{
    public void addStaff(TStaff staff) throws Exception{
        Log.logger.info("StaffDao-->>addStaff()");
        String sql = "insert into tstaff values(?,?,?,?,?)";
        Connection conn = this.openConnection();
        PreparedStatement ps = conn.prepareStatement(sql);
        ps.setString(1, staff.getSno());
        ps.setString(2, staff.getName());
        ps.setDate(3, new java.sql.Date(staff.getBirthday().getTime()));
        ps.setString(4, staff.getAddress());
        ps.setString(5,staff.getTel());
        ps.executeUpdate();
        ps.close();
    }
}
```

5. 事务测试

首先，进行事务提交测试。

测试代码如下。

```java
public static void main(String[] args) {
    TStaff staff = new TStaff();
    staff.setSno("121000131");
    staff.setName("tom");
    SimpleDateFormat sdf = new SimpleDateFormat("yyyy-MM-dd");
    try {
        staff.setBirthday(sdf.parse("1995-10-1"));
    } catch (Exception e) {
    }
    staff.setAddress("北京朝阳区建国门");
    staff.setTel("1352245466221");
    IStaff staffProxy = (IStaff)BeanFactory.getBean("staffBiz");
    try {
        staffProxy.addStaffUser(staff);
        System.out.println(staff.getSno() +"创建成功....");
    } catch (Exception e) {
        e.printStackTrace();
    }
}
```

测试结果如下。

```
INFO - StaffBiz-->>addStaffUser()
INFO - StaffDao-->>addStaff()
INFO - UserDao-->>addUser()
121000131 创建成功....
```

然后，进行事务回滚测试。

测试代码如下。

```java
public void addUser(TUser user) throws Exception {
    Log.logger.info("UserDao-->>addUser()");
    String sql = "insert into tuser values(?,?,?,?)";
    Connection conn = this.openConnection();
    PreparedStatement ps = conn.prepareStatement(sql);
    ps.setString(1, user.getUname());
    ps.setString(2, user.getSno());
    ps.setString(3, user.getPwd());
    ps.setInt(4,user.getRole());
    ps.executeUpdate();
    ps.close();
    throw new RuntimeException("异常测试....");
}
```

测试结果如下。

```
INFO - Loading XML bean definitions from class path resource [beans.xml]
INFO - Loaded JDBC driver: com.mysql.cj.jdbc.Driver
INFO - StaffBiz-->>addStaffUser()
INFO - StaffDao-->>addStaff()
INFO - UserDao-->>addUser()
java.lang.RuntimeException: 异常测试....
    at com.icss.dao.UserDao.addUser(UserDao.java:30)
```

```
    at com.icss.biz.StaffBiz.addStaffUser(StaffBiz.java:33)
```

检查 MySQL 数据库，以确认 TStaff 数据已回滚。

6.5.3 JDBCDaoSupport

不管使用原生的连接，还是使用 JdbcTemplate，都可以通过持久层对象继承 JdbcDaoSupport 的方式编写代码。使用 JdbcDaoSupport 前，需要通过 setDataSource( )方法指明数据源。

```
public abstract class JdbcDaoSupport extends DaoSupport {
    public final void setDataSource(DataSource dataSource) {}
    public final JdbcTemplate getJdbcTemplate() {}
    protected final Connection getConnection()
                throws CannotGetJdbcConnectionException{
        return DataSourceUtils.getConnection(getDataSource());
    }
}
```

调用 getJdbcTemplate( )方法，返回 JdbcTemplate 对象。调用 getConnection( )方法，返回 Spring 上下文中存储的 Connection 对象。

注意： 不能从 DataSource 中直接获取 Connection 对象，如果那样操作，返回的 Connection 对象不受 Spring 事务管理。getConnection( )方法是从 DataSourceUtils 中获取数据库连接的。JDBC 的所有连接操作都封装在 DataSourceUtils 中。

```
public abstract class DataSourceUtils {
    public static Connection getConnection(DataSource dataSource){
        return doGetConnection(dataSource);
    }
    public static void releaseConnection(Connection con, DataSource dataSource) {}
}
```

那么，DataSourceUtils 中的 Connection 对象是从何处来的呢？

在业务操作（如前面的员工项目）中，使用 DataSourceTransactionManager 管理 JDBC 事务，在开启事务时，如果没有数据库连接，它会从 DataSource 中取得一个新连接，然后存储在 TransactionSynchronizationManager 中。而从 DataSourceUtils 中获取数据库连接时，也从 TransactionSynchronizationManager 中查找，因此 TransactionSynchronizationManager 是传递数据库连接的重要载体。而 TransactionSynchronizationManager 在底层使用 ThreadLocal 存储 Connection 对象。

```
public abstract class TransactionSynchronizationManager {
    private static final ThreadLocal<Map<Object, Object>> resources;
    private static final ThreadLocal<Set<TransactionSynchronization>> synchronizations;
    private static final ThreadLocal<String> currentTransactionName ;
    private static final ThreadLocal<Integer> currentTransactionIsolationLevel;
}
```

必须强调一下，对于 Spring 管理的声明性事务，数据库连接是在服务层打开的，然后存储在 ThreadLocal 中，并通过 TransactionSynchronizationManager 作为中间媒体进行传递。从性能优化方面考量，Spring 采用服务层打开数据库连接的方式并不可取，在持久层打开数据库连接才是最佳选择。

6.5.4 通过注解管理声明性事务

除了基于 XML 文件的声明式事务配置外，也可以采用基于注解的事务配置方法。直接在 Java 源代码中声明事务语义的做法，让事务声明和受其影响的代码距离更近了，而且一般来说不会有不恰当耦合的风险。

使用注解管理事务的方式更加简单。分别使用@Service 和@Repository 注解服务层对象与持久层对象。使用@Transactional 注解来标识哪个服务层方法需要事务，需要什么样的事务属性配置。

```
@Target({ElementType.METHOD, ElementType.TYPE})
@Retention(RetentionPolicy.RUNTIME)
@Inherited
@Documented
public @interface Transactional {
    Propagation propagation() default Propagation.REQUIRED;
    Isolation isolation() default Isolation.DEFAULT;
    boolean readOnly() default false;
    Class<? extends Throwable>[] rollbackFor() default {};
}
```

注解@Transactional 的属性（见表 6-2）与 XML 配置中<tx:advice>的属性设置一致。主要配置项是事务传播机制 propagation、异常回滚策略、只读设置等。

表 6-2 注解@Transactional 的属性

属性	类型	描述
propagation	枚举型	可选的传播性设置
isolation	枚举型	可选的隔离性级别（默认值为 ISOLATION_DEFAULT）
readOnly	布尔型	事务类型（读写型事务、只读型事务）设置
timeout	int 型（以秒为单位）	事务超时
rollbackFor	一组类的实例，必须是 Throwable 的子类	一组异常类，遇到时必须进行回滚。默认情况下，对于已检查的异常，不进行回滚；对于未检查的异常（即 RuntimeException 的子类），才进行事务回滚
rollbackForClassname	一组类的名字，必须是 Throwable 的子类	一组异常类名，遇到时必须进行回滚
noRollbackFor	一组类的实例，必须是 Throwable 的子类	一组异常类，遇到时不必回滚
noRollbackForClassname	一组类的名字，必须是 Throwable 的子类	一组异常类，遇到时不必回滚

6.5.5 项目案例：使用注解管理 StaffUser 事务

使用@Transactional 和相关属性设置，实现 StaffUser 系统的事务管理。本节会介绍具体操作步骤。

1. 配置 Spring 核心配置文件

配置 Spring 核心配置文件的步骤如下。

（1）配置 schema，以支持 tx 命名空间。

```
<beans xmlns="http://www.springframework.org/schema/beans"
    xmlns:xsi="http://www.w3.org/2001/XMLSchema-instance"
    xmlns:aop="http://www.springframework.org/schema/aop"
    xmlns:tx="http://www.springframework.org/schema/tx"
    xmlns:context="http://www.springframework.org/schema/context"
    xsi:schemaLocation="http://www.springframework.org/schema/beans
        http://www.springframework.org/schema/beans/spring-beans-4.3.xsd
        http://www.springframework.org/schema/tx
        http://www.springframework.org/schema/tx/spring-tx-4.3.xsd
        http://www.springframework.org/schema/aop
        http://www.springframework.org/schema/aop/spring-aop-4.3.xsd
        http://www.springframework.org/schema/context
        http://www.springframework.org/schema/context/spring-context-4.3.xsd">
</beans>
```

（2）数据源与事务管理器的配置不变。

```
<bean id="dataSource"
     class="org.springframework.jdbc.datasource.DriverManagerDataSource">
    <property name="driverClassName"
            value="com.mysql.cj.jdbc.Driver" />
    <property name="url"
            value="jdbc:mysql://localhost:3306/staff?useSSL=false" />
    <property name="username" value="root" />
    <property name="password" value="123456" />
</bean>
```

（3）配置事务的注解驱动，这与基于 XML 管理事务的配置不同。

```
<tx:annotation-driven transaction-manager="txManager" />
```

（4）配置组件扫描。

```
<context:component-scan base-package="com.icss.biz"/>
    <context:component-scan base-package="com.icss.dao"/>
```

2. 注入服务层和控制层 Bean

通过自动适配的注入模式，给 JdbcDaoSupport 注入数据源。此处使用基于@Autowired 的 set 方法注入模式，注入 dataSource 对象。

```
public abstract class BaseDao extends JdbcDaoSupport {
    @Autowired
    public void setDataSource(org.springframework.jdbc.datasource
                    .DriverManagerDataSource dataSource) {
        super.setDataSource(dataSource);
    }
    public Connection openConnection() {
        return this.getConnection();
    }
}
```

服务层 Bean 使用@Service 注解，持久层 Bean 使用@Repository 注解。服务层依赖持久层对象，使用@Autowired 模式。

```
@Service("staffBiz")
public class StaffBiz implements IStaff{
    @Autowired
    private StaffDao staffDao;
    @Autowired
    private UserDao userDao;
}
@Repository("staffDao")
public class StaffDao extends BaseDao{}
@Repository("userDao")
public class UserDao extends BaseDao{}
```

3. 切面与事务策略

@Transactional 注解应该只应用到 public 方法上。如果在 protected、private 或者 package-visible 方法上使用@Transactional 注解，系统不会报错，但是这个被注解的方法将不会执行已配置的事务。

```
@Transactional(rollbackFor=Throwable.class)
public void addStaffUser(TStaff staff) throws Exception {
    Log.logger.info("StaffBiz-->>addStaffUser()");
    TUser user = new TUser();
    user.setSno(staff.getSno());
    user.setUname(staff.getName());
    user.setRole(2);
    user.setPwd("1234");
    staffDao.addStaff(staff);
    userDao.addUser(user);
}
```

4. 事务测试

首先，进行事务测试。

测试代码如下。

```
public static void main(String[] args) {
    TStaff staff = new TStaff();
    staff.setSno("121000132");
    staff.setName("tom");
    SimpleDateFormat sdf = new SimpleDateFormat("yyyy-MM-dd");
    try {
        staff.setBirthday(sdf.parse("1995-10-1"));
    } catch (Exception e) {

    }
    staff.setAddress("北京朝阳区建国门");
    staff.setTel("1352245466221");
    IStaff staffProxy = (IStaff)BeanFactory.getBean("staffBiz");
    try {
        staffProxy.addStaffUser(staff);
        System.out.println(staff.getSno() + "创建成功....");
    } catch (Exception e) {
        e.printStackTrace();
    }
}
```

测试结果如下。

```
INFO - Loading XML bean definitions from class path resource [beans.xml]
INFO - Loaded JDBC driver: com.mysql.cj.jdbc.Driver
INFO - StaffBiz-->>addStaffUser()
INFO - StaffDao-->>addStaff()
INFO - UserDao-->>addUser()
121000132 创建成功....
```

然后，进行事务回滚测试。

测试代码如下。

```
public void addUser(TUser user) throws Exception {
    Log.logger.info("UserDao-->>addUser()");
    String sql = "insert into tuser values(?,?,?,?)";
    Connection conn = this.openConnection();
    PreparedStatement ps = conn.prepareStatement(sql);
    ps.setString(1, user.getUname());
    ps.setString(2, user.getSno());
    ps.setString(3, user.getPwd());
    ps.setInt(4,user.getRole());
    ps.executeUpdate();
    ps.close();
    throw new RuntimeException("异常测试....");
}
```

测试结果如下。

```
INFO - Loading XML bean definitions from class path resource [beans.xml]
INFO - Loaded JDBC driver: com.mysql.cj.jdbc.Driver
INFO - StaffBiz-->>addStaffUser()
```

```
INFO - StaffDao-->>addStaff()
INFO - UserDao-->>addUser()
java.lang.RuntimeException: 异常测试....
    at com.icss.dao.UserDao.addUser(UserDao.java:32)
    at com.icss.biz.StaffBiz.addStaffUser(StaffBiz.java:34)
```

在 MySQL 数据库中检查，确认 TStaff 数据已回滚。

6.6 Spring 编程式事务

6.6.1 编程式事务的管理

Spring Framework 提供了两种编程式事务管理方法:

- 使用 TransactionTemplate;
- 直接使用 PlatformTransactionManager。

org.springframework.transaction.PlatformTransactionManager 是 Spring 事务管理的统一抽象模型，不管是声明式事务还是编程式事务，都应该使用统一的事务管理器。

Spring 编程式事务的应用并不广泛，与直接调用 JDBC 接口进行编程事务对比，优势仅仅是使用了一致的事务管理抽象。

6.6.2 在 Spring 中通过编程式事务新增员工

要使用编程式事务管理模式新增员工，操作步骤如下。

（1）编写配置文件。事务管理器仍然使用 DataSourceTransactionManager。

```
<context:component-scan     base-package="com.icss.biz" />
<context:component-scan     base-package="com.icss.dao" />
<bean id="dataSource"
      class="org.springframework.jdbc.datasource.DriverManagerDataSource">
    <property name="driverClassName" value="com.mysql.cj.jdbc.Driver" />
    <property name="url"
              value="jdbc:mysql://localhost:3306/staff?useSSL=false" />
    <property name="username" value="root" />
    <property name="password" value="123456" />
</bean>
   <bean id="txManager"
     class="org.springframework.jdbc.datasource.DataSourceTransactionManager">
        <property name="dataSource" ref="dataSource" />
</bean>
```

（2）注入数据源。

```
public abstract class BaseDao extends JdbcDaoSupport {
    public Connection openConnection() throws Exception {
```

```
        return this.getConnection();
    }
    @Autowired
    public void setMyDataSource(org.springframework.jdbc.datasource
                        .DriverManagerDataSource dataSource){
        super.setDataSource(dataSource);
    }
}
```

（3）注入 PlatformTransactionManager。因为在 XML 的配置文件中已经设置了事务管理器，所以在这里注入即可。

```
@Service("staffBiz")
public class StaffBiz implements IStaff{
    @Autowired
    private PlatformTransactionManager txManager;
    @Autowired
    StaffDao staffDao;
    @Autowired
    UserDao userDao;
}
```

（4）实现业务逻辑控制。通过注入事务管理器，采用编程式事务模式控制员工和用户的添加行为。使用 DefaultTransactionDefinition 来实现事务属性的设置，使用 TransactionStatus 控制事务的提交和回滚。

```
public void addStaffUser(Staff staff) throws Exception {
    Log.logger.info(Thread.currentThread().getId() + ":addStaffUser()");
    DefaultTransactionDefinition def = new DefaultTransactionDefinition();
    def.setName("SomeTxName");
    def.setPropagationBehavior(TransactionDefinition.PROPAGATION_REQUIRED);
    TransactionStatus status = txManager.getTransaction(def);
    try {
        staffDao.addStaff(staff);
        User user = new User();
        user.setSno(staff.getSno());
        user.setUname(staff.getName());
        user.setRole(2);
        user.setPwd("1234");
        userDao.addUser(user);
        txManager.commit(status);
    }
    catch (Exception ex) {
       txManager.rollback(status);
       throw ex;
    }
}
```

（5）测试是否可以新增员工。

① 测试代码如下。

```
public static void main(String[] args) {
    Staff staff = new Staff();
    staff.setSno("121000133");
    staff.setName("johson");
    SimpleDateFormat sdf = new SimpleDateFormat("yyyy-MM-dd");
    try {
        staff.setBirthday(sdf.parse("1995-10-1"));
    } catch (Exception e) {
    }
    staff.setAddress("北京朝阳区建国门");
    staff.setTel("13522454666");
    IStaff staffProxy = (IStaff)BeanFactory.getBean(IStaff.class);
    try {
        staffProxy.addStaffUser(staff);
        System.out.println(staff.getSno() + "创建成功....");
    } catch (Exception e) {
        e.printStackTrace();
    }
}
```

② 测试结果如下。

```
INFO - Loading XML bean definitions from class path resource [beans.xml]
INFO - Loaded JDBC driver: com.mysql.cj.jdbc.Driver
INFO - 1:addStaffUser()
INFO - 1:addStaff()
INFO - 1:addUser()
121000133创建成功....
```

（6）完成异常测试。

① 测试代码如下。

```
public void addUser(User user) throws Exception {
    Log.logger.info(Thread.currentThread().getId() + ":addUser()");
    String sql = "insert into tuser values(?,?,?,?)";
    Connection conn = this.openConnection();
    PreparedStatement ps = conn.prepareStatement(sql);
    ps.setString(1, user.getUname());
    ps.setString(2, user.getSno());
    ps.setString(3, user.getPwd());
    ps.setInt(4,user.getRole());
    ps.executeUpdate();
    ps.close();
    throw new RuntimeException("事务回滚测试...");
}
```

② 测试结果如下。

```
INFO - Loading XML bean definitions from class path resource [beans.xml]
INFO - Loaded JDBC driver: com.mysql.cj.jdbc.Driver
INFO - 1:addStaffUser()
INFO - 1:addStaff()
```

```
INFO - 1:addUser()
java.lang.RuntimeException: 事务回滚测试...
    at com.icss.dao.UserDao.addUser(UserDao.java:39)
    at com.icss.biz.impl.StaffBiz.addStaffUser(StaffBiz.java:46)
    at com.icss.ui.TestAddStaffUser.main(TestAddStaffUser.java:24)
```

6.7 声明性事务与编程式事务的选择

到底使用声明性事务，还是编程式事务？Spring 的官方推荐如下。

当只有很少的事务操作时，使用编程式事务通常比较合适。例如，对于一个 Web 应用，其中只有特定的更新操作有事务要求，用户可能不愿使用 Spring 或其他技术设置事务代理。这种情况下，使用 TransactionTemplate 可能是一个好办法。只有编程式事务才能显式地设置事务名称。

如果应用中存在大量事务操作，那么使用声明性事务通常是值得的。它将事务管理与业务逻辑分离。使用 Spring，而不是 EJB 的 CMT，配置成本大大地降低了。

总结：对于绝大多数开发者的业务场景，事务操作的情况并不多，而开发者普遍使用 Spring 声明性事务，这其实是对性能的一种浪费，在高并发环境下尤其要小心。EJB 3.0 后，其 CMT 配置非常简单，在企业级应用中仍然是非常好的选择。

6.8 Spring 事务的传播属性

Spring 事务的传播性类似于 JTA 事务的传播性。JTA 事务的传播属性有 7 种（见表 6-3），Spring 主要支持前 3 种。

表 6-3 JTA 事务的传播属性

JTA 事务的传播属性	说明
PROPAGATION_REQUIRED	• 内部事务与外部事务在逻辑上独立，它们共享同一个物理事务环境。 • 若存在外部事务，则内部事务加入外部事务所在的物理事务环境中。如果不存在外部事务，则创建新事务。 • 在内部事务和外部事务对应的，两个逻辑事务环境中可以独立地设置回滚状态。内部事务的回滚状态会影响外部事务的提交。 • 如果内部事务标记为回滚状态，则仍然提交外部事务；否则，会抛出 UnexpectedRollbackException 异常
PAOPAGATION_REQUIRE_NEW	不论是否存在事务环境，都新建一个事务，新老事务相互独立。内部事务抛出的异常不会影响外部事务的正常提交
PROPAGATION_NESTED	如果当前存在事务，则嵌套在当前事务中执行。如果当前没有事务，则新建一个事务
PROPAGATION_SUPPORTS	支持当前事务，若当前不存在事务，则以非事务方式执行
PROPAGATION_NOT_SUPPORTED	以非事务方式执行，若当前存在事务，则挂起当前事务
PROPAGATION_MANDATORY	强制事务执行，若当前不存在事务，则抛出异常
PROPAGATION_NEVER	以非事务方式执行，如果当前存在事务，则抛出异常

6.8.1 Propagation.REQUIRED

当事务传播模式为 Propagation.REQUIRED 时，始终只有一个物理事务环境（见图 6-7）。

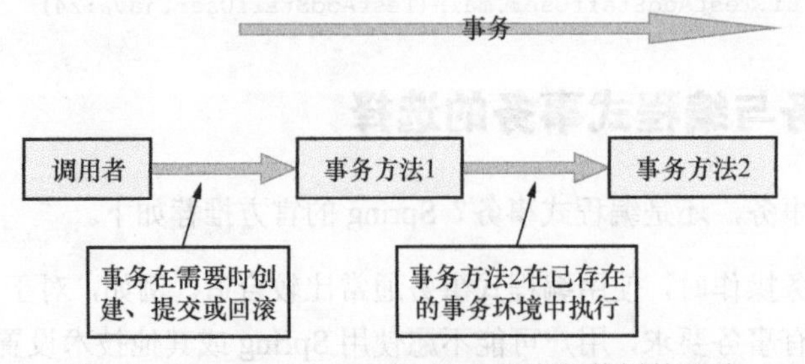

图 6-7　Propagation.REQUIRED 对事务传播的影响

下面通过实例测试一下 Propagation.REQUIRED 对事务的影响。操作步骤如下。

（1）设置业务方法的事务。在方法 operationA( )中调用 operationB( )，设置 operationB( )的事务传播属性为@Transactional(propagation=Propagation.REQUIRED,rollbackFor=Throwable.class)。注意，operationA( )的事务传播属性设置对 operationB( )无影响。

```
@Service("staffBiz")
public class StaffBiz implements IStaff{
    @Transactional(rollbackFor=Throwable.class)
    public void operationA() {
        Log.logger.info("operationA...");
        operationB();
    }
    @Transactional(propagation=Propagation.REQUIRED,
                  rollbackFor=Throwable.class)
    public void operationB() {
        Log.logger.info("operationB...");
    }
}
```

（2）若 operationA( )有事务环境，观察 operationB( )使用什么事务环境。

① 测试代码如下。

```
public static void main(String[] args) {
    IStaff staffProxy = (IStaff)BeanFactory.getBean("staffBiz");
    staffProxy.operationA();
    System.out.println("测试结束...");
}
```

② 测试结果如下。

```
DEBUG - Creating new transaction with name [com.icss.biz.StaffBiz.operationA]:
    PROPAGATION_REQUIRED,ISOLATION_DEFAULT,-java.lang.Throwable
DEBUG - Creating new JDBC DriverManager Connection to
DEBUG - Acquired Connection for JDBC transaction
```

```
INFO - operationA...
DEBUG - Adding transactional method 'com.icss.biz.StaffBiz.operationB' with attribute:
PROPAGATION_REQUIRED,ISOLATION_DEFAULT,-java.lang.Throwable
DEBUG - Participating in existing transaction
INFO - operationB...
DEBUG - Completing transaction for [com.icss.biz.StaffBiz.operationB]
DEBUG - Completing transaction for [com.icss.biz.StaffBiz.operationA]
DEBUG - Committing JDBC transaction on Connection
DEBUG - Releasing JDBC Connection after transaction
DEBUG - Returning JDBC Connection to DataSource
测试结束...
```

如上所示，operationA()与operationB()都有事务设置，operationB()使用了operationA()的事务环境中。

（3）若外部方法无事务设置，观察operationB( )使用什么事务环境。

```
public class StaffBiz implements IStaff{
    public void operationA() {
        Log.logger.info("operationA...");
        operationB();
    }
    @Transactional(propagation=PROPAGATION.REQUIRED,
                    rollbackFor=Throwable.class)
    public void operationB() {
        Log.logger.info("operationB...");
    }
}
```

测试代码不变，测试结果如下。

```
DEBUG - Returning cached instance of singleton bean 'staffBiz'
INFO - operationA...
INFO - operationB...
测试结束...
```

如上所示，operationA( )无事务设置，调用operationB()的结果显示，虽然operationB()设置了事务需求，但是并没有生成物理事务环境，这与官方文档说明相悖，为什么？

（4）继续测试。使用Bean对象调用operationB( )，其他代码不变。

```
@Service("staffBiz")
public class StaffBiz implements IStaff{
    @Autowired
    private IStaff staffBiz;            //注入Bean对象
    public void operationA() {
        Log.logger.info("operationA...");
        staffBiz.operationB();    //Bean对象调用operationB()方法
    }
    @Transactional(propagation=PROPAGATION.REQUIRED,
                   rollbackFor=Throwable.class)
    public void operationB() {
```

```
        Log.logger.info("operationB...");
    }
}
```

测试结果如下。

```
INFO - operationA...
DEBUG - Creating new transaction with name [com.icss.biz.StaffBiz.operationB]:
        PROPAGATION_REQUIRED,ISOLATION_DEFAULT,-java.lang.Throwable
DEBUG - Creating new JDBC DriverManager Connection to
INFO - operationB...
DEBUG - Completing transaction for [com.icss.biz.StaffBiz.operationB]
DEBUG - Committing JDBC transaction on Connection
DEBUG - Releasing JDBC Connection after transaction
DEBUG - Returning JDBC Connection to DataSource
测试结束...
```

总之，设置 operationB( )的事务传播属性为 propagation=PROPAGATION.REQUIRED，如果 operationA( )无事务环境，则在 operationB( )中创建新事务环境（必须是 Bean 对象调用）；如果 operationA( )有事务环境，则 operationB( )使用 operationA( )的事务环境。

（5）进行异常测试。

① 测试代码如下。

```
@Service("staffBiz")
public class StaffBiz implements IStaff{
    @Autowired
    private IStaff staffBiz;
    @Transactional(rollbackFor=Throwable.class)
    public void operationA() {
        Log.logger.info("operationA...");
        try {
            staffBiz.operationB();
        } catch (Exception e) {
        }
    }
    @Transactional(propagation=Propagation.REQUIRES_NEW,
              rollbackFor=Throwable.class)
    public void operationC() {
        Log.logger.info("operationC...");
        throw new RuntimeException("operationC 抛出异常...");
    }
}
```

② 测试结果如下。

```
DEBUG - Creating new transaction with name [com.icss.biz.StaffBiz.operationA]
INFO - operationA...
DEBUG - Participating in existing transaction
DEBUG - Getting transaction for [com.icss.biz.StaffBiz.operationB]
INFO - operationB...
```

```
DEBUG - Completing transaction for [com.icss.biz.StaffBiz.operationB] after exception:
    java.lang.RuntimeException: operationB 抛出异常...
DEBUG - Participating transaction failed - marking existing transaction as rollback-only
DEBUG - Completing transaction for [com.icss.biz.StaffBiz.operationA]
DEBUG - Global transaction is marked as rollback-only
DEBUG - Rolling back JDBC transaction on Connection
DEBUG - Releasing JDBC Connection after transaction
DEBUG - Returning JDBC Connection to DataSource
Exception in thread "main" org.springframework.transaction.UnexpectedRollbackException:
Transaction rolled back because it has been marked as rollback-only
```

总之，operationA( )与 operationB( )使用相同的物理事务环境，当 operationB( )抛出异常后，operationA( )的事务也会回滚。

6.8.2 Propagation.REQUIRES_NEW

如图 6-8 所示，Propagation.REQUIRES_NEW 与 Propagation.REQUIRED 传播模式相反，在其各自的事务范围内，永远使用各自独立的物理事务环境。内部事务永远不会使用外部已存在的事务环境。

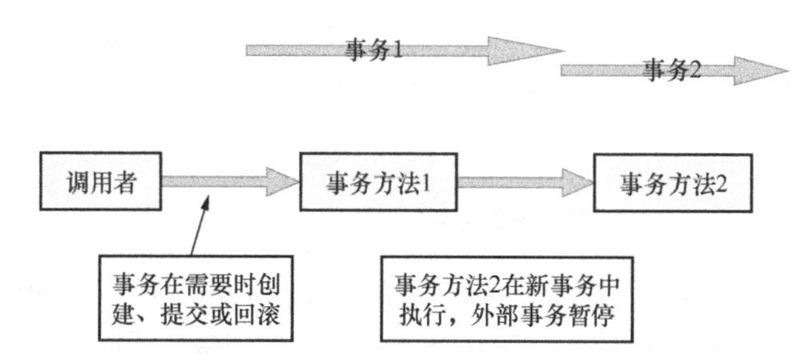

图 6-8 Propagation.REQUIRES_NEW 对事务传播的影响

要测试 Propagation.REQUIRES_NEW 对事务传播的影响，步骤如下。

（1）设置业务方法的事务。外部方法与内部方法都有事务设置。

```
@Service("staffBiz")
public class StaffBiz implements IStaff{
    @Autowired
    private IStaff staffBiz;
    @Transactional(rollbackFor=Throwable.class)
    public void operationA() {
        Log.logger.info("operationA...");
        staffBiz.operationC();
    }
    @Transactional(propagation=Propagation.REQUIRES_NEW,
              rollbackFor=Throwable.class)
    public void operationC() {
        Log.logger.info("operationC...");
    }
}
```

（2）测试 Propagation.REQUIRES_NEW 对事务传播的影响。

① 测试代码如下。

```
public static void main(String[] args) {
    IStaff staffProxy = (IStaff)BeanFactory.getBean("staffBiz");
    staffProxy.operationA();
    System.out.println("测试结束...");
}
```

② 测试结果如下。

```
DEBUG - Creating new transaction with name [com.icss.biz.StaffBiz.operationA]
DEBUG - Creating new JDBC DriverManager Connection
INFO - operationA...
DEBUG - Suspending current transaction, creating new transaction with name
       [com.icss.biz.StaffBiz.operationC]
DEBUG - Creating new JDBC DriverManager Connection to
DEBUG - Getting transaction for [com.icss.biz.StaffBiz.operationC]
INFO - operationC...
DEBUG - Completing transaction for [com.icss.biz.StaffBiz.operationC]
DEBUG - Committing JDBC transaction on Connection
DEBUG - Resuming suspended transaction after completion of inner transaction
DEBUG - Completing transaction for [com.icss.biz.StaffBiz.operationA]
DEBUG - Committing JDBC transaction on Connection
DEBUG - Returning JDBC Connection to DataSource
测试结束...
```

（3）若外部方法无事务设置，测试结果相同。

总之，根据测试结果可知，不管 operationA( )是否存在事务环境，operationC( )都会生成一个新的事务环境。operationA( )与 operationC( )是两个独立的物理事务环境。

注意：在同一个类内 operationA( )与 operationC( )调用会受到限制，两个方法使用同一个事务环境。只有通过 Bean 对象调用 operationC( )，Propagation.REQUIRES_NEW 才有效。因此，这里的测试中 StaffBiz 需要注入自身对象。

（4）进行异常测试。

① 测试代码如下。

```
@Service("staffBiz")
public class StaffBiz implements IStaff{
    @Autowired
    private IStaff staffBiz;
    @Transactional(rollbackFor=Throwable.class)
    public void operationA() {
        Log.logger.info("operationA...");
        try {
            staffBiz.operationC();
        } catch (Exception e) {
```

```
            e.printStackTrace();
        }
        Log.logger.info("operationA...after.........");
    }
    @Transactional(propagation=Propagation.REQUIRES_NEW,
                rollbackFor=Throwable.class)
    public void operationC() {
        Log.logger.info("operationC...");
        throw new RuntimeException("operationC 抛出异常...");
    }
}
```

② 测试结果如下。

```
DEBUG - Creating new transaction with name [com.icss.biz.StaffBiz.operationA]
DEBUG - Getting transaction for [com.icss.biz.StaffBiz.operationA]
INFO - operationA...
DEBUG - Suspending current transaction, creating new transaction with name
        [com.icss.biz.StaffBiz.operationC]
DEBUG - Creating new JDBC DriverManager Connection
DEBUG - Getting transaction for [com.icss.biz.StaffBiz.operationC]
INFO - operationC...
DEBUG - Completing transaction for [com.icss.biz.StaffBiz.operationC] after exception
DEBUG - Rolling back JDBC transaction on Connection
DEBUG - Releasing JDBC Connection
DEBUG - Resuming suspended transaction after completion of inner transaction
java.lang.RuntimeException: operationC 抛出异常...
    at com.icss.biz.StaffBiz.operationC(StaffBiz.java:27)
INFO - operationA...after.........
DEBUG - Completing transaction for [com.icss.biz.StaffBiz.operationA]
DEBUG - Committing JDBC transaction on Connection
DEBUG - Releasing JDBC Connection
测试结束...
```

测试结果显示，operationA()与 operationC()在两个独立的物理事务环境中。当 operationC()异常时，operationA()的事务可以正常提交。

6.8.3 Propagation.NESTED

Propagation.NESTED 使用了单个物理事务，这个事务拥有多个可以回滚的保存点，允许内部事务在它的事务范围内部分回滚，并且外部事务不受影响。这个设置仅仅在 Spring 管理 JDBC 资源时有效，它对应 JDBC 的保存点。

6.9 关于数据库连接管理的总结

Spring 如何管理数据库连接的打开、关闭？

数据库连接是业务系统开发中最宝贵的资源，很容易成为系统的性能瓶颈。如果这个问题处理不好，那么在并发高的环境下系统很容易崩溃。

6.9.1 JdbcDaoSupport

所有的持久层方法都需要继承 JdbcDaoSupport。

```
public abstract class BaseDao extends JdbcDaoSupport {
    public Connection openConnection() {
        return this.getConnection();
    }
}
```

JdbcDaoSupport 的 API 定义如下，其中的连接源于 DataSourceUtils。

```
public abstract class JdbcDaoSupport extends DaoSupport {
    private JdbcTemplate jdbcTemplate;
    public final void setDataSource(DataSource dataSource) {}
    protected final Connection getConnection(){
        return DataSourceUtils.getConnection(getDataSource());
    }
    protected final void releaseConnection(Connection con) {
        DataSourceUtils.releaseConnection(con, getDataSource());
    }
}
```

为了使持久层获得连接，通过原生 API 进行数据操作。

```
public void addStaff(TStaff staff) throws Exception{
    String sql = "insert into tstaff values(?,?,?,?,?)";
    Connection conn = this.openConnection();
    PreparedStatement ps = conn.prepareStatement(sql);
}
public void addUser(TUser user) throws Exception {
    String sql = "insert into tuser values(?,?,?,?)";
    Connection conn = this.openConnection();
    PreparedStatement ps = conn.prepareStatement(sql);
}
```

前面详细讲解了 JdbcDaoSupport 与 DataSourceUtils、DataSourceTransactionManager、TransactionSynchronizationManager 之间的关系。

数据库的连接是在执行有事务（只读事务和写事务）需求的业务方法前开启的，同一个线程后面的所有操作都重用前面这个连接。

6.9.2 数据库连接的控制

作为业务系统最宝贵的资源，数据库连接的数量非常有限。数据库服务器的有效连接数量是受物理硬件约束的，不能无限增加。在一个高并发的系统中，为了允许尽量多的用户同时在线操作，数据库连接必须非常高效地重用才可以。

业务层控制数据库连接的使用。如何合理使用数据库连接非常重要。有些业务方法（如addStaffUser方法）需要事务，有些业务方法（如login方法）需要数据库连接，有些业务方法（如operateNoDb 方法）不需要数据库连接。因此，需要精准地控制数据库连接的打开时间与关闭时间，尽量重用数据库连接。

数据库连接使用的基本原则是晚打开、早关闭，每个用户的使用时间尽量短。如果按照这个原则处理，在持久层打开数据库、在持久层关闭数据库最好。然而，这样做无法在一次请求内重用数据库连接。

综上所述，推荐模式是在持久层打开数据连接，在服务层关闭数据库连接，特殊情况（如懒加载）下，则在视图层关闭数据库连接。

1. 在业务类中使用@Transactional

在业务类中使用@Transactional(readOnly=true)是否合适？

无数据库操作的业务方法是否会占用数据库连接？示例代码如下。

```
@Transactional(readOnly=true)
@Service("staffBiz")
public class StaffBiz implements IStaff{
    @Autowired
    private StaffDao staffDao;
    @Autowired
    private UserDao userDao;
    /**
     * 无数据库操作的业务方法
     */
    public void operateNoDb() {
        Log.logger.info("StaffBiz-->>operateNoDb()");
    }
}
```

测试代码如下。

```
public static void main(String[] args) {
    IStaff biz = (IStaff)BeanFactory.getBean("staffBiz");
    biz.operateNoDb();
}
```

测试结果如下。

```
DEBUG - Creating new transaction with name [com.icss.biz.StaffBiz.operateNoDb]:
        PROPAGATION_REQUIRED,ISOLATION_DEFAULT,readOnly
DEBUG - Creating new JDBC DriverManager Connection
DEBUG - Getting transaction for [com.icss.biz.StaffBiz.operateNoDb]
INFO - StaffBiz-->>operateNoDb()
DEBUG - Completing transaction for [com.icss.biz.StaffBiz.operateNoDb]
DEBUG - Committing JDBC transaction on Connection
DEBUG - Resetting read-only flag of JDBC Connection
DEBUG - Releasing JDBC Connection
```

总之，如果在业务类 StaffBiz 中使用@Transactional(readOnly=true)注解，即使没有数据库操作，类中的业务方法（如很多 get( )方法）也会打开一个数据库连接，这将造成极大的浪费。在业务类中使用@Transactional(readOnly=true)要慎重。

2. 在读操作中使用@Transactional

在业务类 StaffBiz 与方法 login( )中都不使用@Transactional 注解，下面进行用户登录测试。

（1）业务层代码无事务配置。

```
@Service("staffBiz")
public class StaffBiz implements IStaff{
    @Autowired
    private UserDao userDao;
    public TUser login(String uname, String pwd) throws Exception {
        if(uname == null || pwd == null
            || uname.trim().equals("") || pwd.trim().equals("")) {
            throw new Exception("用户名或密码不能为空......");
        }
        TUser user = userDao.login(uname, pwd);
        return user;
    }
}
```

（2）在持久层获取数据库连接。

```
@Repository("userDao")
public class UserDao extends BaseDao{
    public TUser login(String uname, String pwd) throws Exception {
        Log.logger.info("UserDao-->>login()");
        String sql = "select * from tuser where uname=? and pwd=?";
        Connection conn = this.openConnection();
        PreparedStatement ps = conn.prepareStatement(sql);
        ......
}
```

（3）测试用户是否可以登录。

① 测试代码如下。

```
public static void main(String[] args) {
    IStaff biz = (IStaff)BeanFactory.getBean("staffBiz");
    try {
        TUser user = biz.login("tom","1234");
        if(user != null) {
            System.out.println(user.getUname() + "登录成功");
        }else {
            System.out.println("登录失败...");
        }
    } catch (Exception e) {
    }
}
```

② 测试结果如下。

```
DEBUG - Returning cached instance of singleton bean 'staffBiz'
INFO - UserDao-->>login()
DEBUG - Fetching JDBC Connection from DataSource
DEBUG - Creating new JDBC DriverManager Connection
tom 登录成功
```

总之，如上操作会在持久层打开一个新的数据库连接，但是没有释放数据库连接。这样的操作非常危险，很容易导致系统崩溃。

（4）推荐在读方法上使用@Transactional(readOnly=true)。

```
@Service("staffBiz")
public class StaffBiz implements IStaff{
    @Autowired
    private UserDao userDao;
    @Transactional(readOnly=true)
    public TUser login(String uname, String pwd) throws Exception {
        if(uname == null || pwd == null
            || uname.trim().equals("") || pwd.trim().equals("")) {
            throw new Exception("用户名或密码不能为空......");
        }
        TUser user = userDao.login(uname, pwd);
        return user;
    }
}
```

测试结果如下。

```
DEBUG - Creating new transaction with name [com.icss.biz.StaffBiz.login]
        :PROPAGATION_REQUIRED,ISOLATION_DEFAULT,readOnly
DEBUG - Getting transaction for [com.icss.biz.StaffBiz.login]
INFO - UserDao-->>login()
DEBUG - Completing transaction for [com.icss.biz.StaffBiz.login]
DEBUG - Committing JDBC transaction on Connection
DEBUG - Resetting read-only flag of JDBC Connection
DEBUG - Releasing JDBC Connection after transaction
DEBUG - Returning JDBC Connection to DataSource
tom 登录成功
```

总之，如果业务类中读库的方法（如 login( )）用@Transactional(readOnly=true)注解，可以正确释放数据库连接。但是这样的操作不是最佳的，原因有两个：第一，数据库在事务开始时就过早打开了；第二，即使是只读事务控制，也会浪费性能。摆脱 Spring 的事务管理，在服务层手动关闭连接是最好的选择。

3. 使用@Transactional 控制事务

测试写操作的事务控制。

```
@Transactional(rollbackFor=Throwable.class)
public void addStaffUser(TStaff staff) throws Exception {
    .....
    staffDao.addStaff(staff);
    userDao.addUser(user);
}
```

使用了事务注解后，测试结果如下。

```
DEBUG - Creating new transaction with name [com.icss.biz.StaffBiz.addStaffUser]:
        PROPAGATION_REQUIRED,ISOLATION_DEFAULT,-java.lang.Throwable
DEBUG - Creating new JDBC DriverManager Connection
DEBUG - Getting transaction for [com.icss.biz.StaffBiz.addStaffUser]
INFO - StaffBiz-->>addStaffUser()
INFO - StaffDao-->>addStaff()
INFO - UserDao-->>addUser()
DEBUG - Completing transaction for [com.icss.biz.StaffBiz.addStaffUser]
DEBUG - Committing JDBC transaction on Connection
DEBUG - Triggering afterCommit synchronization
DEBUG - Clearing transaction synchronization
DEBUG - Releasing JDBC Connection after transaction
DEBUG - Returning JDBC Connection to DataSource
121000134 创建成功....
```

总之，使用@Transactional(rollbackFor=Throwable.class)控制事务，数据库连接唯一，事务环境唯一，数据库资源在业务方法结束后正确释放。

第 7 章 Spring MVC

7.1 Spring MVC 介绍

Spring MVC 是当前流行的 MVC 框架，它功能强大，用法简单，可扩展性极强。用 Spring MVC 框架做开发的特点是上手容易。为了掌握 Spring MVC 的运行机制，要深入了解 Spring MVC 的运行原理，这是一项艰巨的任务。

不要重复发明轮子，是 Rod Johnson 奉行的至理名言，也是 Spring Framework 的基石。但是 Spring MVC 是唯一的例外，原因是 Rod Johnson 觉得市场上的所有 MVC 框架都太糟糕了，没有整合的必要性，因此干脆新造了一个轮子。

Spring MVC 框架围绕 DispatcherServlet 设计而成。DispatcherServlet 的主要作用是分发请求到不同的 Action，根据 Action 返回的结果转向不同的视图。

Action 的处理基于@Controller 和@RequestMapping 等注解，还可以使用@PathVariable 注解，并与@Controller 联合使用，创建 RESTful 风格的 Web 站点。

Spring MVC 框架设计的核心理念是著名的开闭原则——软件实体（包括类、模板、功能）等应对扩展开放，但对修改关闭。Spring MVC 框架的很多核心类的重要方法都被标记为 final，即不允许通过重写的方式改变其行为，这也是满足开闭原则的体现。

Spring MVC 的视图解决方案非常灵活。视图名字和 Action 返回的 Model 数据组成

ModelAndView 对象。一个 ModelAndView 实例包含一个视图名字和一个类型为 Map 的 Model 对象。视图名字的解析由可配置的视图解析器完成。

Map 类型的 Model 是高度抽象的，适用于各种表现层技术。这些视图技术有 JSP、Velocity、Freemarker、JSON、XML 等。Map 可以根据不同的视图，选择合适的格式，如把 JSP 页面转为 request 属性格式，把 Velocity 转为模板格式。

7.1.1 视图与控制层技术

在 Java EE 6 之前，使用 Servlet、Filter、Listener 等，都必须要在 web.xml 中配置，非常麻烦；自定义的 Servlet 控制器需要继承自 HttpServlet，在里面只能写 doGet( )、doPost( )等方法，无法采用面向对象编程，非常死板。另外，要把接收的 HTTP 参数转为 POJO 对象也很麻烦。这些问题都可以采用优化方案进行解决。

针对 Java EE 传统方案的缺陷，许多 MVC 框架纷纷诞生，如 Struts 2、WebWorks 等，当前流行的框架是 Spring MVC。这些 MVC 框架最大的特点是控制器为面向对象编程而开发。

针对 JSP 响应速度慢、数据修改不灵活的问题，可以采用模板技术进行优化，如 Velocity 和 FreeMarker 等。Spring MVC 不但支持 JSP，而且支持在视图层使用模板。

现在视图层的主要趋势是使用 HTML5+AJAX+Vue 等前端框架，这样前端页面响应速度快，变化灵活。Spring MVC 支持与 AJAX 交互，支持在视图层使用模板，支持报表技术 JasperReport，支持输出 PDF 或 Excel，支持 XSLT，支持 WebSocket 等，非常灵活。

7.1.2 Spring MVC 支持的特性

Spring 的 Web 模块提供了以下独特的功能。

- 清晰的角色划分：包括控制器（controller）、验证器（validator）、命令对象（command object）、表单对象（form object）、模型对象（model object）、分发器（dispatcher）、处理程序映射（handler mapping）、视图解析器（view resolver）等角色。每一个角色都可以由一个专门的对象来实现。
- 强大而直接的配置方式：将框架类和业务类都作为 Bean 配置，支持跨多个上下文的引用，例如，在 Web 控制器中对业务对象和验证器的引用。
- 可适配性：使用一个简单的带参数注解，即可定义控制器方法的签名。在参数传入中用 @RequestParam、@RequestHeader、@PathVariable 等。
- 可重用的业务代码：可以使用现有的业务对象作为命令或表单对象，而不需要去扩展某个特定框架的基类。

- 可定制的绑定（binding）和验证（validation）：例如，将类型不匹配作为应用级的验证错误，本地化的日期和数字绑定等。在其他某些框架中，只能使用字符串表单对象，需要手动解析它并转换为业务对象。

- 控制映射和视图解析：控制映射转换，从最简单的 URL 映射到复杂的、专用的定制策略映射均可。与某些 Web MVC 框架强制开发人员使用单一特定技术相比，Spring 显得更加灵活。

- 灵活的模型转换：在 Spring MVC 框架中，使用基于 Map 的键/值对来轻易地与各种视图技术集成。

- 简单而强大的 JSP 标签库（Tag Library）：支持包括诸如数据绑定和主题之类的许多功能。

7.2 HelloMVC 项目

在本节中，我们先使用 Servlet 实现一个简单的入门项目，然后用 Spring MVC 框架改造这个 HelloMVC 项目，最后通过对比两个项目，看看 Spring MVC 带来了哪些变化。

7.2.1 Eclipse 和 Tomcat 8 的环境配置

开发工具使用 eclipse-jee-photon、eclipse-jee-oxygen，或版本更高的 Eclipse EE 版。配置开发环境的步骤如下。

（1）新建服务器（见图 7-1）。

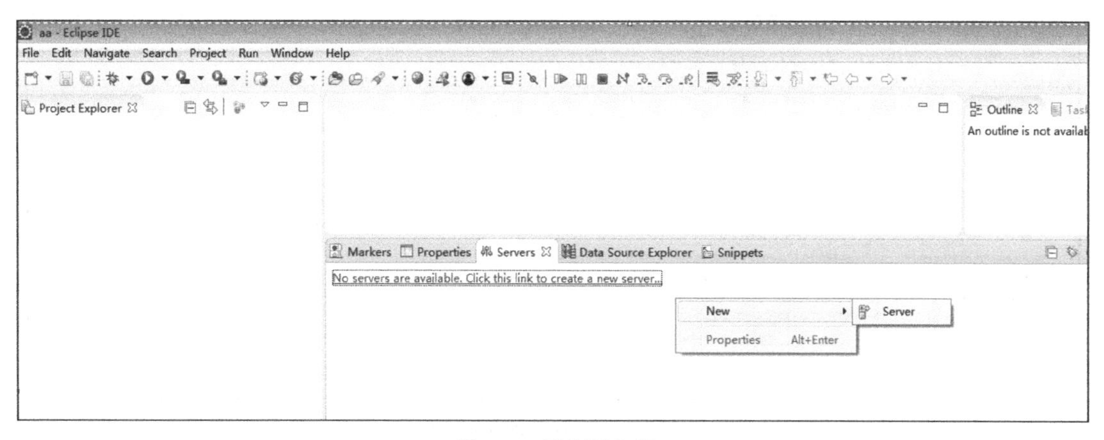

图 7-1　新建服务器

（2）选择 Apache 的 Tomcat v8.0 Server（见图 7-2）或 Tomcat v 8.5 Server。

（3）双击新建的服务器，单击 Use Tomcat installation (takes control of Tomcat installation)单选按钮，把 Deploy path 设置为 Tomcat 安装目录下的 webapp（见图 7-3）。

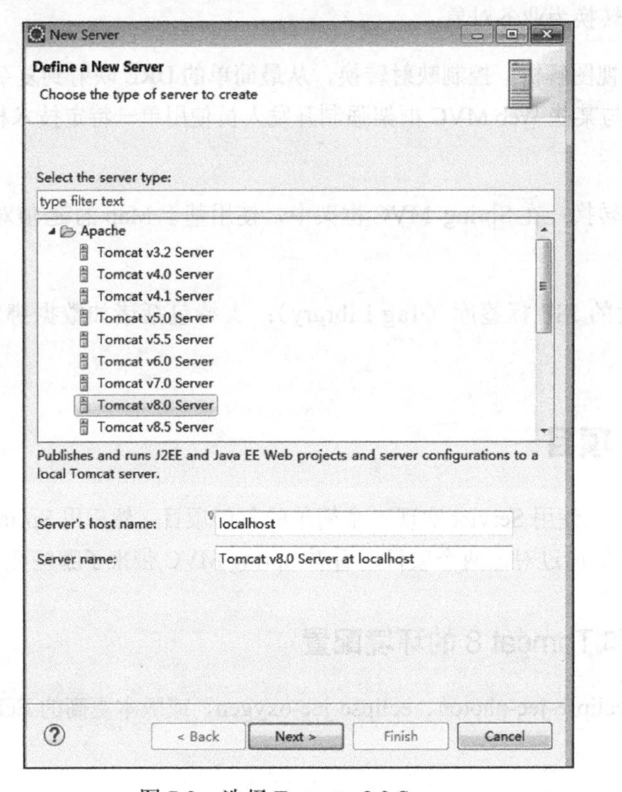

图 7-2　选择 Tomcat v8.0 Server

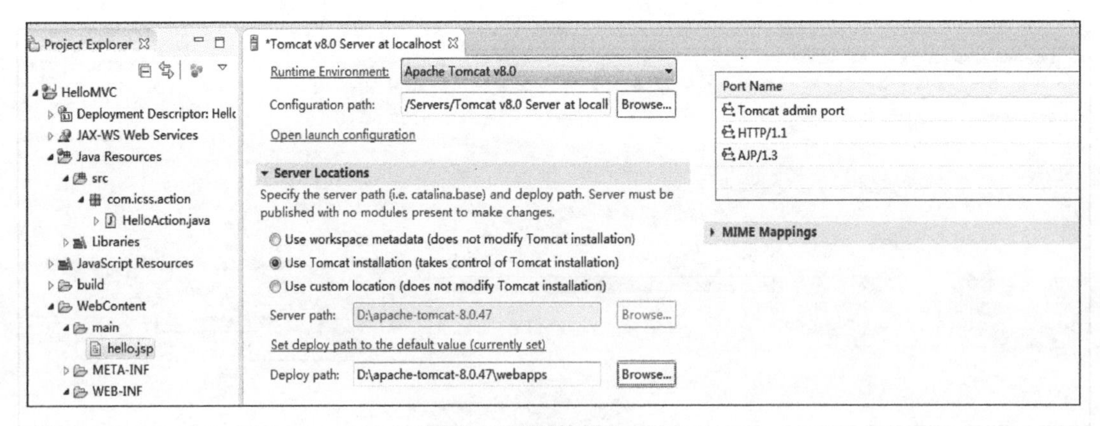

图 7-3　配置外部 Tomcat

（4）新建动态 Web 项目（见图 7-4）。

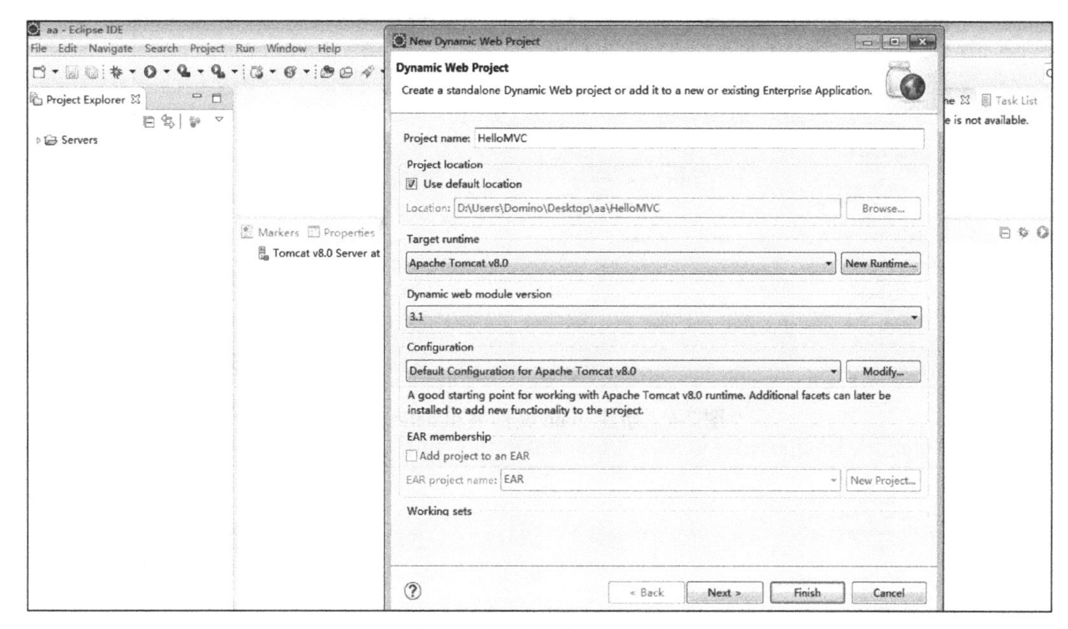

图 7-4　新建动态 Web 项目

注意：在 Target runtime 下拉列表中选择 Apache Tomcat v8.0，在 Dynamic web module version 下拉列表中选择 3.1。

（5）新建 Servlet（见图 7-5），命名为 HelloMVC。

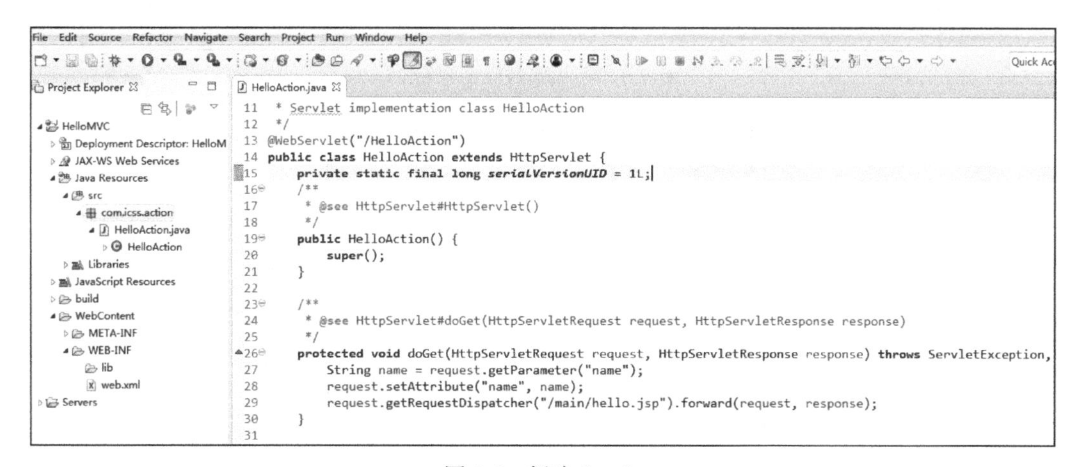

图 7-5　新建 Servlet

（6）新建 main 文件夹和 hello.jsp（见图 7-6）。

（7）右击服务器，选择 Add and Remove，添加和移除服务器（见图 7-7）。添加 HelloMVC 项目（见图 7-8）。

```
<%@ page language="java" contentType="text/html; charset=utf-8"
    pageEncoding="utf-8"%>
<!DOCTYPE html>
<html>
<head>
<meta charset="ISO-8859-1">
<title>Insert title here</title>
</head>
<body>

  hello Mr. ${name}

</body>
</html>
```

图 7-6　新建 main 文件夹和 hello.jsp

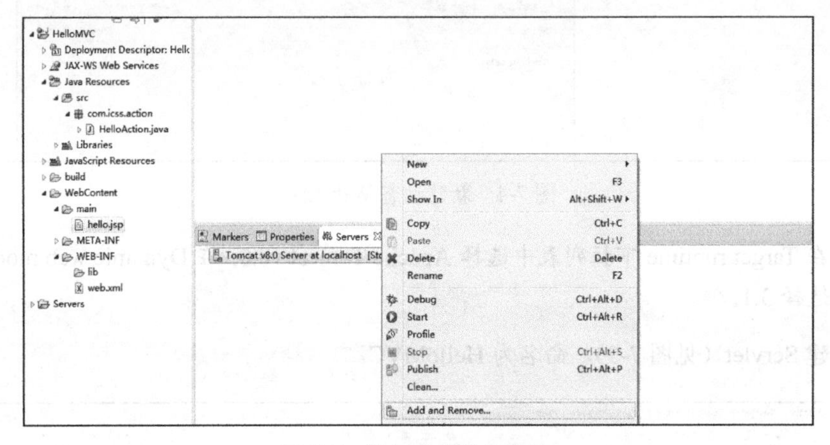

图 7-7　添加和移除服务器

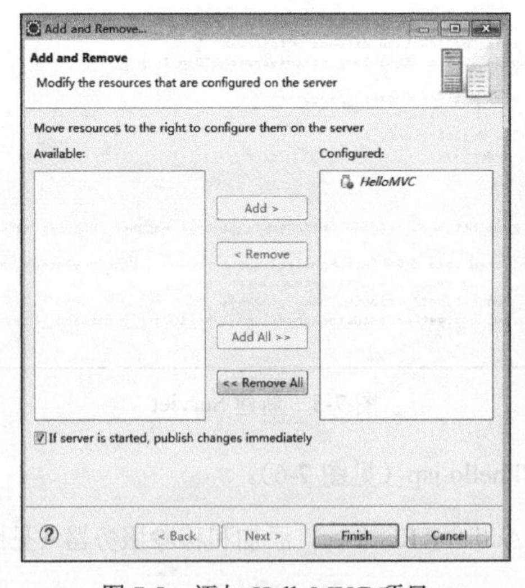

图 7-8　添加 HelloMVC 项目

（8）右击服务器，选择 Publish（见图 7-9）。

提示：在开发过程中，代码会变化，因此要经常发布，要保持工作区中的代码与 Tomcat 中发布的内容一致。

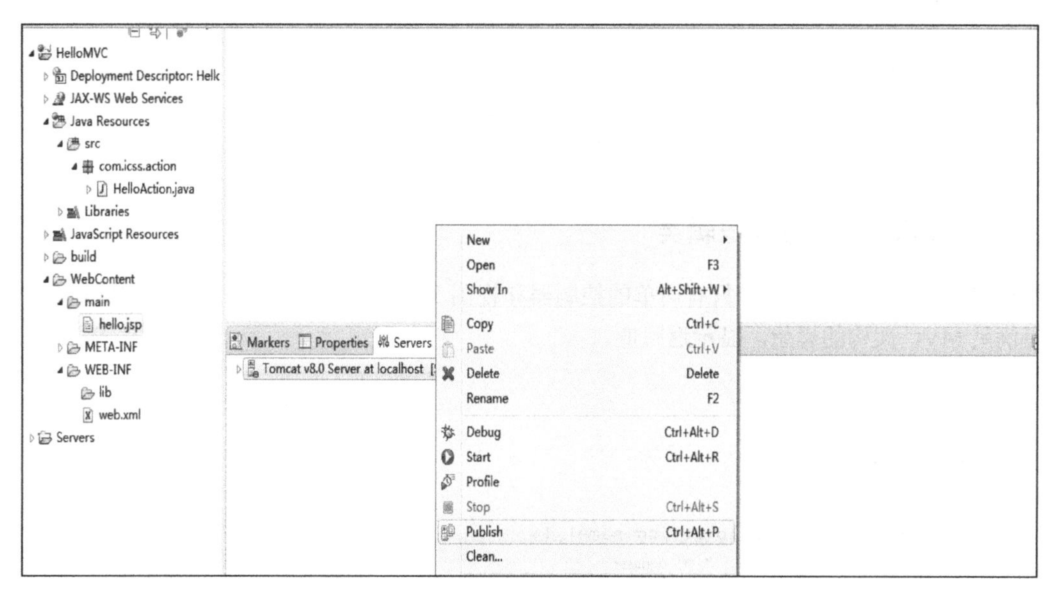

图 7-9　发布项目

（9）右击服务器，选择 Start，启动 Tomcat（见图 7-10）。

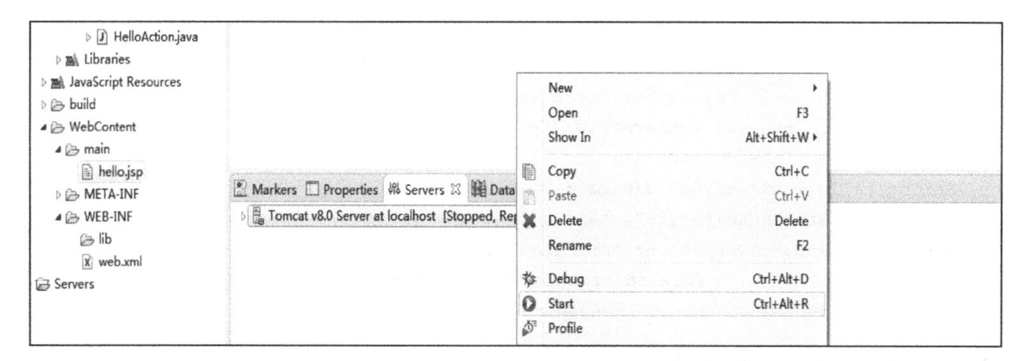

图 7-10　启动 Tomcat

Tomcat 启动成功，提示如下。

```
org.apache.coyote.AbstractProtocol start
信息: Starting ProtocolHandler ["ajp-nio-8009"]
org.apache.catalina.startup.Catalina start
信息: Server startup in 49051 ms
```

（10）启动外部浏览器，进行测试（见图 7-11）。Tomcat 成功启动后，即可通过浏览器访问 Servlet。

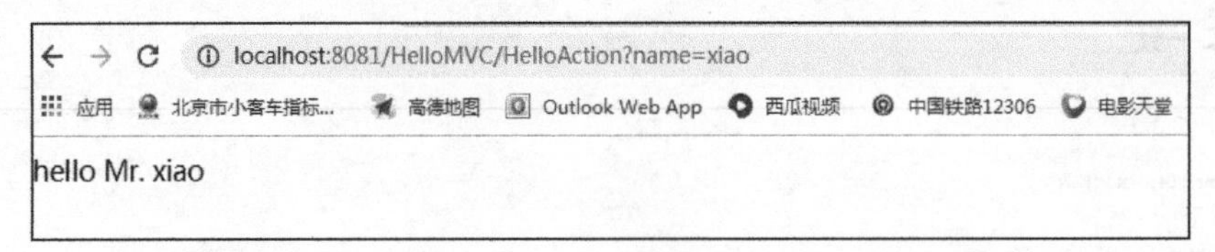

图 7-11　测试

7.2.2　Servlet 控制器与逻辑类

在前面的 HelloMVC 中，只有简单的控制器和视图 JSP，在这一节中，我们增加逻辑类，让项目满足 MVC 架构的思想，显示返回值。

操作步骤如下。

（1）新增业务逻辑类。

```
public class HelloBiz {
    public String sayHello(String name) {
        return "hello,Mr. " + name;
    }
}
```

（2）修改 HelloAction 的代码。在 HelloAction 中调用业务逻辑对象 HelloBiz。

```
protected void doGet(HttpServletRequest request,
                      HttpServletResponse response)
                      throws ServletException, IOException {
    String name = request.getParameter("name");
    HelloBiz biz = new HelloBiz();
    String hello = biz.sayHello(name);
    request.setAttribute("hello", hello);
    request.getRequestDispatcher("/main/hello.jsp")
                        .forward(request, response);
}
```

（3）在 hello.jsp 中显示打招呼的信息。

```
<!DOCTYPE html>
<html>
<body>
    张三说：${hello}
</body>
</html>
```

（4）测试。测试结果见图 7-12。

localhost:8080/HelloMVC/HelloAction?name=xiao

张三说：hello, Mr. xiao

图 7-12　测试结果

7.2.3　MVC 架构

图 7-13 是标准的 MVC 架构，不管控制器使用 Servlet，还是 Struts 2/Spring MVC，MVC 架构的基本思想不变。控制器的作用是接收客户端传来的 HTTP 请求，转交给业务逻辑对象处理。业务逻辑层处理完成后，结果信息通过 Model 返回。控制器根据逻辑层返回结果的不同，转向不同的视图，同时把 Model 对象传给视图。

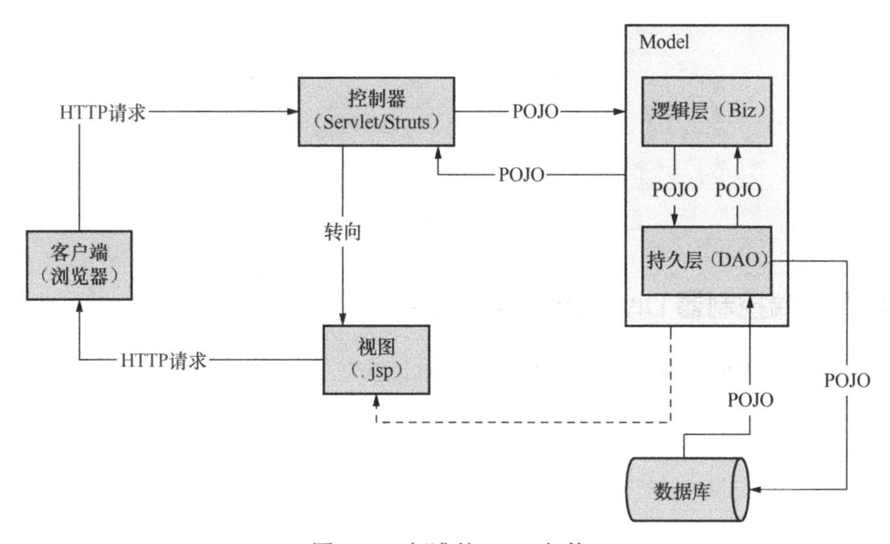

图 7-13　标准的 MVC 架构

7.3　HelloSpringMVC 示例

在本节中，我们使用 Spring MVC 改造前面的 HelloMVC 项目。这是 Spring MVC 的入门示例，包含了 Spring MVC 开发的所有重要步骤。

7.3.1　导入模块和包

在图 7-14 所示的 Spring 模块中，Servlet、Web、AOP、切面、Bean、Core、上下文和 SpEL 模块为 Web 开发需要导入的模块。

- 导入核心包——spring-beans-4.3.19.jar、spring-core-4.3.19.jar、spring-context-4.3.19.jar、spring-context-support-4.3.19.jar、spring-expression-4.3.19.jar。
- 导入 AOP 包——spring-aop-4.3.19.jar、spring-aspects-4.3.19.jar。
- 导入 Web 包——spring-web-4.3.19.jar、spring-webmvc-.jar。
- 导入依赖包——commons.logging-1.1.jar、org.aspectj.weaver-1.6.8.jar。

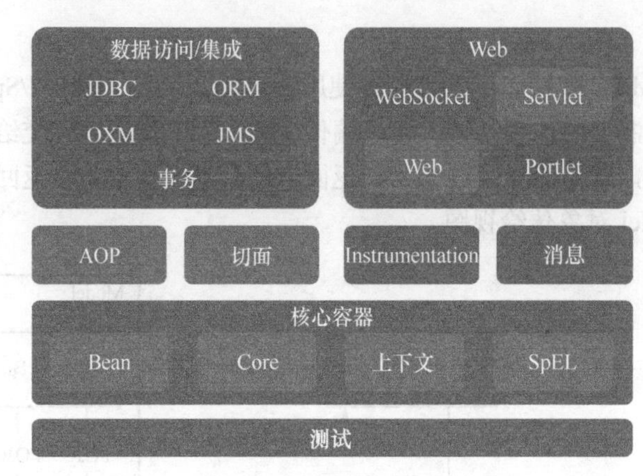

图 7-14 Spring 模块

7.3.2 配置前端控制器 DispatcherServlet

在 web.xml 中配置前端控制器 DispatcherServlet。

```
<servlet>
    <servlet-name>aa</servlet-name>
    <servlet-class>org.springframework.web.servlet.DispatcherServlet</servlet-class>
    <init-param>
        <param-name>contextConfigLocation</param-name>
        <param-value>/WEB-INF/spring-mvc.xml</param-value>
    </init-param>
    <load-on-startup>1</load-on-startup>
</servlet>
<servlet-mapping>
    <servlet-name>aa</servlet-name>
    <url-pattern>*.do</url-pattern>
</servlet-mapping>
```

7.3.3 配置 spring-mvc.xml

在 WEB-INF 目录下，新建 spring-mvc.xml 文件（其他文件名也可以）。新建的 spring-mvc.xml 与 web.xml 位置相同。

配置 spring-mvc.xml 的步骤如下。

（1）配置 schema。

```
<beans xmlns="http://www.springframework.org/schema/beans"
    xmlns:xsi="http://www.w3.org/2001/XMLSchema-instance"
    xmlns:context="http://www.springframework.org/schema/context"
    xsi:schemaLocation="
    http://www.springframework.org/schema/beans
    http://www.springframework.org/schema/beans/spring-beans.xsd
    http://www.springframework.org/schema/context
    http://www.springframework.org/schema/context/spring-context.xsd">
</beans>
```

（2）配置组件扫描。

```
<context:component-scan base-package="com.icss.action"/>
```

（3）配置视图解析器。所有视图 JSP 文件都放在/WEB-INF/views 下。

```
<bean class="org.springframework.web.servlet.view.InternalResourceViewResolver">
    <property name="prefix"  value="/WEB-INF/views/" />
</bean>
```

7.3.4 编写 HelloAction

使用@Controller 和@RequestMapping 注解，分别标注 Action 类与方法。

```
@Controller("helloAction")
public class HelloAction {
    @RequestMapping("/hello")
    public String sayHello(String name,Model model) {
        model.addAttribute("name",name);
        return "/main/hello.jsp";
    }
}
```

7.3.5 编写视图

在 WEB-INF 文件夹下新建 views 文件夹，在 views 下新建 main 文件夹，在 main 下新建 hello.jsp。把视图放到 WEB-INF 文件夹下，是为了提高 Web 项目的安全性，不允许通过浏览器直接访问 JSP 页面。所有的 JSP 页面必须通过控制器对象访问。

```
<html>
    <body>
        hello Mr. ${name}
    </body>
</html>
```

7.3.6 浏览器测试

为了测试浏览器，发出请求信息 http://localhost:8080/HelloSpringMVC/hello.do?name=xiao（见图 7-15）。注意，请求必须是*.do 的格式（要与 web.xml 中的配置一致），这样 Spring MVC 的前端控制器才能拦截请求。

localhost:8080/HelloSpringMvc/hello.do?name=xiao

hello Mr. xiao

图 7-15　发送请求信息

7.3.7 配置 log4j 日志

通过配置 log4j 日志，并且与 Spring MVC 结合，可以检查控制器是否配置成功，这在开发实践中非常重要。配置 log4j 日志的步骤如下。

（1）导入 log4j.jar。

（2）在项目 src 下新建 log4j.properties 配置文件。

```
log4j.rootLogger=INFO,BB,AA
log4j.appender.AA=org.apache.log4j.ConsoleAppender
log4j.appender.AA.layout=org.apache.log4j.SimpleLayout
log4j.appender.BB=org.apache.log4j.FileAppender
log4j.appender.BB.File=proxy.log
log4j.appender.BB.layout=org.apache.log4j.PatternLayout
log4j.appender.BB.layout.ConversionPattern
                         =%d{yyyy-MM-dd HH:mm:ss} %l %F %p %m%n
```

（3）Tomcat 启动成功后，要检查如下输出信息，这对于调试非常有帮助。如果出现 Mapper URL path [/hello]，说明 HelloAction 配置正确。

```
信息: Initializing Spring FrameworkServlet 'aa'
INFO - FrameworkServlet 'aa': initialization started
INFO - Refreshing WebApplicationContext for namespace 'aa-servlet': startup
INFO - Loading XML bean definitions from ServletContext resource
[/WEB-INF/spring-mvc.xml]
INFO - Mapped URL path [/hello] onto handler 'helloAction'
INFO - Mapped URL path [/hello.*] onto handler 'helloAction'
INFO - Mapped URL path [/hello/] onto handler 'helloAction'
INFO - FrameworkServlet 'aa': initialization completed in 1446 ms
1月03, 2020 11:47:51上午 org.apache.coyote.AbstractProtocol start
```

7.4 前端控制器 DispatcherServlet

7.4.1 Spring Web MVC 架构

图 7-16 所示是 Spring MVC 的高层架构，其中主要包括前端控制器、用户自定义控制器和视

图模板。

前端控制器负责接收浏览器或其他客户端发出的 HTTP 请求，并把处理结果中的数据流返回客户端。前端控制器将代理请求转发给用户自定义控制器，自定义控制器把模型数据转发给前端控制器。此外，用户自定义控制器还处理客户端请求，调用服务层对象，创建模型。前端控制器提交模型数据给视图模板。视图模板解析模型和视图模板，把结果返回给前端控制器。

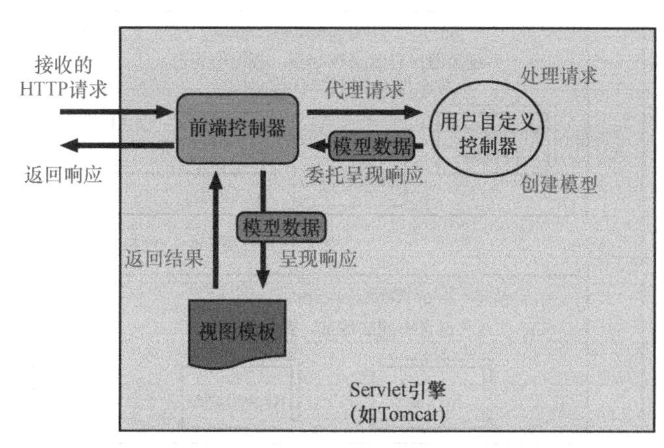

图 7-16　Spring MVC 的高层架构

客户端请求的处理流程如下。

（1）客户端向前端控制器发出请求。

（2）前端控制器接收客户端请求信息。

（3）前端控制器调用代理对象，把请求转发给用户自定义控制器处理。

（4）用户自定义控制器处理请求，并把处理结果通过模型对象返回给前端控制器。

（5）前端控制器把模型数据发给视图模板进行处理。

（6）视图模板把结果返回给前端控制器。

（7）前端控制器把最终的视图信息通过流返回给客户端。

7.4.2　DispatcherServlet 与 IoC 容器的关系

DispatcherServlet 与 IoC 容器的关系如图 7-17 所示。前端控制器 DispatcherServlet 中有一个 IoC 容器（Servlet WebApplicationContext），这个 IoC 容器可能指向它的父容器——Root WebApplicationContext。

- DispatcherServlet：表示前端控制器。

- Servlet WebApplicationContext：表示 Web 层的 IoC 容器，需要依赖 Servlet 环境。其中，HandlerMapping 表示处理视图映射的接口，用于返回拦截处理链。
- Root WebApplicationContext：表示根 IoC 容器，包含中间层业务对象、数据源对象等。其中，Services 表示服务层 Bean 对象；Repositories 表示持久层 Bean 对象。

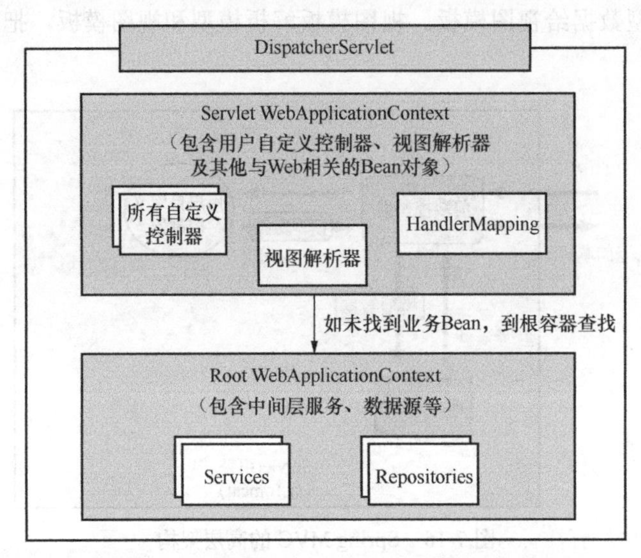

图 7-17　DispatcherServlet 与 IoC 容器的关系

习惯上，我们把自定义控制器、视图解析器、映射器等当成 Spring 的 Bean，配置在 Servlet WebApplicationContext 中。

服务层对象和持久层对象也是 IoC 容器中的 Bean，它们习惯上配置在 Root WebApplication Context 容器中。

当 Tomcat 启动时，DispatcherServlet 的初始化参数如下。

```
<servlet>
    <servlet-name>aa</servlet-name>
    <servlet-class>org.springframework.web.servlet.DispatcherServlet</servlet-class>
    <init-param>
        <param-name>contextConfigLocation</param-name>
        <param-value>/WEB-INF/spring-mvc.xml</param-value>
    </init-param>
    <load-on-startup>1</load-on-startup>
</servlet>
```

如果发现<load-on-startup>为 1 的配置，Tomcat 会在启动时创建 DispatcherServlet 对象实例，同时加载 IoC 容器，加载 spring-mvc.xml 文件中的 Bean。

7.4.3 DispatcherServlet 的功能

DispatcherServlet 继承自 HttpServlet，实现了 javax.servlet.Servlet 接口。注意，所有的 Servlet 对象都是 Java EE 容器中的组件，它受 Tomcat 容器管理。DispatcherServlet 的详细功能参见 API 描述。

```
org.springframework.web.servlet
Class DispatcherServlet
java.lang.Object
     javax.servlet.GenericServlet
     javax.servlet.http.HttpServlet
         org.springframework.web.servlet.HttpServletBean
             org.springframework.web.servlet.FrameworkServlet
                org.springframework.web.servlet.DispatcherServlet
```

所有已实现的接口包括 java.io.Serializable、Servlet、ServletConfig、Aware、ApplicationContextAware、EnvironmentAware 和 EnvironmentCapable。

与其他请求驱动的 MVC 框架明显不同，DispatcherServlet 提供了如下功能。

- 基于 JavaBeans 的配置机制。
- 使用 HandlerMapping 的实现类，提前处理对控制器的请求路由。默认使用的是 BeanNameUrlHandlerMapping 和 DefaultAnnotationHandlerMapping。HandlerMapping 作为 Bean 对象由 IoC 容器统一管理。
- 使用 HandlerAdapter 处理请求。对于 Spring 的 HttpRequestHandler 和 Controller 接口，默认使用 HttpRequestHandlerAdapter 和 SimpleControllerHandlerAdapter 适配器处理。同时，将注册 AnnotationMethodHandlerAdapter。HandlerAdapter 也作为 Bean 对象由 IoC 容器统一管理。
- 前端控制器中的异常由 HandlerExceptionResolver 解析。可以映射异常到指定的错误显示页。默认使用 AnnotationMethodHandlerExceptionResolver、ResponseStatusExceptionResolver 和 DefaultHandlerExceptionResolver 等异常解析器。
- 视图解析策略由配置的 ViewResolver 类来处理，它可以把视图名字解析成视图对象。默认使用 InternalResourceViewResolver 视图解析器。所有视图解析器由 IoC 容器统一管理。
- 如果用户没有提供视图对象或视图名字，使用 RequestToViewNameTranslator 配置，可以把请求转换到相应的视图对象，默认视图名字为 viewNameTranslator。默认解析器是 DefaultRequestToViewNameTranslator。
- 前端控制器使用 MultipartResolver 的实现类处理实体文件的上传。使用的典型解析器为

CommonsMultipartResolver，它基于 Apache 的 CommonsFileUpload。

- 国际化地使用 LocaleResolver，它的默认实现类是 AcceptHeaderLocaleResolver。
- 主题解析策略使用 ThemeResolver，默认实现类是 FixedThemeResolver。

7.5 通过源代码解析 DispatcherServlet 的工作流程

本节采用源代码跟踪的方式，详细解析前端控制器 DispatcherServlet 的工作流程。通过分析源代码，更容易理解前端控制 DispatcherServlet 的工作机制。

7.5.1 添加源代码

要添加源代码，如图 7-18 所示，在 Eclipse IDE 的左侧面板中，选择 Web App Libraries 下的 spring-webmvc-4.3.19.RELEASE.jar，右击并选择 properties，在弹出的窗口中选择 Java Source Attachment，单击 External location 单选按钮，从 Spring 官网下载源代码并指明存放位置（spring-webmvc-4.3.19.RELEASE- sources.jar），单击 Apply 按钮。

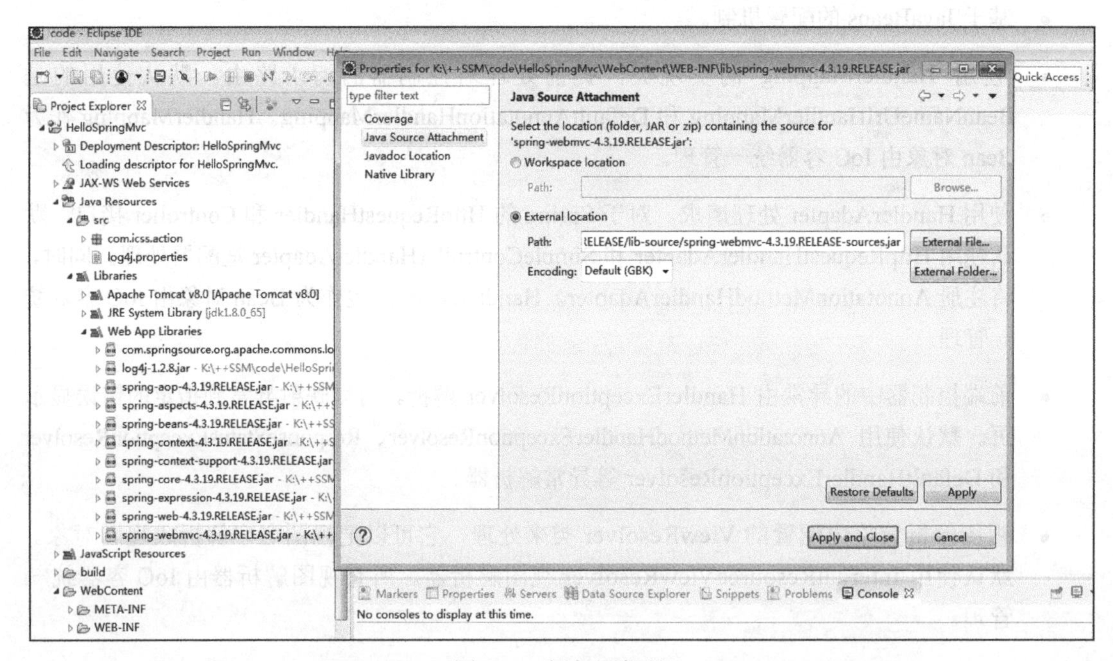

图 7-18 添加源代码

如果成功添加源代码，双击 DispatcherServlet，即可看到源代码（见图 7-19）。同理，添加其他包的源代码，如 Spring 的核心包。

```
154  * @see org.springframework.web.context.ContextLoaderListener
155  */
156 @SuppressWarnings("serial")
157 public class DispatcherServlet extends FrameworkServlet {
158
159     /** Well-known name for the MultipartResolver object in the bean factory for thi
160     public static final String MULTIPART_RESOLVER_BEAN_NAME = "multipartResolver";
161
162     /** Well-known name for the LocaleResolver object in the bean factory for this n
163     public static final String LOCALE_RESOLVER_BEAN_NAME = "localeResolver";
164
165     /** Well-known name for the ThemeResolver object in the bean factory for this na
166     public static final String THEME_RESOLVER_BEAN_NAME = "themeResolver";
167
168     /**
169      * Well-known name for the HandlerMapping object in the bean factory for this na
170      * Only used when "detectAllHandlerMappings" is turned off.
171      * @see #setDetectAllHandlerMappings
172      */
173     public static final String HANDLER_MAPPING_BEAN_NAME = "handlerMapping";
174
175     /**
```

图 7-19　源代码

7.5.2　通过断点跟踪观察 DispatcherServlet 的工作流程

把发送请求给 UserAction 的 login()，触发图 7-20 所示的工作流程。

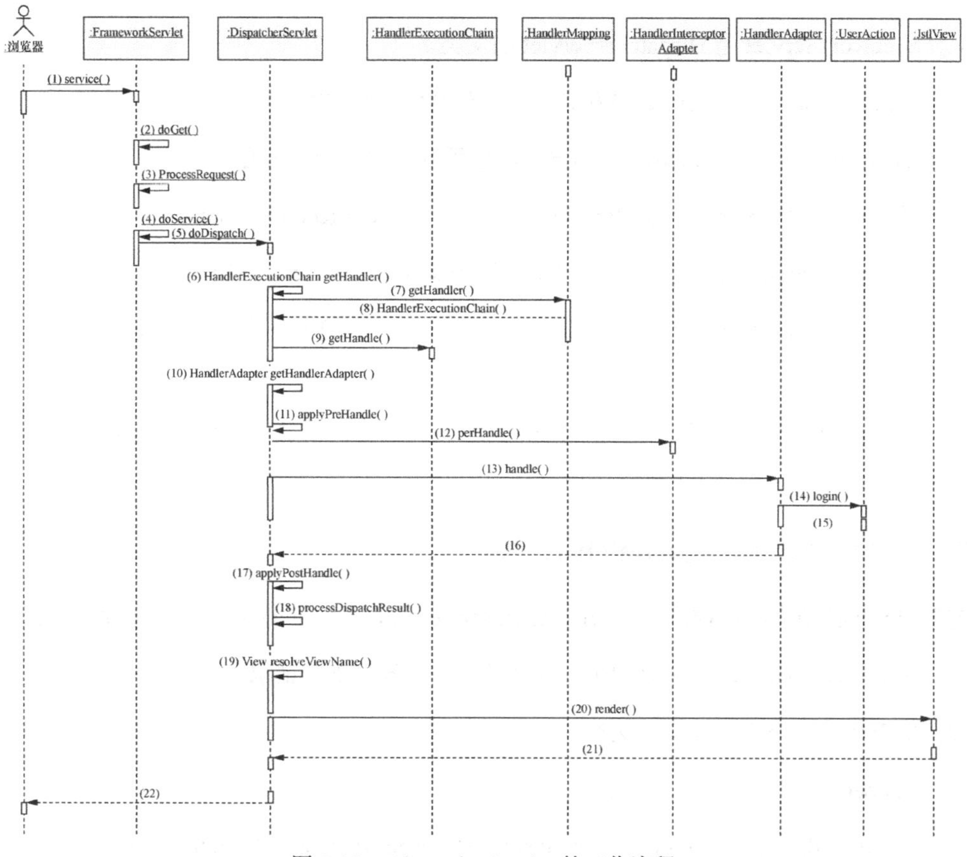

图 7-20　DispatcherServlet 的工作流程

下面的这些类和接口是 DispatcherServlet 工作流程时序图中的重点交互部分。

```
public class DispatcherServlet extends FrameworkServlet {}
public abstract class FrameworkServlet
            extends HttpServletBean implements ApplicationContextAware {
    protected void service(HttpServletRequest request, HttpServletResponse response)
        throws ServletException, IOException {}
}
public class HandlerExecutionChain {
    private final Object handler; //处理请求的 Action，如 HelloAction 对象
    private HandlerInterceptor[] interceptors; //多个拦截器组成链
}
public interface HandlerMapping {
    HandlerExecutionChain getHandler(PortletRequest request) throws Exception;
}
public interface HandlerAdapter {
    ModelAndView handle(HttpServletRequest request,
          HttpServletResponse response, Object handler) throws Exception;
}
```

通过源代码中的断点跟踪，了解 HTTP 请求从前端控制器到 UserAction 的处理流程。

- FrameworkServlet 是 DispatcherServlet 的父类。
- FrameworkServlet 的 service( )方法，是所有 HTTP 请求的入口。
- DispatcherServlet 的 doDispatch( )方法，是所有流程控制的核心。
- 调用 HandlerMapping 的 getHandler( )，返回 HandlerExecutionChain。
- HandlerExecutionChain 包含用户自定义控制器（UserAction）和多个拦截器。
- 调用 HandlerAdapter 的 handle()方法，前端控制器把客户端请求转发给用户自定义控制器。
- doDispatch 接收 UserAction 的返回结果 ModelAndView，然后调用 processDispatchResult 方法处理视图。

7.5.3 前端控制器的 doDispatch()方法

前端控制器的所有核心控制都在 DispatcherServlet 的 doDispatch( )中，主要处理步骤如下。

（1）WebAsyncManager 判断是同步处理请求，还是异步处理请求。

（2）检查是否为 multipart 实体文件上传。

（3）mappedHandler 拦截器处理请求。

（4）HandlerAdapter：：handle(processedRequest, response, mappedHandler.getHandler( ))激活自

定义 Action 的实际方法，以处理请求。

（5）返回 ModelAndView 对象。

（6）不管是否存在异常，统一由 processDispatchResult( )处理。

如下为 doDispatch( )方法的核心代码。

```
protected void doDispatch(HttpServletRequest request,
                  HttpServletResponse response) throws Exception {
    HttpServletRequest processedRequest = request;
    HandlerExecutionChain mappedHandler = null;  //处理程序链
    try {
        ModelAndView mv = null;
        try {
            mappedHandler = getHandler(processedRequest);.
            HandlerAdapter ha = getHandlerAdapter(mappedHandler.getHandler());
            if (!mappedHandler.applyPreHandle(processedRequest, response)) {
                //拦截器前置处理
                return;
            }
            // 适配器调用自定义 Action 方法处理请求，返回 ModelAndView
            mv = ha.handle(processedRequest, response, mappedHandler.getHandler());
            //拦截器后置处理
            mappedHandler.applyPostHandle(processedRequest, response, mv);
        }
        catch (Exception ex) {
            dispatchException = ex;
        }
        //调用视图的 render 方法，返回视图给用户
        processDispatchResult(processedRequest, response,
                                  mappedHandler, mv, dispatchException);
    }finally {
    }
}
```

7.5.4 创建 IoC 容器

配置 DispatcherServlet 的初始化参数，把 Spring 的配置信息 spring-mvc.xml 传递给 IoC 容器。此处的 IoC 容器是 WebApplicationContext，它通常为单例模式。IoC 容器的创建流程见图 7-21。

HttpServletBean 是前端控制器 DispatcherServlet 的父类。

```
public abstract class HttpServletBean extends HttpServlet
implements EnvironmentCapable, EnvironmentAware {}
```

在 Servlet 的 init( )方法中，可以使用 ServletConfig 读取 spring-mvc.xml 配置信息，然后把这个信息传递给 WebApplicationContext。

```
<servlet>
    <servlet-name>aa</servlet-name>
```

```
        <servlet-class>org.springframework.web.servlet.DispatcherServlet</servlet-class>
        <init-param>
            <param-name>contextConfigLocation</param-name>
            <param-value>/WEB-INF/spring-mvc.xml</param-value>
        </init-param>
        <load-on-startup>1</load-on-startup>
    </servlet>
```

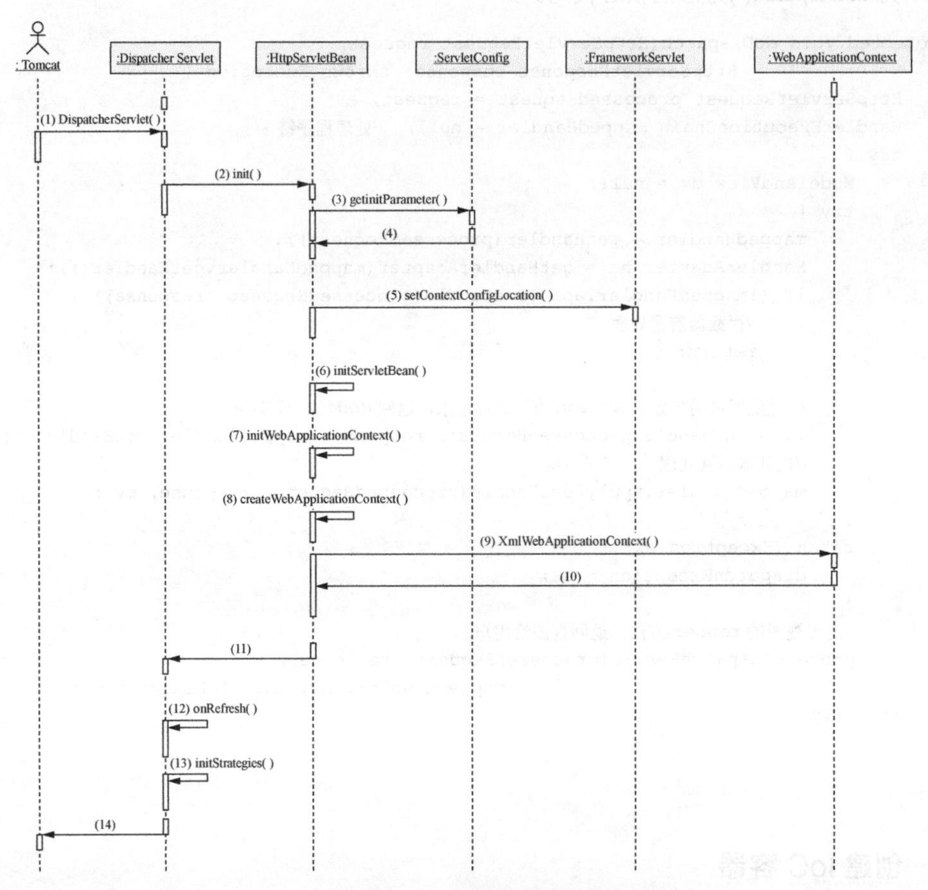

图 7-21　IoC 容器的创建流程

在 initWebApplicationContext()方法中读取配置信息，创建 IoC 容器。

```
public abstract class FrameworkServlet extends HttpServletBean
                                implements ApplicationContextAware {
    protected final void initServletBean() throws ServletException {
        this.webApplicationContext = initWebApplicationContext();
        initFrameworkServlet();
    }
    protected WebApplicationContext initWebApplicationContext() {
        WebApplicationContext rootContext =
            WebApplicationContextUtils.getWebApplicationContext(getServletContext());
```

```
            if (this.webApplicationContext != null) {
                wac = this.webApplicationContext;
            }
        }
    }
```

也可以使用监听器创建 IoC 容器。ContextLoaderListener 实现了 ServletContextListener 接口，这个监听器在 Tomcat 启动时触发，即 Tomcat 启动时加载配置文件 root-context.xml。

```
public class ContextLoaderListener extends ContextLoader
implements ServletContextListener {}
```

使用 ServletContext 的 getInitParameter( )方法读取配置信息，然后把读取的配置信息传给 WebApplicationContext 容器。

```
<context-param>
    <param-name>contextConfigLocation</param-name>
    <param-value>/WEB-INF/root-context.xml</param-value>
</context-param>
<servlet>
    <servlet-name>dispatcher</servlet-name>
    <servlet-class>
        org.springframework.web.servlet.DispatcherServlet
    </servlet-class>
    <init-param>
        <param-name>contextConfigLocation</param-name>
        <param-value></param-value>
    </init-param>
    <load-on-startup>1</load-on-startup>
</servlet>
<servlet-mapping>
    <servlet-name>dispatcher</servlet-name>
    <url-pattern>/*</url-pattern>
</servlet-mapping>
<listener>
    <listener-class>
        org.springframework.web.context.ContextLoaderListener
    </listener-class>
</listener>
```

7.6 控制器@Controller

7.6.1 @Controller 概述

1. @Controller 与 Controller 接口

控制器是指 MVC 中的 C（Controller），它的典型操作是调用服务层接口，返回 Model 数据。

Spring 以一种抽象的方式实现了控制器概念，这样可以支持不同类型的控制器。Spring 控制器的实现无须继承指定的基类或实现某些接口。另外，控制器无须依赖Servlet API或Portlet API。

@Controller 控制器与 Controller 接口是两个完全不同的概念。参见如下控制器接口的定义，用户自定义的控制器无须直接实现这些接口。

org.springframework.web.servlet.mvc.Controller 接口用于 Servlet 环境。

```
public interface Controller {
    ModelAndView handleRequest(HttpServletRequest request,
                    HttpServletResponse response) throws Exception;
}
public abstract class AbstractController
        extends WebContentGenerator implements Controller {
    protected abstract ModelAndView handleRequestInternal(HttpServletRequest request,
                    HttpServletResponse response)throws Exception;
    public ModelAndView handleRequest(HttpServletRequest request,
                    HttpServletResponse response)throws Exception {}
}
```

前端控制器 DispatcherServlet 与 Controller 接口有着非常紧密的关系。如果浏览器发送一个 HTTP 请求（如 http://localhost:8080/HelloSpringMVC/hello.do），它的处理流程如下。

（1）Java EE 服务器（如 Tomcat）接收请求。

（2）Tomcat 根据请求的*.do 后缀，创建 ServletRequest 对象和 ServletResponse 对象。

（3）Tomcat 调用 DispatcherServlet 的 service(ServletRequest req, ServletResponse res)，把创建的 ServletRequest 对象引用和 ServletResponse 对象引用传递给 DispatcherServlet。

（4）DispatcherServlet 调用 Controller 接口的 handleRequest，传入 ServletRequest 对象引用和 ServletResponse 对象引用，返回 ModelAndView。

（5）DispatcherServlet 调用 ModelAndView 中 View 对象的 render 方法，生成数据流。

```
void render(Map<String, ?> model, HttpServletRequest req, HttpServletResponse res)
```

（6）Tomcat 把数据流返回给浏览器。

使用@Controller 注解标注控制器类，是配置自定义控制器的常用方式（很少使用 XML 方式配置控制器）。@Controller 注解是@Component 注解在控制层的具体应用。在普通类上使用@Controller 即可，扫描 Spring 组件时，会创建一个控制器实例。

```
<context:component-scan
        base-package="com.icss.action"></context:component-scan>
@Controller
public class HelloAction {
    @RequestMapping("/hello")
    public String sayHello(String name,Model model) {
        HelloBiz biz = new HelloBiz();
```

```
        String msg = biz.sayHello(name);
        model.addAttribute("msg",msg);
        return "/main/hello.jsp";
    }
}
```

HelloAction 中的 sayHello( )返回值为视图的名字，它与 spring-mvc.xml 中配置的视图解析器 InternalResourceViewResolver 组合，即可创建视图对象。

```
<bean id="jspViewResolver"
    class="org.springframework.web.servlet.view.InternalResourceViewResolver">
    <property name="prefix" value="/WEB-INF/views/" />
</bean>
```

控制器调用逻辑层方法，根据返回结果生成 model 数据。在 Servlet 环境中，model 数据被作为 request 对象的 attribute 信息，在 request 对象域中传递。如下两句代码等效。

```
model.addAttribute("msg",msg);
request.setAttribute("msg", msg);
```

2. @Controller 的作用域

对@Controller 尽量不要使用属性信息，这样就可以使用单例模式（在控制器中注入其他单例对象，不受影响）。Struts 2 的特点是在 Action 中用属性传递信息，这导致 Struts 的 Action 只能使用 prototype 模式，严重影响了性能。@Controller 还可以使用其他作用域，如 request、session、application 等，具体使用哪个作用域更好，由具体的业务环境决定。

下面通过示例介绍@Controller 的作用域的定义方式。

示例 7-1：默认的单例作用域。

```
@Controller
public class HelloAction {}
```

示例 7-2：原型作用域。

```
@Controller
@Scope("prototype")
public class HelloAction {}
```

示例 7-3：request 作用域。

```
@Controller
@RequestScope
public class HelloAction {}
```

示例 7-4：session 作用域。

```
@Controller
@SessionScope
public class HelloAction {}
```

示例 7-5：application 作用域。

```
@Controller
@ApplicationScope
public class HelloAction {}
```

7.6.2 @RequestMapping

1. HTTP 请求映射

使用@RequestMapping 注解去映射 HTTP 的 URL，如@RequestMapping("/hello")，这映射的 URL 就是 http://localhost:8081/HelloSpringMvc/hello.do。

@RequestMapping 可以用于注解 class 和 method。在类级映射，可以把 URL 映射到表单控制器；在方法级映射，可以映射 HTTP 请求中的具体方法，如 GET、POST、PUT、DELETE 等。

@RequestMapping 注解位于类后，请求的 URL 应该同时包含类和方法的@RequestMapping 值。

以下示例中的 URL 为 http://localhost:8080/HelloSpringMvc/back/buyinfo.do。

```
@Controller
@RequestMapping("/back")
public class BackAction {
    @RequestMapping("/buyinfo")
    public String lookBuyinfo() {
        return null;
    }
}
```

@RequestMapping 的 value 属性和 path 属性在 Servlet 环境下是等效的。method 属性与 HTTP 方法对应。如果在@RequestMapping 中没有指定 HTTP 的方法类型，即默认映射到所有 HTTP 方法。示例代码如下。

```
@Controller
public class UserAction {
    @GetMapping("/login")
    public String login() {
        return "/main/login.jsp";
    }
    @PostMapping("/login")
    public String login(String uname, String pwd,
                    Model model, HttpSession session) throws Exception {
    }
}
```

在这个示例中，浏览器发送请求 http://localhost:8081/HelloSpringMvc/user/login.do，这里对 GET 请求与 POST 请求分别进行了处理。GET 请求通常用于页面转向，而 POST 请求用于表单数据提交。

2. @RequestParam

@RequestParam 的作用是绑定 HTTP 请求参数到控制器的方法参数。

当 HTTP 请求中的参数与控制器中的映射方法参数名不一致时，必须使用@RequestParam 进行转换。例如，http://localhost:8080/HelloSpringMvc/hello.do?name=xiao（见图 7-22）这个 HTTP 请求中的参数名为 uname，而以下代码中映射方法 sayHello( )里的参数名为 name，如果没有进行参数映射，则正确的参数就无法接收到。

```
@RequestMapping("/hello")
public String sayHello(String name,Model model) {
    HelloBiz biz = new HelloBiz();
    String msg = biz.sayHello(name);
    model.addAttribute("msg",msg);
    return "/main/hello.jsp";
}
```

通过@RequestParam 映射参数后，HTTP 请求中的 uname 值会映射到 sayHello()方法中的参数 name 上。

```
@RequestMapping("/hello")
public String sayHello(@RequestParam("uname")String name,
                       Model model) {
    HelloBiz biz = new HelloBiz();
    String msg = biz.sayHello(name);
    model.addAttribute("msg",msg);
    return "/main/hello.jsp";
}
```

以上代码的执行结果如图 7-22 所示。

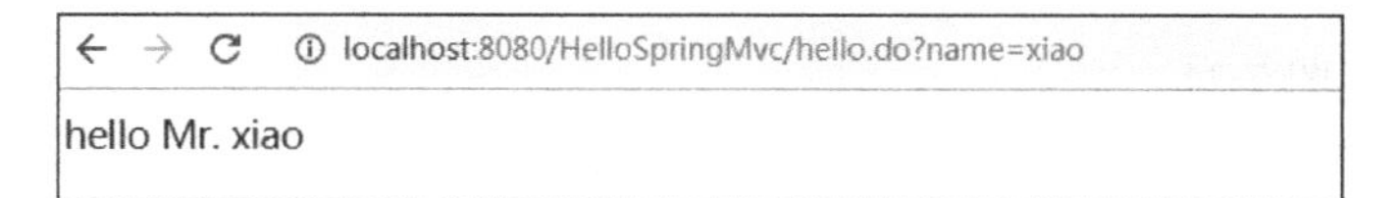

图 7-22　执行结果

注意，@RequestParam 注解的参数在 URL 中必须存在，否则报错（见图 7-23）。

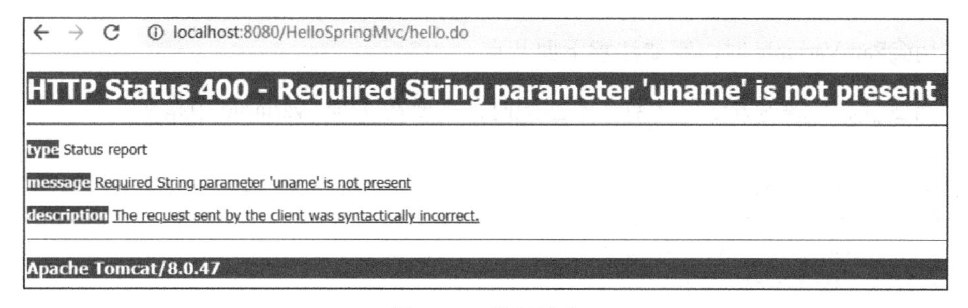

图 7-23　错误消息

如果参数是可选的，需要设置 required 属性为 false，如@RequestParam(name="uname", required=false)，这样即使 HTTP 请求中没有这个参数，也不会报错（required 属性默认为 true）。

在@RequestParam 中还可以设置默认值，当请求中没有这个参数时，就使用默认值。在传递翻页参数时，经常设置参数 page 的默认值为 1。

示例代码如下。

```
@Controller
@RequestMapping("/back")
public class OrderAction {
    @RequestMapping("/buyinfo")
    public String readUserBuyRecord(
                @RequestParam(name = "page", defaultValue = "1") int page,
                String uname, Model model) throws Exception {}
}
```

3. @PathVariable

使用@PathVariable 注解，可以创建 RESTFul 风格的 Web 站点。在解析@PathVariable 时使用了 URI 模板。传统的参数传递方法是通过 HTTP 的方法参数。@PathVariable 通过 URL 中存放的变量传递参数，这种方法并不流行。

```
@Controller("helloAction")
public class HelloAction {
    @RequestMapping("/hello/{name}")
    public String sayHello(@PathVariable String name,Model model) {
        model.addAttribute("name",name);
        return "/main/hello.jsp";
    }
}
```

请求的 URL 见图 7-24。

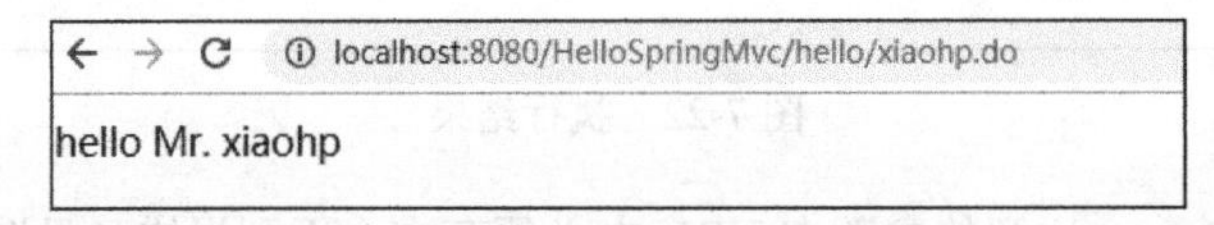

图 7-24　请求的 URL

在使用@PathVariable 时，需要注意下面几点。

（1）当变量名与映射方法的参数名不匹配时，需要在@PathVariable 中指明。

```
@GetMapping("/owners/{ownerId}")
public String findOwner(@PathVariable("ownerId") String theOwner, Model model) {
    Owner owner = ownerService.findOwner(ownerId);
    model.addAttribute("owner", owner);
    return "displayOwner";
}
```

（2）在同一个方法参数中，可以有多个@PathVariable 注解。

```
@GetMapping("/owners/{ownerId}/pets/{petId}")
public String findPet(@PathVariable String ownerId,
        @PathVariable String petId, Model model) {
    Owner owner = ownerService.findOwner(ownerId);
    Pet pet = owner.getPet(petId);
    model.addAttribute("pet", pet);
    return "displayPet";
}
```

（3）为了装配 URL 模板，使用@RequestMapping 注解类和方法均可。

```
@Controller
@RequestMapping("/owners/{ownerId}")
public class RelativePathUriTemplateController {
    @RequestMapping("/pets/{petId}")
    public void findPet(@PathVariable String ownerId,
                        @PathVariable String petId, Model model) {
    }
}
```

4. 映射处理程序类

从 Spring Framework 3.1 起，@RequestMapping 映射方法的处理程序用 RequestMappingHandlerMapping 和 RequestMappingHandlerAdapter 解析。RequestMappingHandlerMapping 会实现 HandlerMapping 接口，RequestMappingHandlerAdapter 会实现 HandlerAdapter。HandlerMapping 接口和 HandlerAdapter 接口非常重要，它们是处理@RequestMapping 的关键。

RequestMappingHandlerMapping 的父类 AbstractHandlerMapping 实现了接口 HandlerMapping。

```
public class RequestMappingHandlerMapping
    extends RequestMappingInfoHandlerMapping
    implements MatchableHandlerMapping, EmbeddedValueResolverAware {}

public abstract class AbstractHandlerMapping
    extends WebApplicationObjectSupport
    implements  HandlerMapping, Ordered {}
```

HandlerMapping 接口的核心方法是 getHandler，调用该方法会返回 HandlerExecutionChain。

```
public interface HandlerMapping {
    HandlerExecutionChain getHandler
          (HttpServletRequest request) throws Exception;
}
```

HandlerExecutionChain 包含核心处理对象 handler，它是处理 HTTP 请求的用户自定义 Action，如 HelloAction。handler 还包含拦截器集合，即在处理请求前，Action 需要先处理拦截器行为。

```
public class HandlerExecutionChain {
    private final Object handler;
    private HandlerInterceptor[] interceptors;
}
```

RequestMappingHandlerAdapter的父类AbstractHandlerMethodAdapter实现了HandlerAdapter接口。

```
public class RequestMappingHandlerAdapter
    extends AbstractHandlerMethodAdapter
    implements BeanFactoryAware, InitializingBean {}

public interface HandlerAdapter {
  ModelAndView handle(HttpServletRequest request,
         HttpServletResponse response, Object handler)throws Exception;
}
```

@RequestMapping 的处理流程参见 DispatcherServlet 中的 doDispatch( )方法，该方法的核心代码如下。

```
//调用 HandlerMapping 的 getHandler 返回处理程序链
HandlerExecutionChain mappedHandler = getHandler(processedRequest);
//创建适配器，激活 helloAction 的方法
//mappedHandler.getHandler()返回的是 HelloAction 对象
HandlerAdapter ha = getHandlerAdapter(mappedHandler.getHandler());
ModelAndView mv =
    ha.handle(processedRequest, response, mappedHandler.getHandler());
```

5. @GetMapping 与@PostMapping

Spring Framework 4.3 之后，可以使用如下注解简化@RequestMapping 的写法：

- @GetMapping；
- @PostMapping；
- @PutMapping；
- @DeleteMapping；
- @PatchMapping。

为了使@GetMapping 生效，必须增加<mvc:annotation-driven />配置。

```
<beans xmlns="http://www.springframework.org/schema/beans"
 xmlns:xsi="http://www.w3.org/2001/XMLSchema-instance"
    xmlns:mvc="http://www.springframework.org/schema/mvc"
    xsi:schemaLocation="
    http://www.springframework.org/schema/mvc
    http://www.springframework.org/schema/mvc/spring-mvc.xsd">
            <mvc:annotation-driven />
</beans>
```

示例代码如下。

```
@Controller
public class UserAction {
    @GetMapping("/login")
    public String login() {
```

```
        return "/main/login.jsp";
    }
    @PostMapping(value = "/login")
    @ResponseBody
    public String login(@RequestBody String user) {
    }
}
```

6. 使用媒体类型与@RequestBody

在网络请求中，使用 Content-Type 标识访问的资源类型。常用的 Content-Type 如下。

- 要访问 HTML 或 Servlet，可以使用 text/html、text/plain。
- 要访问 JavaScript 和 CSS 文件，可以使用 text/javascript、text/css。
- 要访问图片，可以使用 image/jpeg、image/png、image/gif。
- 要表示表单提交的默认类型，可以使用 application/x-www-form-urlencoded。
- 要通过表单上传实体文件，可以使用 multipart/form-data。
- 要通过 AJAX 提交数据，可以使用 application/json、application/xml。

@RequestMapping 有一个 consumes 属性，它表示映射方法使用的媒体类型。

只有 HTTP 请求头中的 Content-Type 与@RequestMapping 中规定的媒体类型一致，控制器的方法才能响应。注意，Content-Type 中可以使用通配符，如"Content-Type=text/*"会匹配"text/plain"和"text/html"。

在以下示例代码中，用户异步登录，在 JavaScript 脚本中指明 contentType:"application/json"。

```
function tijiao(){
       var uname = $('#uname').val();
       var pwd = $('#pwd').val();
       var user =  {uname: uname, pwd: pwd};
       if(uname != ""){
           var destUrl = "<%=basePath%>login.do";
             $.ajax({
                     type : "post",
                     url : destUrl,
                     data:user,
                     dataType:"json",
                     contentType :"application/json",
                     success : function(msg) {
                         if (msg == 0) {
                              alert("登录失败");
                         } else if (msg == 1) {
                              alert("登录成功");
                         } else if (msg == -1) {
                              alert("登录异常");
```

```
                    }
                },
                error : function() {
                    alert("异常，请检查");
                }
            });
        }
    }
```

下面处理 UserAction 中的 login()方法。

```
@Controller
public class UserAction {
    @GetMapping(value="/login",consumes="")
    public String login() {
        return "/main/login.jsp";
    }
    @PostMapping(value = "/login",consumes = "application/json")
    @ResponseBody
    public String login(@RequestBody String user) {
        System.out.println("收到：" + user);
        return "1";
    }
}
```

如上代码中，@PostMapping(value = "/login",consumes = "application/json")表明了请求数据的提交格式只能为 JSON 字符串。

@RequestBody 注解指明了方法参数与 HTTP 请求体数据绑定。在上面的示例中，@RequestBody 用来接收前端传递给后端的 JSON 字符串，数据不是源于 HTTP 请求中的参数，而是源于 HTTP 请求体。

@ResponseBody 用于 AJAX 的数据响应，不再使用 ModelAndView 转向 MVC，这个注解后面会详细讲解。

如果修改客户端的代码为 contentType :"text/html"，则会提示如下错误。

```
Resolved [org.springframework.web.HttpMediaTypeNotSupportedException: Content type 'text/html'
    not supported]
```

7. 生产媒体类型

@RequestMapping 有一个 produces 属性，它表示生产媒体类型。如果在映射方法中指明了生产媒体类型，则 HTTP 请求的 Accept 必须与生产媒体类型一致才可以。

在 AJAX 中指定 dataType:"json"后，HTTP 请求头的格式如下。

```
Accept: application/json, text/javascript, */*
```

示例代码如下。

```
function tijiao(){
        var uname = $('#uname').val();
        var pwd = $('#pwd').val();
        var user =  {uname: uname, pwd: pwd};
        if(uname != ""){
            var destUrl = "<%=basePath%>login.do";
              $.ajax({
                        type : "post",
                        url : destUrl,
                        data:user,
                        dataType:"json",
                        success : function(msg) {
                            var user = eval(msg);
                            if (user == null) {
                                 alert("登录失败");
                            } else {
                                 alert(user.uname + "登录成功");
                            }
                        },
                        error : function() {
                            alert("异常，请检查");
                        }
                  });
          }
     }
```

在 UserAction 中设置 produces=MediaType.APPLICATION_JSON_UTF8_VALUE，表示只能接收 JSON 类型的请求。

```
@Controller
public class UserAction {
     @GetMapping("/login")
     public String login() {
          return "/main/login.jsp";
     }
     @PostMapping(value = "/login",produces=MediaType.APPLICATION_JSON_UTF8_VALUE)
     @ResponseBody
     public User login(String uname,String pwd) {
          User user = new User();
          user.setUname(uname);
          user.setRole(1);
          return user;
     }
}
```

如果修改@PostMapping 的 produces 属性为文本类型，与上传的 JSON 类型不符，就会报错。

```
@PostMapping(value = "/login",produces=MediaType.TEXT_HTML_VALUE)
```

错误提示如下。

```
Resolved [org.springframework.web.HttpMediaTypeNotAcceptableException: Could not find
acceptable representation]
```

8. params 属性

params 属性表示只有 HTTP 请求包含指定的参数值，映射方法才会处理请求；否则，报错。在如下示例代码中，要求在 HTTP 请求中必须包含 userName 参数。

```
@PostMapping(value = "/login2", params = {"userName"})
public User login2(String uname, String pwd) {
    System.out.println("login post....");
    return user;
}
```

若参数信息不匹配，会报如下错误。

```
WARN - Resolved
[org.springframework.web.bind.UnsatisfiedServletRequestParameterException:
Parameter conditions "username" not met for actual request
parameters: uname={admin},  pwd={123}]
```

9. headers 属性

只有指定 HTTP 请求包含指定的 header 属性值，才能让映射方法处理请求。HTTP 请求的 header 属性值可以携带很多信息，因此更加灵活。

示例代码（在百度主页抓包的请求头）如下。

```
Accept: text/html,application/xhtml+xml,application/xml;
        q=0.9,image/webp,image/apng,*/*;q=0.8,application/signed-exchange;v=b3
Accept-Encoding: gzip, deflate, br
Accept-Language: zh-CN,zh;q=0.9
Cache-Control: max-age=0
Connection: keep-alive
Cookie: BIDUPSID=8D58C3E297D1EA655AFB2866FD3F7FDB;
      PSTM=1517223831; Host: www.baidu.com
Upgrade-Insecure-Requests: 1
```

使用 headers 约束 Content-Type 和 Accept，起到了约束 consumes 和 produces 参数的效果。当然，还可以增加其他 headers 属性约束。

```
@PostMapping(value = "/login",
        headers = {"Content-Type=application/json", "Accept=application/json" })
@ResponseBody
public User login(@RequestBody String userString) {
    User user = new User();
    String uname = userString.split("&")[0].split("=")[1];
    user.setUname(uname);
    user.setRole(2);
    return user;
}
```

10. path 属性与 value 属性

@RequestMapping 的 path 属性与 value 属性容易混淆。在 Servlet 环境下，@RequestMapping("/foo")、@RequestMapping(path="/foo")、@RequestMapping(value="/foo")等效。

参考 RequestMapping 源代码中的属性定义，value 属性的别名是 path，path 属性的别名是 value，它们相互等效。

```
@Target({ElementType.METHOD, ElementType.TYPE})
@Retention(RetentionPolicy.RUNTIME)
public @interface RequestMapping {
    @AliasFor("path")
    String[] value() default {};

    @AliasFor("value")
    String[] path() default {};
}
```

value 属性的格式是 URI，也可以是 URI 模板。value 属性可以应用于类和方法，当应用于类时，方法映射的 value 属性都继承自类。

```
@Controller
@RequestMapping(value="/user")
public class UserAction {
    @GetMapping(value="/login")
    public String login() {
        return "/main/login.jsp";
    }
}
```

11. 控制器方法的参数

控制器方法支持表 7-1 所示的参数类型。也就是说，表 7-1 中第 1 列的对象可以直接注入 Spring MVC 的控制器环境中。

在业务 Bean 中，要获取 IoC 容器的环境，通常使用的方法是实现 Aware 接口。在控制器方法中直接注入表 7-1 所示的参数类型，效果等同于使用 Aware 接口注入对象，而且前一种注入方式比后一种注入方式更加方便、灵活。

表 7-1　　控制器方法的参数类型

参数类型	描述
HttpServletRequest HttpServletResponse HttpSession	Servlet API 对象
org.springframework.web.context .request.WebRequest 和 NativeWebRequest.	允许使用泛化的请求参数，不和本地 Servlet/Portlet API 耦合

续表

参数类型	描述
java.util.Locale	在 MVC 环境中，使用 LocaleResolver/LocaleContextResolver 等本地化解析器
java.time.ZoneId	时区
java.io.InputStream/java.io.Reader	从 Servlet API 中提取字节和字符输入流的原生对象
java.io.OutputStream/java.io.Writer	从 Servlet API 中提取字节和字符输出流的原生对象
org.springframework.http.HttpMethod	HTTP 请求中的方法
ava.security.Principal	包含当前授权用户信息
@PathVariable	请求路径包含参数
@MatrixVariable	请求中包含多个 name-value 对
@RequestParam	从 HTTP 请求中提取参数
@RequestHeader	从 HTTP 请求头中提取信息
@RequestBody	从 HTTP 请求体中提取信息
@RequestPart	请求包含 multipart/form-data 数据
@SessionAttribute	从 session 对象中提取数据
@RequestAttribute	从 request 对象中提取数据
HttpEntity<?>	包含请求头和请求体的信息
java.util.Map org.springframework.ui.Model java.util.Map org.springframework.ui.ModelMap	模型数据
org.springframework.web.servlet .mvc.support.RedirectAttributes	重定向时传递的属性数据
Java Bean	使用属性编辑器，把 HTTP 参数自动绑定到 Java Bean 对象
org.springframework.validation.Errors org.springframework.validation.BindingResult	数据校验结果
org.springframework.web.bind .support.SessionStatus	会话状态
org.springframework.web .util.UriComponentsBuilder	在 Servlet 映射中，提取 host、port、scheme、context path 等信息

示例代码如下。

```
@RequestMapping("/test")
public void test(HttpServletRequest request , HttpServletResponse response ,
                 HttpSession session , Writer out ,Reader reader ,Model model ) {
}
```

在高并发环境下使用 HttpSession 对象时，Spring MVC 需要考虑并发访问的安全性。可以设置 RequestMappingHandlerAdapter 的 synchronizeOnSession 标记为 true，这样可以通过锁增加并发访问的安全性。

```
public class RequestMappingHandlerAdapter extends AbstractHandlerMethodAdapter
                implements BeanFactoryAware, InitializingBean {
```

```
    private boolean synchronizeOnSession = true;
    public void setSynchronizeOnSession(boolean synchronizeOnSession) {
        this.synchronizeOnSession = synchronizeOnSession;
    }
}
```

HttpSession 的底层数据结构是 Map<String,Object>，即按照 sessionid 来操作数据。HttpSession 没有线程的并发控制。当高并发访问 HttpSession 对象时，可以在 Bean 启动时设置并发访问的同步约束。参见如下示例，通过实现 BeanPostProcessor 接口，在 IoC 容器启动时查找 RequestMapping HandlerAdapter 对象，并设置它的 synchronizeOnSession 属性为 true。

示例代码如下。

```
@Component
public class MyProcessor implements BeanPostProcessor {
    @Override
    public Object postProcessAfterInitialization(Object bean, String arg1) throws
        BeansException {
        if (bean instanceof RequestMappingHandlerAdapter) {
                RequestMappingHandlerAdapter adapter = (RequestMappingHandlerAdapter) bean;
                adapter.setSynchronizeOnSession(true);
                Log.logger.info("MyProcessor,setSynchronizeOnSession(true)");
            }
            return bean;
    }
}
```

12. 控制器方法的返回类型

Spring MVC 的控制器方法支持表 7-2 所示的返回类型。控制器方法最常用的返回类型是 String，它表示返回视图的名字。前端控制器激活用户自定义控制器，返回的类型是 ModelAndView。如果通过 AJAX 访问控制器，则返回类型可以任意指定。

表 7-2　　控制器方法的返回类型

返回类型	说明
ModelAndView	包含模型和视图两部分数据
Model	模型数据
Map	键-值类型的数据
View	视图
String	对于 MVC 模式，解析为视图名字； 对于 AJAX 模式，解析为字符串
void	空类型，可以使用 ServletResponse 直接写数据
Any	用@ResponseBody 注解的方法可以返回任意类型的数据，返回信息由 HttpMessageConverters 合理转换
HttpEntity<?> ResponseEntity<?>	包含 HTTP 头和 HTTP 体的响应信息

续表

返回类型	说明
HttpHeaders	响应信息中无正文
Callable<?>	Spring MVC 的异步处理
DeferredResult<?>	用于耗时的异步处理
ListenableFuture<?> CompletableFuture<?> CompletionStage<?>	返回线程池的异步处理结果
ResponseBodyEmitter	返回用于产生流对象的值
SseEmitter	用于定制响应的状态和标头
StreamingResponseBody	返回 OutputStream 的异步处理结果
others	在方法上使用@ModelAttribute，结果存于 model 中

示例 7-6：控制器方法的多种返回值。

```
@Controller
@RequestMapping(value="/user")
public class UserAction {
    @GetMapping(value="/login")
    public String login() {
        return "/main/login.jsp";
    }
    @PostMapping(value = "/login")
    public ModelAndView login(String uname,String pwd) {
        ModelAndView mv = new ModelAndView();
        return mv;
    }
    @PostMapping(value = "/login2")
    @ResponseBody
    public User login2(String uname,String pwd) {
        return user;
    }
    @PostMapping(value = "/login3")
    @ResponseBody
    public void login3(Writer out) {
        out.write("1");
        out.flush();
        out.close();
    }
}
```

13. @RequestBody 与消息转换器

@RequestBody 注解指明方法参数与 HTTP 请求体数据绑定。前面讲过用户异步登录的示例，@RequestBody 用来接收前端传递给后端的 JSON 字符串，数据不是源于 HTTP 的参数，而是源于 HTTP 请求体。

示例代码如下。

```
@Controller
public class UserAction {
    @PostMapping(value = "/login",consumes = "application/json")
    @ResponseBody
    public String login(@RequestBody String user) {
        System.out.println("收到：" + user);
        return "1";
    }
}
```

为了转换 HTTP 请求体信息为控制器方法的参数，需要使用 HttpMessageConverter。消息转换器还负责把方法参数转换为 HTTP 响应体（response body）。

RequestMappingHandlerAdapter 支持@RequestBody 注解，使用如下默认消息转换器。

- ByteArrayHttpMessageConverter：转换字节数组。
- StringHttpMessageConverter：转换字符串。
- FormHttpMessageConverter：转换表单数据为 MultiValueMap<String, String>。
- SourceHttpMessageConverter：与 XML 数据源之间进行数据转换。

为了在 Spring MVC 中使用消息转换器，需要在 spring-mvc.xml 中进行配置。

消息转换器的配置方法 1 如下。

```
<mvc:annotation-driven>
  <mvc:message-converters register-defaults="true"
    <bean  class="org.springframework.http
              .converter.StringHttpMessageConverter" >
    <property name = "supportedMediaTypes">
        <list>
              <value>application/json;charset=utf-8</value>
              <value>text/html;charset=utf-8</value>
            </list>
         </property>
     </bean>
    </mvc:message-converters>
</mvc:annotation-driven>
```

消息转换器的配置方法 2 如下。

```
<bean class="org.springframework.web.servlet.mvc.
             method.annotation.RequestMappingHandlerAdapter">
  <property name="messageConverters">
      <util:list id="beanList">
          <ref bean="stringHttpMessageConverter" />
          <ref bean="marshallingHttpMessageConverter" />
      </util:list>
```

```
        </property>
    </bean>
    <bean id="stringHttpMessageConverter"
            class="org.springframework.http.converter.StringHttpMessageConverter" />
    <bean id="marshallingHttpMessageConverter"
    class="org.springframework.http.converter.xml.MarshallingHttpMessageConverter">
        <property name="marshaller" ref="castorMarshaller" />
        <property name="unmarshaller" ref="castorMarshaller" />
    </bean>
    <bean id="castorMarshaller"
            class="org.springframework.oxm.castor.CastorMarshaller" />
```

消息转换器的相关内容参见后面章节。

14. @ResponseBody 注解

@ResponseBody 注解用于方法，表明将方法返回的信息直接写入 HTTP 响应体。因为使用响应体传递数据，所以映射方法不再返回视图数据。

示例代码如下。

```
@Controller
public class HelloAction {
    @RequestMapping("/hello")
    public String sayHello(String name,Model model) {
        model.addAttribute("name",name);
        return "/main/hello.jsp";
    }
    @RequestMapping("/hello2")
    @ResponseBody
    public String sayHello(String name) {
        return "hello Mr." +  name;
    }
}
```

其中，sayHello( )方法返回 String，如果没有@ResponseBody 注解，则返回视图名字。有了@ResponseBody 注解后，String 表示把方法的返回值直接返回给用户。

浏览器分别发出请求，在有@ResponseBody 注解和没有@ResponseBody 注解的情况下，代码的执行结果（见图 7-25）相同，但是后台运行机制完全不一样。

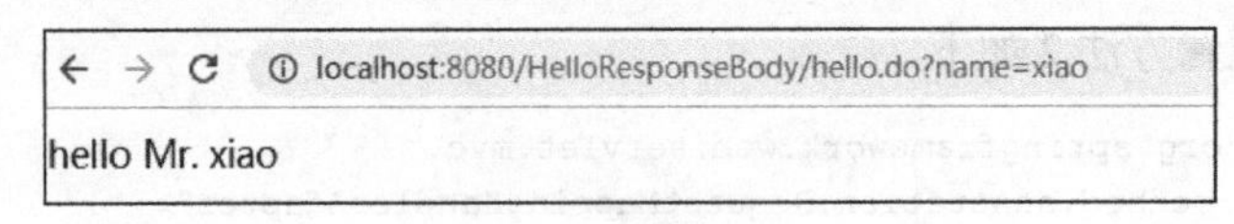

图 7-25　执行结果

浏览器抓包的示例见图 7-26（a）与（b）。

（a）

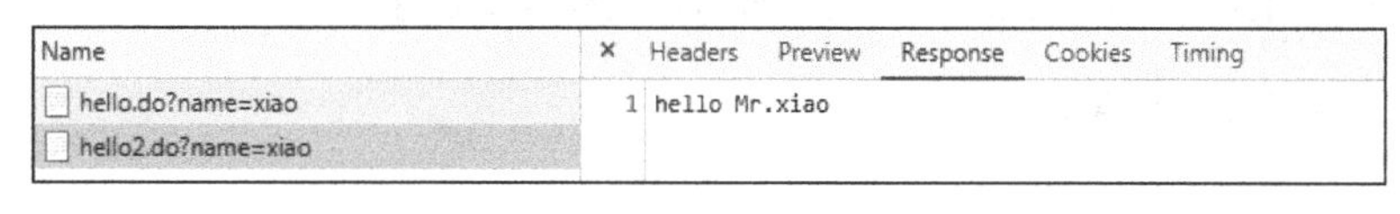

（b）

图 7-26　浏览器抓包的示例

如上示例中，没有@ResponseBody 注解，返回的数据流为 HTML 信息，内容很多。使用@ResponseBody 后，只返回最终的数据结果。

同@RequestBody 一样，@ResponseBody 也使用 HttpMessageConverter 把方法返回的数据转换为 HTTP 响应体。如果访问参数中包含中文（如 http://localhost:8080/HelloResponseBody/hello2.do?name=张三），则会出现乱码。

解决方法是配置合适的消息转换器，如下所示。

```
   <mvc:annotation-driven>
       <mvc:message-converters register-defaults="true">
       <bean  class="org.springframework.http
                     .converter.StringHttpMessageConverter" >
             <property name = "supportedMediaTypes">
                     <list>
                          <value>application/json;charset=utf-8</value>
                          <value>text/html;charset=utf-8</value>
                     </list>
             </property>
       </bean>
   </mvc:message-converters>
</mvc:annotation-driven>
```

案例 7-1：用户异步登录

客户端浏览器采用 AJAX 异步方式发出登录请求，然后在 JavaScript 中接收服务器返回的值，并用 alert 提示。

（1）编写控制器代码，使用@ResponseBody 将登录的处理结果返回给客户端。

```
@Controller
public class UserAction {
```

```
    @PostMapping(value = "/login")
    @ResponseBody
    public int login2(String uname,String pwd) {
        if(uname.equals("tom") && pwd.equals("123")) {
            return 1;                //登录成功
        }else {
            return 0;                //登录失败
        }
    }
}
```

（2）配置 HttpMessageConverter，参见前面的消息转换器配置。

（3）在客户端实现异步登录调用。

```
function tijiao(){
    var uname = $('#uname').val();
    var pwd = $('#pwd').val();
    if(uname != ""){
        var destUrl = "<%=basePath%>login.do";
        $.ajax({
                    type : "post",
                    url : destUrl,
                    data: "uname=" + uname + "&pwd=" + pwd,
                    success : function(msg) {
                        if (msg == 0) {
                            alert("登录失败");
                        } else if (msg == 1) {
                            alert("登录成功");
                        } else {
                            alert("登录异常");
                        }
                    },
                    error : function() {
                        alert("异常，请检查");
                    }
            });
    }
}
```

案例 7-2：省市区三级联动

把 Java EE 项目 CityThree 改造成 Spring MVC 模式的省、市、区三级联动，比较两个项目在代码实现方法方面的区别。项目显示效果如图 7-27 所示。

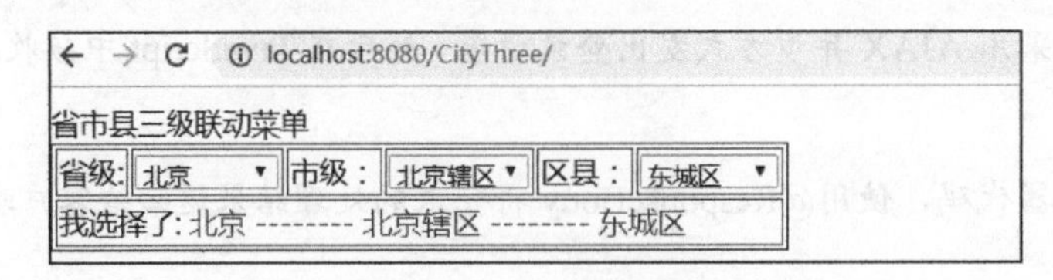

图 7-27　项目显示效果

省的数据在第一次发出 HTTP 请求时使用集合返回（MVC 模式）。市和区的数据采用 AJAX 方式请求，需要把集合转换成 JSON 字符串并返回。

当选择省时，<select>下拉框的 onchange 方法会触发 AJAX 请求，服务器根据请求参数中的省 id 提取该省下的市数据，并通过 JSON 对象返回客户端。

当选择市的数据时，<select>下拉框的 onchange 方法会触发 AJAX 请求，服务器根据请求参数中的市 id 会提取区县的数据，并通过 JSON 对象返回客户端。

在 Java EE 的项目 CityThree 中，为了把集合对象转换为 JSON 字符串，需要使用 json-lib-2.3-jdk15.jar 包，并手动编码。把集合对象转换为 JSON 字符串的样例参见如下代码。

```
List<City> cityList = testBiz.getShiBySheng(shengID);
JSONArray jsonArray = JSONArray.fromObject(cityList);
response.getWriter().print(jsonArray.toString());
```

在 Spring MVC 中，使用 jackson-core-2.6.3.jar 包把集合对象转换为 JSON 字符串。这个工作是 Spring 框架调用 jackson 包自动完成的，无须使用任何显式代码转换，非常简洁。

项目 CityThreeSpring 的核心代码示例如下。

（1）控制器的代码如下。

```
@Controller
public class CityAction {
    //读取所有省的数据，以 MVC 方式返回
    @RequestMapping("/GetShengSvl")
    public String getAllSheng(Model model) {
        TestBiz testBiz = new TestBiz();
        List<Province> arrSheng = testBiz.getAllSheng();
        if(arrSheng != null){
            model.addAttribute("arrSheng", arrSheng);
        }
        return "/main/city.jsp";
    }
    //根据省的编号，读取该省下的所有市数据，Spring 自动把数据转换为 JSON 字符串
    @RequestMapping("/GetShiSvl")
    @ResponseBody
    public List<City> getCitys(String sheng) {
        TestBiz testBiz = new TestBiz();
        return testBiz.getShiBySheng(sheng);
    }
    //根据省和市的编号，返回区县数据，Spring 自动把数据转换为 JSON 字符串
    @RequestMapping("/GetQvSvl")
    @ResponseBody
    public List<Country> getQvList(String sheng,String shi) {
        TestBiz biz = new TestBiz();
        return biz.getCountrys(sheng, shi);
    }
}
```

（2）使用 StringHttpMessageConverter 转换器和 MappingJackson2HttpMessageConverter 转换器，配置消息转换器。

```
<mvc:annotation-driven>
    <mvc:message-converters register-defaults="true">
            <bean class="org.springframework.http.
                    converter.StringHttpMessageConverter" >
            <property name = "supportedMediaTypes">
                <list>
                    <value>application/json;charset=utf-8</value>
                    <value>text/html;charset=utf-8</value>
                </list>
            </property>
            </bean>
            <bean class="org.springframework.http.
                converter.json.MappingJackson2HttpMessageConverter"></bean>
        </mvc:message-converters>
    </mvc:annotation-driven>
```

（3）实现 city.jsp 页面的 AJAX 操作。

```
function getShiList() {
   $.getJSON("<%=basePath%>GetShiSvl.do",{sheng:$("#shengID").val()},
                                                    function callback(data) {
        $("#shiID").empty();
        $(data).each( function(i){
            $("<option value=" + data[i].id + ">" + data[i].name
                            + "</option>").appendTo("#shiID");;
                });
            $("#shiID").change();
        });
   }
}
function getQvList(){
      var s = document.form1.shiSelect.options[document.form1.shiSelect.selectedIndex].text;
      var shi = document.getElementById("cityNameSpan");
      shi.innerHTML = "<font color=\"red\">" + s + "</font>";
      $.getJSON("<%=basePath%>GetQvSvl.do",
              {shi:$("#shiID").val(),sheng:$("#shengID").val()}, function callback(data) {
                        $("#qvID").empty();
                        $(data).each( function(i){
                           $("<option value=" + data[i].id + ">"
                                            + data[i].name +"</option>").appendTo("#qvID");;
                     });
                        $("#qvID").click();
               });
}
```

（4）实现 city.jsp 页面的布局。

```
<table border="1">
     <tr>
          <td>省级:</td>
```

```
        <td>
            <select name="shengSelect" onchange="getShiList();"
                        id="shengID" onclick = "getShengName()">
                <option value="nullSelect">     --请选择--</option>
                    <c:forEach var="sheng" items="${arrSheng}">
                        <option value="${sheng.id}">${sheng.name}</option>
                    </c:forEach>
            </select>
        </td>
        <td>市级</td>
            <td>
                <select name="shiSelect" id="shiID" onchange = "getQvList()">
                    <option>--请选择--</option>
                </select>
            </td>
            <td>区县：</td>
            <td>
                <select name="qvSelect" id="qvID" onclick="getQvName()">
                     <option>--请选择--</option>
                </select>
            </td>
    </tr>
</table>
```

15. @RestController 注解

如果控制器只服务于 JSON、XML、多媒体数据等，是 REST 风格，就适合使用@RestController 注解。@RestController 是一种简化写法，相当于控制器中所有@RequestMapping 方法都使用了@ResponseBody 注解，即@RestController = @ResponseBody + @Controller。

示例 7-7：在微服务开发中，所有对外的服务都使用@RestController 注解。

```
@RestController
public class HelloController {
    @RequestMapping("/hello")
    public String say() {
        return "hello xiaohp";
    }
}
@SpringBootApplication
public class App {
    public static void main( String[] args ){
       SpringApplication.run(App.class, args);
    }
}
```

示例 7-8：在 MVC 的 Web 项目中，使用@RestController。

（1）编写控制器代码。

```
@RestController
public class HelloAction {
    @RequestMapping("/hello")
    public String sayHello(String name,Model model) {
        model.addAttribute("name",name);
        return "/main/hello.jsp";
    }
    @RequestMapping("/hello2")
    public String sayHello(String name) {
         return "Hello Mr. " + name;
    }
}
```

（2）配置下面这个参数后，@RestController才能生效。

```
<mvc:annotation-driven></mvc:annotation-driven>
```

（3）测试 http://localhost:8080/HelloRest/login.do，返回信息"/main/hello.jsp"。因为配置了@RestController注解，所以所有控制器的方法都直接返回数据，不再通过视图解析器返回结果。

16. HttpEntity

HttpEntity表示HTTP的request和response实体，它由消息头和消息体组成。

从HttpEntity中可以获取HTTP请求头和响应头，也可以获取HTTP请求体和响应体信息。

HttpEntity的使用方法与@RequestBody、@ResponseBody类似。

```
package org.springframework.http;
public class HttpEntity<T> {}
```

HttpEntity的已知子类为RequestEntity、ResponseEntity。

HttpEntity的典型应用是在微服务中携带很多额外信息，如状态码、提示信息。

示例代码如下。

```
HttpHeaders headers = new HttpHeaders();
headers.setContentType(MediaType.TEXT_PLAIN);
HttpEntity<String> entity = new HttpEntity<String>(helloWorld,headers);
URI location = template.postForLocation("http://example.com", entity);
```

或者：

```
HttpEntity<String> entity = template.getForEntity("http://example.com", String.class);
String body = entity.getBody();
MediaType contentType = entity.getHeaders().getContentType();
```

在Spring MVC中也可以直接使用HttpEntity的子类RequestEntity和ResponseEntity。

用户登录示例如下。

（1）在 Action 中提取 RequestEntity 中的请求头信息，并用 ResponseEntity 响应。

```
@RequestMapping("/login")
public ResponseEntity<User> handle(@RequestBody String userString,
                                    RequestEntity requestEntity){
    System.out.println(requestEntity.getHeaders().getContentLength());
    System.out.println(requestEntity.getHeaders().getContentType().toString());
    System.out.println(requestEntity.getHeaders().getAccept().toString());
    System.out.println(requestEntity.getHeaders().getOrigin());
    String requestHeader = requestEntity.getHeaders().getFirst("token");
    System. out .println(requestHeader);
    System.out.println(requestEntity.getUrl());
    //设置响应信息
    HttpHeaders responseHeaders = new HttpHeaders();
    responseHeaders.set("myResponseHeader", "myValue");
    User user = new User();
    String uname = userString.split("&")[0].split("=")[1];
    user.setUname(uname);
    user.setRole(2);
    return new ResponseEntity<User>(user,responseHeaders,HttpStatus.CREATED);
}
```

（2）客户端发出登录请求，HTTP 请求头携带了一个用于安全校验的令牌。

```
function tijiao(){
    var uname = $('#uname').val();
    var pwd = $('#pwd').val();
    var user = {uname: uname, pwd: pwd};
    if(uname != ""){
        var destUrl = "<%=basePath%>login.do";
        $.ajax({
          type : "post",
          url : destUrl,
          data:user,
          dataType:"json",
          contentType :"application/json",
          beforeSend: function (xhr) {
             xhr.setRequestHeader("token", "eyJhbGciOJzdWIiOiIxOD.....");
          },
          success : function(msg) {
             var user = eval(msg);
             if (user == null) {
                 alert("登录失败");
             } else {
                 alert(user.uname + "登录成功");
             }
         },
        error : function() {
            alert("异常，请检查");
        }
    });
  }
}
```

（3）测试。

浏览器发出请求 http://localhost:8081/HelloHttpEntity/login.do。

测试结果如下。

```
20
application/json
[application/json, text/javascript, */*]
http://localhost:8080
eyJhbGciOJzdWIiOiIxOD.....
http://localhost:8080/HelloHttpEntity/login.do
```

17. @ModelAttribute

@ModelAttribute 注解可以用于方法或方法参数，两种情况下的含义不同。

1）用@ModelAttribute 注解方法

用@ModelAttribute 注解方法表明这个方法主要用于添加一个或多个 model 属性。用@ModelAttribute 注解的方法会在此控制器的所有@RequestMapping 方法前执行，一个控制器中允许存在多个用@ModelAttribute 注解的方法。

示例代码如下。

```
@Controller
public class AccountAction {
    //在@RequestMapping 方法前执行，查找 Account 对象并自动存储在 model 属性中
    @ModelAttribute
    public Account addAccount(@RequestParam String number) {
        return accountManager.findAccount(number);
    }
}
```

如果@ModelAttribute 中没有指定 model 的名字，则使用默认名，属性名可以显式指定。

```
@ModelAttribute ("account")
public Account addAccount(@RequestParam String number) {
    return accountManager.findAccount(number);
}
```

@ModelAttribute 可应用于有返回值的方法和无返回值的方法。无返回值的方法需要使用 Model 对象显式地设置属性。

```
@ModelAttribute
public void populateModel(@RequestParam String number, Model model) {
    model.addAttribute("account",accountManager.findAccount(number));
}
```

案例 7-3：账户查询

在每次 HTTP 请求中，根据参数 number 读取相应账户的信息。

（1）编写服务层的代码（模拟系统中的所有账户数据）。

```
@Service
public class AccountManager {
    private static Map<String,Account> allAccount;
    static {
        allAccount = new HashMap<>();
        allAccount.put("001", new Account("001",1000));
        allAccount.put("002", new Account("002",2000));
        allAccount.put("003", new Account("003",3000));
        allAccount.put("004", new Account("004",4000));
        allAccount.put("005", new Account("005",5000));
    }
    public Account findAccount(String number) {
        return allAccount.get(number);
    }
}
```

（2）编写控制器代码。根据输入参数中的账户编号，查找用户的账户信息（http://localhost:8080/HelloModelAttribute/info.do?number=002），getAccountInfo( )方法执行前，addAccount( )先执行，获得当前控制器要操作的账户对象。

```
@Controller
public class AccountAction {
    @Autowired
    private AccountManager accountManager;
    @ModelAttribute
    public Account addAccount(@RequestParam String number) {
        return accountManager.findAccount(number);
    }
    @RequestMapping("/info")
    public String getAccountInfo(String number) {
        return "/main/account.jsp";
    }
}
```

（3）在并发环境中，AccountAction 控制器默认是单例模式，这样后面的请求会把前面的请求存放在 Model 中的 account 对象替换掉，因此必须要修改作用域为@RequestScope。

```
@Controller
@RequestScope
public class AccountAction { }
```

（4）编写视图层代码。

```
<html>
    <body>
        账户编号：${account.number}  <br>
        余额：${account.money}
    </body>
</html>
```

（5）测试。测试结果见图 7-28。

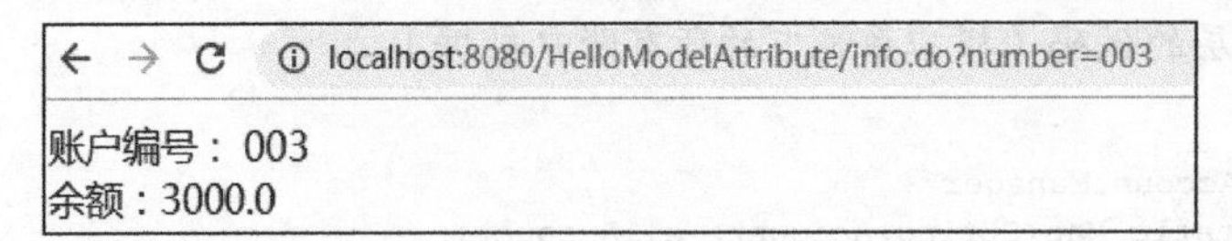

图 7-28　测试结果

如上示例使用了@ModelAttribute 后，AccountAction 与要处理的账户 AccountManager 捆绑得更加紧密，它表示 AccountAction 中的所有请求都必须与一个当前账户有关。

2）用@ModelAttribute 注解方法参数

当把@ModelAttribute 应用于方法参数时，表明这个参数的值应该从 Model 中获取。如果在 Model 中没有找到对应名称的对象，则创建一个对象并添加到 Model 中。如果在 Model 中找到了对应名称的对象，这个对象的属性值应该从请求参数中按名字匹配、查找。

示例代码如下。

```
@Controller
public class PetAction {
    @GetMapping("/owner/{ownerId}/pet/{petId}/info")
    public String processSubmit(@ModelAttribute Pet pet) {
        return "/main/pet.jsp";
    }
}
```

如上代码在第一次访问时会自动创建一个 pet 对象，然后存放于 model 中。pet 对象的属性值从请求中提取。

```
<html>
    <body>
        ${pet.ownerId} <br>
        ${pet.petId}
    </body>
</html>
```

测试结果如图 7-29 所示。

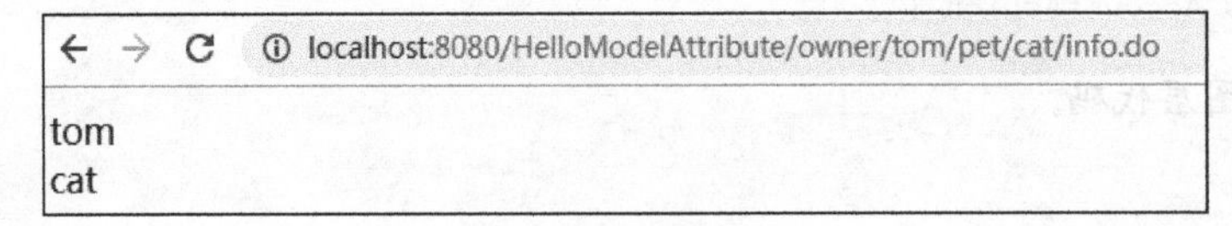

图 7-29　测试结果

总之，如果把@ModelAttribute 应用于方法参数，代码更加简洁，而且可读性更强。这是一种非常重要的应用，在 Spring MVC 项目中经常使用。

18. @SessionAttributes 与@SessionAttribute

@SessionAttributes 注解用于在会话中存储 Model 中指定名称的属性，用在类的级别。它可以同时存储多个对象。

@SessionAttribute 注解用于方法参数，获取已经存储的会话数据。注意，@SessionAttribute 与@ModelAttribute 的使用方法有显著不同。

以下代码用于从 Model 中提取 user 属性和 shopcar 对象，然后存储在会话中。

```
@SessionAttributes(value = {"user","shopcar"})
public class TestAction {...}
```

以下代码用于从会话中查找 account 对象并使用，若找不到，就给出错误提示。

```
@GetMapping("/buypet")
public String processSubmit(@SessionAttribute Account account) {...}
```

下面实现一个案例（基于前面的账户查询）。

（1）发送请求 http://localhost:8080/HelloSessionAttribute/account.do?number=003，提取指定账户的数据，存储在会话中。方法 addAccount( )会在 getAccountInfo( )之前执行。getAccountInfo( )提取指定账户的数据，存放在 Model 中。框架自动提取 Model 中的属性对象 account，然后存储在会话中。

```
@Controller
@SessionAttributes("account")
public class AccountAction {
    @Autowired
    private AccountManager accountManager;
    @ModelAttribute("account")
    public Account addAccount(@RequestParam String number) {
        return accountManager.findAccount(number);
    }
    @RequestMapping("/account")
    public String getAccountInfo(String number) {
        return "/main/account.jsp";
    }
}
```

（2）购买宠物，从当前账户扣款。当用于方法参数时，@SessionAttribute 注解类似于@ModelAttribute。不同的是，当 Model 中没有这个对象时，@ModelAttribute 会自动创建一个对象，而@SessionAttribute 只是从会话中查找对象，不会自动创建对象。先调用 account.do 获取当前账户信息，再调用 buypet.do 从当前账户扣款。

```
http://localhost:8080/HelloSessionAttribute/account.do?number=003
http://localhost:8080/HelloSessionAttribute/buypet.do?money=200
@GetMapping("/buypet")
public String processSubmit(@SessionAttribute Account account,
                            double money,Model model) {
```

```
    System.out.println(account.getNumber() + ":" + account.getMoney());
    model.addAttribute("oldMoney", account.getMoney());
    model.addAttribute("money", money);
    double pay = account.getMoney() - money;
    account.setMoney(pay);
    return "/main/pet.jsp";
}
```

如果会话中没有 account 对象，则会报 400 错误（见图 7-30）。

HTTP Status 400 - Missing session attribute 'account' of type Account

type Status report

message Missing session attribute 'account' of type Account

description The request sent by the client was syntactically incorrect.

图 7-30　400 错误

（3）编写页面 pet.jsp 的代码。

```
<html>
    <body>
        原有金额：${oldMoney}<br>
        付款：${money}<br>
        余额：${account.money}
    </body>
</html>
```

（4）测试。测试结果见图 7-31。

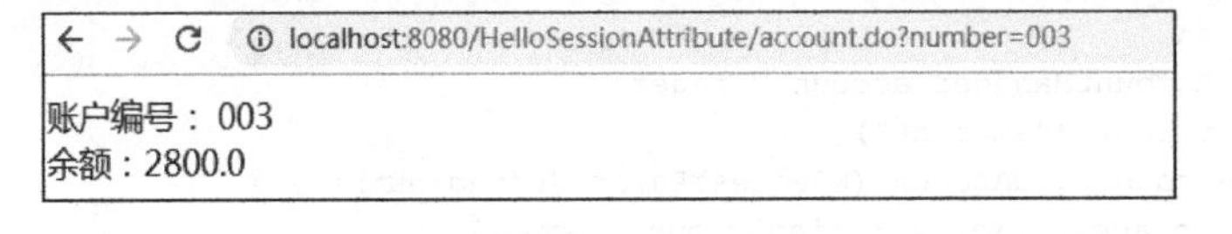

localhost:8080/HelloSessionAttribute/buypet.do?money=200

原有金额：2800.0
付款：200.0
余额：2600.0

图 7-31　测试结果

19. @RequestAttribute

使用@RequestAttribute 注解，可以提取 HttpServletRequest 的属性数据。

下面给出一个示例。

（1）在过滤器中校验用户是否登录，登录成功后，获得令牌。

```
public void doFilter(ServletRequest request, ServletResponse response,
        FilterChain chain) throws IOException, ServletException {
```

```java
        if(request instanceof HttpServletRequest){
            HttpServletRequest req = (HttpServletRequest)request;
            Object object = req.getSession().getAttribute("user");
            if(object != null){
                //创建一个令牌
                String token = new String("tk" + new Date().getTime());
                req.setAttribute("token", token);
                chain.doFilter(request, response);
            }else{
                request.setAttribute("msg","请先登录");
                req.getRequestDispatcher("/WEB-INF/views/main/login.jsp")
                                        .forward(request, response);
            }
        }
    }
```

（2）用户登录成功，在业务操作中先收取令牌，用于增强安全校验。

```java
@Controller
@RequestMapping("/user")
public class AccountAction {
    @RequestMapping("/account")
    public String getAccountInfo(String number,
                                @RequestAttribute String token) {
        System.out.println("收到令牌：" + token);
        return "/main/account.jsp";
    }
}
```

20．PUT 和 PATCH 请求

Servlet 规范要求，使用 ServletRequest.getParameter*( )接收 GET、POST 请求参数，不支持 PUT、PATCH 请求。

在浏览器中，使用 HTTP GET 或 HTTP POST 提交 form 数据；在非浏览器中，可以使用 HTTP PUT 提交数据（通过微服务提交请求不需要浏览器）。

为了支持 HTTP PUT、PATCH 请求，Spring MVC 提供了 HttpPutFormContentFilter。

web.xml 的配置如下。

```xml
<filter>
    <filter-name>httpPutFormcontentFilter</filter-name>
    <filter-class>
        org.springframework.web.filter.HttpPutFormContentFilter
    </filter-class>
</filter>
<filter-mapping>
    <filter-name>httpPutFormContentFilter</filter-name>
    <url-pattern>/*</url-pattern>
</filter-mapping>
```

在 Restful 风格的项目中，通常使用 GET、POST、PUT、DELETE，分别对应查找、增加、修改、删除操作。

示例代码如下。

```
RestTemplate restTemplate = new RestTemplate();
HttpHeaders header = new HttpHeaders();
header.setContentType(MediaType.APPLICATION_JSON_UTF8);
Map<String, Object> m = new HashMap<String, Object>();
m.put("t1", "xx");
m.put("flag", "1");
ObjectMapper mapper = new ObjectMapper();
String value = mapper.writeValueAsString(m);
HttpEntity<String> entity = new HttpEntity<String>(value,header);
restTemplate.put("http://localhost:8080/orders/update.do", entity);
```

在以下代码中，使用 PUT 请求，实现用户登录。

```
function tijiao(){
    var uname = $('#uname').val();
    var pwd = $('#pwd').val();
    if(uname != ""){
        var destUrl = "<%=basePath%>login.do?uname=" + uname + "&pwd=" + pwd;
        $.ajax({
                type : "put",
                url : destUrl,
                success : function(msg) {
                    if (msg == 0) {
                        alert("登录失败");
                    } else if (msg == 1) {
                        alert("登录成功");
                    } else if (msg == -1) {
                        alert("登录异常");
                    }
                },
                error : function() {
                    alert("异常，请检查");
                }
        });
    }
}
```

控制器使用@PutMapping 接收客户端的 PUT 请求。

```
@Controller
public class UserAction {
    @PutMapping(value = "/login")
    @ResponseBody
    public String login(String uname,String pwd) {
        return "1";
    }
}
```

21. @CookieValue

@CookieValue 用于提取 HTTP 请求中指定名称的 Cookie。

例如，HTTP 的每次会话都使用唯一的 sessionid 标识这次会话，这个 sessionid 一般会存储于 Cookie 中。下面的例子提取了存放于 Cookie 中的 sessionid。如果自己创建 Cookie 对象，通过 @CookieValue 同样可以提取到。

下面给出一个示例。

（1）第一次访问 hello.do，写入 Cookie 数据。

```
@Controller
public class HelloAction {
    @RequestMapping("/hello")
    public String sayHello(String name,Model model,
                    HttpServletResponse response) {
        model.addAttribute("name",name);
        Cookie ck = new Cookie("user", "xiaohp");
        response.addCookie(ck);
        return "/main/hello.jsp";
}
```

（2）访问 login.do，读取 Cookie 数据。

```
@Controller
public class UserAction {
    @GetMapping("/login")
    public String login(@CookieValue("JSESSIONID") String cookie,
                    @CookieValue String user) {
        System.out.println("会话 ID: " + cookie);
        System.out.println("用户: " + cookie);
        return "/main/login.jsp";
    }
}
```

（3）通过浏览器抓包，可以查看当前请求携带了哪些 Cookie（见图 7-32）。

```
▾ Request Headers     view source
    Accept: text/html,application/xhtml+xml,application/xml;q=0.9,image/webp,image/apng,*/*;q=0.8,application/s
    Accept-Encoding: gzip, deflate, br
    Accept-Language: zh-CN,zh;q=0.9
    Cache-Control: max-age=0
    Connection: keep-alive
    Cookie: user=xiaohp; JSESSIONID=3EFB7E810B01BAB8B4851746F5C83F76
    Host: localhost:8080
```

图 7-32　抓包分析

22. @RequestHeader

@RequestHeader 用于提取 HTTP 请求头的信息。请求头有很多信息，可以按照 key 值提取指定键的 value 信息。

以下代码用于读取请求头的信息。

```
@GetMapping("/login")
public String login(@RequestHeader("Accept-Encoding") String encoding,
                    @RequestHeader Map<String, String> headers) {
    System.out.println(encoding);
    for(String key : headers.keySet()) {
        System.out.println(key);
    }
    return "/main/login.jsp";
}
```

输出结果如下。

```
信息: Server startup in 52999 ms
gzip, deflate, br
host
connection
cache-control
upgrade-insecure-requests
user-agent
accept
accept-encoding
accept-language
cookie
```

23. @InitBinder

当对 HTTP 请求中的参数与控制器方法中的对象进行数据绑定时，使用@InitBinder 定制绑定行为。

WebDataBinder 的作用是选择合适的属性绑定器，把 HTTP 请求参数与 javaBean 对象的属性进行绑定。常规的操作模式是用 ServletRequest.getParameter*( )接收参数，然后用 set 方法给 javaBean 对象的属性赋值。当参数很多时，这个操作相当麻烦。使用了 WebDataBinder 后，绑定参数对象的操作变得非常简单。WebDataBinder 用于 Web 环境，但不依赖于 Servlet API。

API 描述如下。

```
org.springframework.web.bind
Class WebDataBinder
java.lang.Object
    org.springframework.validation.DataBinder
        org.springframework.web.bind.WebDataBinder
```

所有实现的接口包括 PropertyEditorRegistry、TypeConverter。

直接子类包括 PortletRequestDataBinder、ServletRequestDataBinder、WebRequestDataBinder。

案例 7-4：添加员工

要求如下。

- 保证输入的字符型日期格式正确转换为 Date 对象。
- 当出生日期为空时，转换不能出错。
- 身高为数值型，类型转换不能出错，当身高为空时，应该转换为 0。

操作步骤如下。

（1）如图 7-33 所示，当用户注册时，使用 jQuery-easyui 的两个控件，输入出生日期和身高。另外，还可以使用正则表达式、JavaScript 脚本等进行客户端校验。代码如下。

```
<table>
    <tr><td>出生日期</td>
        <td>
            <input type="text" name="birthday" class="easyui-datebox"/>
        </td>
    </tr>
    <tr><td>身高</td>
        <td>
            <input class="easyui-numberbox" value="0" ame="height"></input>
        </td>
    </tr>
</table>
```

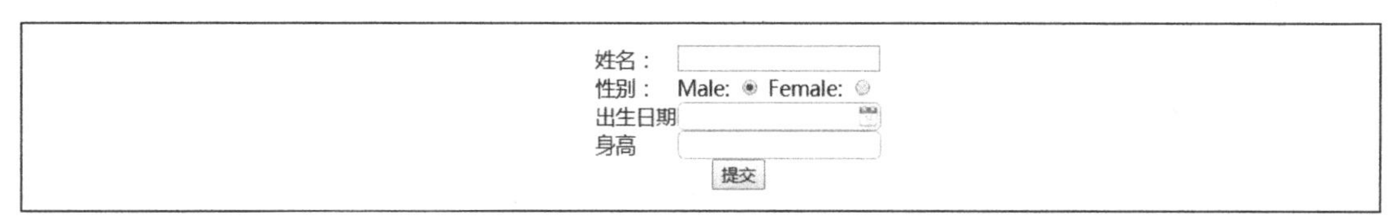

图 7-33　用户注册

（2）把客户端其他的 HTTP 参数自动绑定到实体 Staff 的属性上。

```
public class Staff {
    private String name;
    private String sex;
    private Date birthday;
    private int height;
}
```

（3）绑定控制器数据。

```
@Controller
public class StaffAction {
    @InitBinder
    public void initBinder(WebDataBinder binder,String height) {
```

```java
                    binder.registerCustomEditor(String.class,
                    new StringTrimmerEditor(true));
            binder.registerCustomEditor(Date.class,
            new CustomDateEditor(new SimpleDateFormat("MM/dd/yyyy"), true));
            if(height==null || height.equals("")) {
                binder.setDisallowedFields("height");
            }
        }
        @GetMapping("/add")
        public String add() {
            return "/main/addStaff.jsp";
        }
        @PostMapping("/add")
        public String add(Staff staff) {
            System.out.println("add ok");
            return "/main/addStaff.jsp";
        }
}
```

总之，@InitBinder 注解的方法在@RequestMapping 映射方法前执行。

调用 WebDataBinder 的父接口 PropertyEditorRegistry::registerCustomEditor( )方法，可以定制属性编辑器的行为。例如，CustomDateEditor 是把字符串转换为日期对象时需要使用的编辑器。下面的代码定义了日期格式，并允许绑定属性时日期为空。

```java
new CustomDateEditor(new SimpleDateFormat("MM/dd/yyyy"), true)
```

调用 WebDataBinder::setDisallowedFields( )方法，可以在绑定属性时把某些属性排除在外，不进行属性绑定。

24. @JsonView

使用@JsonView 在@ResponseBody 返回的 JSON 数据中标识显示哪些数据。敏感数据应该忽略，如密码，即响应数据不包含密码信息。

下面给出一个示例。

（1）用@JsonView 标识 User 属性。

```java
public class User {
    public interface WithoutPasswordView {};
    public interface WithPasswordView extends WithoutPasswordView {
    };
    private String username;
    private String password;
    public User(String username, String password) {
        this.username = username;
        this.password = password;
    }
    @JsonView(WithoutPasswordView.class)
```

```
    public String getUsername() {
        return this.username;
    }
    @JsonView(WithPasswordView.class)
    public String getPassword() {
        return this.password;
    }
}
```

（2）在 Action 的方法中标识返回数据的类型。

```
@RestController
public class UserAction {
    @GetMapping("/user")
    @JsonView(User.WithoutPasswordView.class)
    public User getUser() {
        return new User("tom", "7!jd#h282");
    }
    @GetMapping("/userDetail")
    @JsonView(User.WithPasswordView.class)
    public User getUser2() {
        return new User("tom", "7!jd#h282");
    }
}
```

（3）测试。请求用户信息，密码被屏蔽（见图 7-34）。

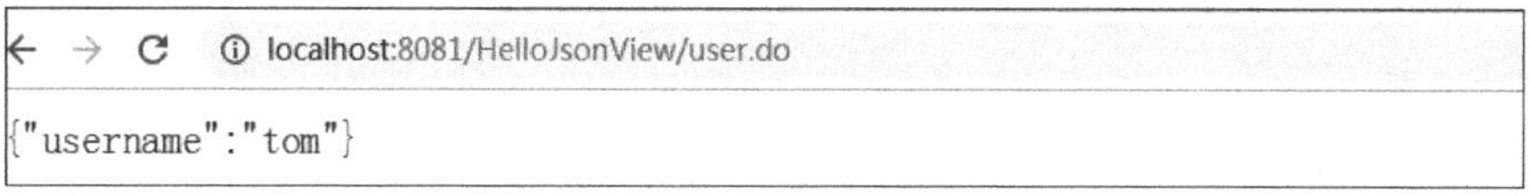

```
localhost:8081/HelloJsonView/user.do
{"username":"tom"}
```

图 7-34　无密码

（4）请求用户详细信息，会显示密码（见图 7-35）。

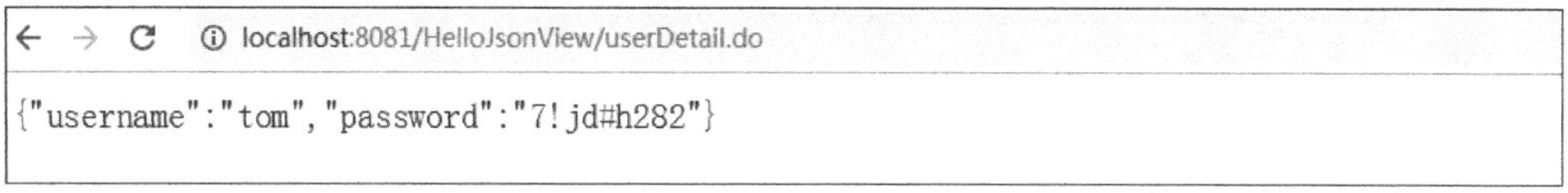

```
localhost:8081/HelloJsonView/userDetail.do
{"username":"tom","password":"7!jd#h282"}
```

图 7-35　显示了密码

25. @ControllerAdvice

@ControllerAdvice 可以简单理解为全局控制器，它比@Component 注解的控制器有更多功能。通过@ControllerAdvice 声明一个类为组件后，表示当前控制器中用@ExceptionHandler、@InitBinder、@ModelAttribute 注解的方法在所有组件中共享。

@ControllerAdvice 也是 IoC 容器的 Bean，在容器启动时需要统一加载。

对于 Spring MVC 项目，所有未捕获的异常最后都会抛给全局异常处理程序来统一处理。这里可以统一设置各种异常类型的提醒页面。对于同一类型的异常，若转到不同的页面处理，不能在全局异常处理程序中设置。

下面给出一个示例。

（1）定义全局异常处理程序。

```
@ControllerAdvice
public class GlobalExceptionHandler {
    @ExceptionHandler(Exception.class)
    public ModelAndView otherException(Exception e) {
        ModelAndView mv = new ModelAndView();
        mv.addObject("msg", e.getMessage());
        mv.setViewName("/error/error.jsp");
        return mv;
    }
    @ExceptionHandler(MaxUploadSizeExceededException.class)
    public ModelAndView fileMaxException(MaxUploadSizeExceededException e) {
        ModelAndView mv = new ModelAndView();
        mv.addObject("msg", "上传文件超过了允许范围");
        mv.setViewName("/error/OverMaxUploadSize.jsp");
        return mv;
    }
}
```

（2）测试。测试代码如下。

```
@RestController
public class UserAction {
    @GetMapping("/user")
    public User getUser() {
        throw new RuntimeException("getUser 异常....");
    }
    @GetMapping("/login")
    public String login(){
        throw new MaxUploadSizeExceededException(1024);
    }
}
```

7.6.3 控制器的异步处理

从 Java EE 6 开始，服务器端支持异步处理。从 Spring MVC 3.2 开始，Spring 支持 Java EE 的异步处理规范。当遇到耗时的处理时，Servlet 容器的主线程可以先退出，用于处理其他请求。这时启动异步线程，专门用于处理耗时的任务。控制器异步处理的返回结果一般使用 java.util.concurrent.Callable 返回。

1. Java EE 7 与 AsyncContext

Spring MVC 与 Java EE 7 的异步都是基于 AsyncContext 来实现的。参见 Servlet API。

```
public interface AsyncContext{
    void addListener(AsyncListener listener);
    void complete()
    void dispatch(String path)
    ServletResponse getResponse()
    void start(Runnable run)
}
public interface ServletRequest{
    AsyncContext startAsync(ServletRequest servletRequest,
            ServletResponse servletResponse) throws IllegalStateException
}
```

AsyncContext 表示异步执行环境，它的对象在 ServletRequest 中实例化。在 Java EE 容器的主线程中，调用 ServletRequest.startAsync( )或 ServletRequest.startAsync(ServletRequest, ServletResponse) 创建 AsyncContext 的实例。重复调用 startAsync( )，返回的是相同的 AsyncContext 实例。

案例 7-5：通过 Java EE 异步抢票

模拟 12306 进行网上抢票。由于在春节等高峰时期，车票资源非常紧张，因此在约定时间放票时，会存在大量抢票行为。在服务器端，为了缓解并发高峰时的压力，抢票请求会用消息中间件进行排队处理。出票在服务器的独立线程中完成，这样主线程主要用于接收购票请求，而不是出票。

通过以下步骤实现抢票。

（1）设置 Servlet 支持异步模式。

```
@WebServlet(asyncSupported=true,urlPatterns="/AuctionSvl")
public class AuctionSvl extends HttpServlet {}
```

（2）Servlet 的 service 方法用于高并发环境。每个 HTTP 请求都使用独立的 request 对象。调用 request.startAsync( )，为每个 HTTP 请求创建一个异步响应的环境。

```
public void service(HttpServletRequest request,
        HttpServletResponse response)     throws ServletException, IOException {
    System.out.println("servlet 主线程: " + Thread.currentThread().getId() );
    String linenum = request.getParameter("lineNum");    //lineNum 为行号
    AsyncContext act = request.startAsync();
    act.getRequest().setAttribute("linenum", linenum);
    AuctionListener.add(act);
}
```

（3）所有异步响应在独立线程中统一处理。

```
public class AuctionListener implements ServletContextListener{
    private static final BlockingQueue<AsyncContext> queue
                    = new LinkedBlockingQueue<AsyncContext>();
        private volatile Thread thread;
        public static void add(AsyncContext c) {
```

```
        queue.add(c);
        }
        contextInitialized();      //参见下面的代码实现
}
```

ServletContextListener 的 contextInitialized( )方法的代码如下。

```
public void contextInitialized(ServletContextEvent servletContextEvent) {
        thread = new Thread(new Runnable() {
         public void run() {
           while (true) {
              AsyncContext acontext = null;
              while (queue.peek() != null) {
                try {
                       acontext = (AsyncContext)queue.poll();
                       ServletResponse response = acontext.getResponse();
                       response.setContentType("text/html;charset=utf-8");
                        PrintWriter out = response.getWriter();
                        Thread.sleep(200);
                        String name = "异步线程: " + Thread.currentThread().getId();
                        long duration = System.currentTimeMillis();
                        //提取前面输入的行号，并输出
                        out.println(acontext.getRequest().getAttribute("linenum")
                                + " "+ name +" "+ duration);
                        out.close();
                } catch (Exception e) {
                  throw new RuntimeException(e.getMessage(), e);
                } finally {
                    if(acontext != null)
                          acontext.complete();
                }
              }
           }
         }
        });
        thread.start();
}
```

独立的后台线程从队列中顺序提取 AsyncContext 对象，然后分别调用每个 AsyncContext 的 response 对象，给不同的客户端发送响应数据。

（4）客户端模拟多用户并发抢票。循环发出 20 个异步请求。

```
$.ajax({
    url: "AuctionSvl",
    type:"post",
    dataType:"html",
    data: data,
    timeout:50000,
    cache:false,
    dataFilter:function (data, type) {
```

```
            return data;
    },
    success:function(data,testStatus){
        var dataArray = Array();
        dataArray = data.split(" ");
        /*填充表格*/
        $("#table1").append("<tr id='tr"+i+"' class='mytr'></tr>");
        for(var j=0; j<dataArray.length; j++){
            $("#tr"+i).append("<td>"+dataArray[j]+"</td>");
        }
    },
    error:function(msg) {
    }
});
```

（5）对比服务器异步响应和同步响应的效果（见图 7-36）。

在这个示例中，所有用户的并发购票请求都会先存放在一个阻塞队列中，然后由一个后台异步线程顺序从阻塞队列中提取请求，出票后，响应客户端。这样做的好处是服务器的压力很小，虽然用户的等待时间较长，但是这保证了用户购票请求的正确处理。

userno	threadName	duration
	start	
0	异步线程：14	1574324654214
1	异步线程：14	1574324655216
2	异步线程：14	1574324656219
4	异步线程：14	1574324657223
3	异步线程：14	1574324658229
5	异步线程：14	1574324659233
6	异步线程：14	1574324660235
7	异步线程：14	1574324661238
8	异步线程：14	1574324662239
9	异步线程：14	1574324663241
10	异步线程：14	1574324664245
11	异步线程：14	1574324665246
12	异步线程：14	1574324666252
13	异步线程：14	1574324667253
14	异步线程：14	1574324668265
15	异步线程：14	1574324669266
16	异步线程：14	1574324670267

图 7-36　服务器异步响应和同步响应的效果

2. Callable 接口与服务器异步

Spring MVC 包装了 Java EE 的异步机制，其本质完全相同。

- request.startAsync( )在 Servlet 中启动后，Servlet 的主线程退出当前 Servlet 和过滤器，只

有 response 对象持续打开，直到结束处理。

- Spring MVC 在 Callable 接口的 call( )中隐式调用 request.startAsync( )，启动一个新的线程，提交 Callable 接口任务到 TaskExecutor 并在新线程中执行。
- 启动异步机制后，DispatcherServlet 和所有的过滤器都退出当前的 Servlet 容器线程，只有 response 对象保持打开。

参见 JDK 1.8 中 Callable 接口的描述。

```
java.util.concurrent
public interface Callable<V>{
    V  call();                    //V 表示异步执行的任务返回的类型
}
```

这段代码表示一个可以返回结果并可能抛出异常的任务。实现者定义了一个没有参数的方法，称为 call。

Callable 接口类似于 Runnable 接口，它们都表示对其他线程要执行的任务。然而，Runnable 接口不返回结果，也不能抛出已检查的异常。

案例 7-6：通过 Spring MVC 异步抢票

使用 Spring MVC 的服务器异步机制，实现异步抢票的功能。步骤如下。

（1）配置前端控制器的异步支持。在 web.xml 中配置<async-supported>true</async-supported>。

```
<servlet>
    <servlet-name>Example</servlet-name>
    <servlet-class>
        org.springframework.web.servlet.DispatcherServlet
    </servlet-class>
    <init-param>
        <param-name>contextConfigLocation</param-name>
        <param-value>/WEB-INF/spring-mvc.xml</param-value>
    </init-param>
    <load-on-startup>1</load-on-startup>
    <async-supported>true</async-supported>
</servlet>
```

（2）异步处理客户端的并发请求。控制器是一个并发环境，允许很多客户端同时访问 /auction.do，客户端的每个请求在控制器方法中都对应一个服务器线程。每个线程在进入 Callable 接口的 call( )方法后，都会隐式启动一个新的线程，原来的主线程则退出。

```
@PostMapping("/auction")
@ResponseBody
public Callable<String> buy(String lineNum) {
    System.out.println("servlet 主线程：" + Thread.currentThread().getId());
```

```
        return new Callable<String>() {
            public String call() throws Exception {
                Thread.sleep(200);
                String name = "异步线程：" + Thread.currentThread().getId();
                long duration = System.currentTimeMillis();
                String retStr = lineNum + " " + name + " " + duration;
                return retStr;
            }
        };
    }
```

（3）对比服务器异步响应和同步响应的效果（见图 7-37）。

userno	threadName	duration
start		
1	异步线程：83	1574322167033
0	异步线程：82	1574322167035
3	异步线程：86	1574322167063
4	异步线程：85	1574322167069
2	异步线程：84	1574322167050
5	异步线程：87	1574322167069
7	异步线程：88	1574322167069
6	异步线程：89	1574322167073
8	异步线程：92	1574322167109
10	异步线程：91	1574322167093
9	异步线程：90	1574322167092
16	异步线程：98	1574322167148
11	异步线程：93	1574322167131
14	异步线程：96	1574322167145
15	异步线程：97	1574322167147
12	异步线程：95	1574322167145
13	异步线程：94	1574322167132
17	异步线程：99	1574322167169
18	异步线程：100	1574322167181

图 7-37　服务器异步响应和同步响应的效果

总之，采用 Spring MVC 的服务器异步机制后，没有采用阻塞队列，代码更加简洁。所有出票的请求都是在异步线程池中进行的操作，对主线程没有影响。在实际的项目中，会使用集群的消息中间件代替阻塞队列，这样异步线程就可以并发调用消息中间件，进行出票处理，这样的效率明显更高。

7.7 拦截器

7.7.1 HandlerMapping 接口

HandlerMapping 接口的声明如下。

```
public interface HandlerMapping {
        HandlerExecutionChain getHandler(HttpServletRequest request)
                                                throws Exception;
}
```

HandlerMapping 接口的实现类有 AbstractUrlHandlerMapping、RequestMappingHandlerMapping、SimpleUrlHandlerMapping 等。

HandlerMapping 接口主要用于实现 HTTP 请求与处理请求的控制器对象之间的映射关系。在 Spring MVC 中，最常用的控制器方法与 HTTP 请求之间的关系就是通过 HandlerMapping 接口的子类 RequestMappingHandlerMapping 实现的。

```
public abstract class AbstractHandlerMapping
                        extends ApplicationObjectSupport
                        implements HandlerMapping, Ordered {

    private Object defaultHandler;
    private int order = Integer.MAX_VALUE;
    private final List<Object> interceptors = new ArrayList<Object>();
    private HandlerInterceptor[] adaptedInterceptors;
}
```

所有的 HandlerMapping 类都继承自 AbstractHandlerMapping，它有如下属性（用于定制映射行为）。

- interceptors 列表：定制这个 HandlerMapping 使用哪些拦截器。
- defaultHandler：当 handlerMapping 的结果中没有匹配的处理程序时，使用默认处理程序。
- order：设置拦截器排序。
- alwaysUseFullPath：配置使用相对路径还是绝对路径。
- urlDecode：配置 Servlet 路径是 encoded 模式还是 decoded 模式。

要自定义拦截器，必须通过 HandlerMapping 来实现。

7.7.2 项目案例：在非工作时间拒绝服务

自定义一个拦截器，用于配置系统的工作时间。如果在正常工作时间，则接收请求；否则，拒绝请求。具体步骤如下。

（1）新建拦截器。

```
public class TimeBasedAccessInterceptor extends HandlerInterceptorAdapter{
    private int openingTime;
    private int closingTime;
    public void setOpeningTime(int openingTime) {
        this.openingTime = openingTime;
    }
    public void setClosingTime(int closingTime) {
```

```
        this.closingTime = closingTime;
    }
    public boolean preHandle(HttpServletRequest request,
                            HttpServletResponse response,
                            Object handler) throws Exception {
        System.out.println("TimeBasedAccessInterceptor doing ...........");
        Calendar cal = Calendar.getInstance();
        int hour = cal.get(Calendar.HOUR_OF_DAY);
        if (openingTime <= hour && hour < closingTime) {
            return true;
        } else {
            request.getRequestDispatcher(
                    "/WEB-INF/views/error/outsideOfficeHours.html")
                    .forward(request, response);
            return false;
        }
    }
}
```

（2）配置拦截器。在 RequestMappingHandlerMapping 中使用拦截器。

```
<bean class="org.springframework.web.servlet.mvc
           .method.annotation.RequestMappingHandlerMapping">
        <property name="interceptors">
            <list>
                <ref bean="officeHoursInterceptor" />
            </list>
        </property>
    </bean>
    <bean id="officeHoursInterceptor"
          class="com.icss.intercept.TimeBasedAccessInterceptor">
        <property name="openingTime" value="9" />
        <property name="closingTime" value="20" />
    </bean>
```

（3）由于拦截器与<mvc:annotation-driven>发生了冲突，因此系统中的消息转换器需要通过 RequestMappingHandlerAdapter 属性设置，不能使用<mvc:annotation-driven>方式，否则拦截器会失效。

```
<bean class="org.springframework.web.servlet.mvc
              .method.annotation.RequestMappingHandlerAdapter">
        <property name="messageConverters">
            <list>
                <bean class="org.springframework.http
                        .converter.StringHttpMessageConverter">
                    <constructor-arg value="utf-8" />
                    <property name="writeAcceptCharset" value="false" />
                </bean>
                <bean class="org.springframework.http
                        .converter.ByteArrayHttpMessageConverter" />
                <bean class="org.springframework.http
                        .converter.json.MappingJackson2HttpMessageConverter" />
```

```
        </list>
    </property>
</bean>
```

（4）进行测试，所有请求都会被拦截器提前捕获。

7.7.3 拦截器运行流程分析

拦截器的运行流程见图 7-38。

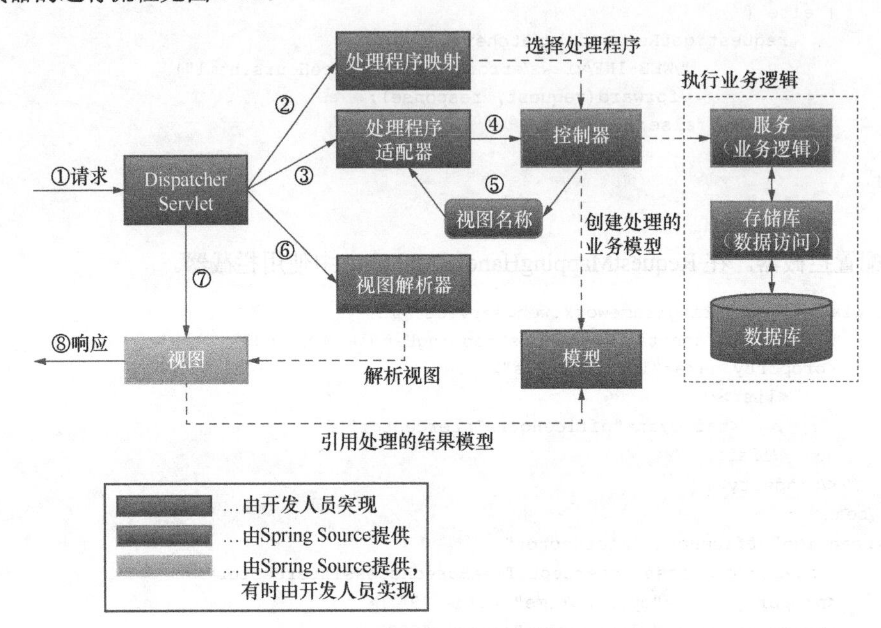

图 7-38　拦截器的运行流程

DispatcherServlet 的 doDispatch( )方法如下，此处可以采用断点调试的方式观察拦截器的运行。

```
protected void doDispatch(HttpServletRequest request,
                HttpServletResponse response) throws Exception {
        HttpServletRequest processedRequest = request;
        HandlerExecutionChain mappedHandler = null; //处理程序链
        try {
            ModelAndView mv = null;
            try {
                mappedHandler = getHandler(processedRequest);.
                HandlerAdapter ha = getHandlerAdapter(mappedHandler.getHandler());
                if (!mappedHandler.applyPreHandle(processedRequest, response)) {
                    //拦截器前置处理
                    return;
                }
                //适配器调用自定义 Action 方法处理请求
                mv = ha.handle(processedRequest, response, mappedHandler.getHandler());
                //拦截器后置处理
```

```java
                mappedHandler.applyPostHandle(processedRequest, response, mv);
            }
            catch (Exception ex) {
                dispatchException = ex;
            }
            processDispatchResult(processedRequest, response,
                                  mappedHandler, mv, dispatchException);
        }finally {
        }
}
```

关键流程如下。

（1）HTTP 请求首先被前端控制器 DispatchServlet 的 service( )方法接收。

（2）service( )调用 doDispatch( )方法。

（3）调用 HandlerMapping 的 getHandler( )方法，返回 HandlerExecutionChain。

（4）HandlerExecutionChain 提取成员 interceptors，找到拦截器。

（5）激活拦截器的 preHandle( )方法，对 HTTP 请求进行前置处理。

（6）HandlerAdapter 调用 handle( )方法，激活自定义控制器，处理 HTTP 请求。

（7）激活拦截器的 postHandle ( )方法，对 HTTP 请求进行后置处理。

7.8 视图解析

Spring MVC 没有把模型与具体的视图技术绑定，它提供了很大的灵活性，允许使用不同的视图技术解析模型，如 JSP、Velocity 模板和 XSLT 等。

7.8.1 视图解析的主要接口

Spring MVC 主要使用 ViewResolver 和 View 这两个接口管理视图。

```java
public interface ViewResolver {
    View resolveViewName(String viewName, Locale locale) throws Exception;
}
```

ViewResolver 接口通过调用 resolveViewName( )方法实现，可以通过视图的名字解析视图对象。在应用的运行过程中，视图状态不能改变，因此视图解析器对象可以缓存视图数据。使用视图解析器，可以支持国际化。

```java
public interface View {
    String  getContentType();
    void render(Map<String, ?> model, HttpServletRequest request,
            HttpServletResponse response) throws Exception;
}
```

在 Web 交互中，MVC 的视图负责提交内容、展示模型数据。单个视图可以展示很多模型的属性数据。视图的实现有很多种，典型实现方式是 JSP，其他还有 XSLT、HTML 等。

视图对象由视图解析器创建。

View 接口是无状态的，View 接口的实现应该保证线程安全。View 接口最主要的作用是调用 render( )方法，提交模型的数据并处理。

render( )方法的主要作用是提交模型数据和给定的视图。以 JSP 视图为例，render( )方法首先把模型数据转换为 request.setAttribute( )模式，然后使用 RequestDispatcher 提交 JSP 视图。

视图提交本质上是把视图数据从前端控制器提交给 Java EE 服务器，如 Tomcat。

7.8.2 JSP 视图

在 Web 应用下，InternalResourceView 包装了 JSP 或其他资源，解析 Model 属性数据为 request 对象的属性值，使用 RequestDispatcher 将请求提交到指定的 URL 地址。

```
public class InternalResourceView extends AbstractUrlBasedView {
    void render(...);
    RequestDispatcher getRequestDispatcher(HttpServletRequest request, String path);
}
```

如下示例中，控制方法返回“/main/hello.jsp”，InternalResourceViewResolver 负责解析模型数据，并存储到 request 属性中，然后提交到 hello.jsp 下。

示例代码如下。

```
<bean
    class="org.springframework.web.servlet.view.InternalResourceViewResolver">
    <property name="prefix" value="/WEB-INF/views/" />
</bean>
@Controller
public class HelloAction {
    @RequestMapping("/hello")
    public String sayHello(String name,Model model) {
        model.addAttribute("name",name);
        return "/main/hello.jsp";
    }
}
```

7.8.3 通过 ViewResolver 解析视图

视图解析器通过逻辑视图名字或 ModelAndView 解析视图。

常见的视图解析器见表 7-3。

表 7-3　　　　常见的视图解析器

常见的视图解析器	描述
AbstractCachingViewResolver	实现了 ViewResolver 接口。 通常，在使用视图之前需要准备继承这个解析器，用于提供缓存功能
XmlViewResolver	继承自 AbstractCachingViewResolver。 实现了 ViewResolver 接口，接受相同 DTD 定义的 XML 配置文件作为 Spring 的 XMLBean 工厂。也就是说，通过 XML 指定逻辑名称与真实视图间的关系，它从 XML 配置文件中查找视图实现（默认 XML 配置文件为 /WEB-INF/views.xml）
ResourceBundleViewResolver	继承自 AbstractCachingViewResolver。 和 XmlViewResolver 一样，它也需要有一个配置文件，以定义逻辑视图名称和真正的 View 对象的对应关系。不同的是，ResourceBundleViewResolver 的配置文件是一个属性文件，而且必须是放在 classpath 下面的。默认情况下，这个配置文件是在 classpath 根目录下的 views.properties 文件
UrlBasedViewResolver	继承自 AbstractCachingViewResolver。 无须显示映射定义，它直接把逻辑视图的名字映射到 URL 资源
InternalResourceViewResolver	继承自 UrlBasedViewResolver，用于解析 InternalResourceView
VelocityViewResolver / FreeMarkerViewResolver	继承自 UrlBasedViewResolver，用于解析 VelocityView 和 FreeMarkerView
ContentNegotiatingViewResolver	直接实现 ViewResolver 接口。 它不能直接解析视图，而由其他视图解析器代理

关于 JSP 视图的解析示例如下。

```
<bean id="viewResolver"
    class="org.springframework.web.servlet.view.UrlBasedViewResolver">
    <property name="viewClass"
            value="org.springframework.web.servlet.view.JstlView"/>
    <property name="prefix" value="/WEB-INF/jsp/"/>
    <property name="suffix" value=".jsp"/>
</bean>
```

注意：AbstractCachingViewResolver 的子类缓存了解析的视图实例，用于提高性能。可以设置 cache 属性为 false，关闭缓存功能；也可以调用 removeFromCache( )方法，刷新视图。

7.8.4　视图解析器链

Spring 支持多种视图解析器，可以同时配置多个解析器，形成解析器链。根据需要，可以通过 order 属性设置解析顺序。order 数值越高，解析越晚。

示例代码如下。

```
<bean id="jspViewResolver"
        class="org.springframework.web.servlet
```

```
            .view.InternalResourceViewResolver">
    <property name="viewClass"
            value="org.springframework.web.servlet.view.JstlView"/>
    <property name="prefix" value="/WEB-INF/jsp/"/>
    <property name="suffix" value=".jsp"/>
</bean>
<bean id="excelViewResolver"
    class="org.springframework.web.servlet.view.XmlViewResolver">
    <property name="order" value="1"/>
    <property name="location" value="/WEB-INF/views.xml"/>
</bean>
```

上面的示例配置了两个视图解析器。InternalResourceViewResolver 用于解析 JSP，它总是自动定位在解析器链的最后。XmlViewResolver 用于解析 Excel 视图，InternalResourceViewResolver 不支持这种视图。

7.8.5 重定向到视图

1. 重定向

通常情况下，控制器返回一个逻辑视图名字，视图解析器使用具体的视图技术进行视图解析。例如，JSP 视图的解析就需要 JSP 或 Servlet 引擎 InternalResourceViewResolver 和 InternalResourceView 联合处理，在其内部调用 Servlet API 的 RequestDispatcher.forward()方法或 RequestDispatcher.include( )方法进行处理，这个称为转发模式。JSP 视图的解析过程还可以使用重定向模式。

使用重定向模式的好处如下。

- 防止客户端提交的数据在多个控制器间共享（内部转发意味着发送的数据可以在其他控制间共享，有很大的安全隐患）。
- 防止表单数据重复提交（重定向会修改浏览器的地址栏）。

2. RedirectView

RedirectView 的 API 描述如下。

```
public class RedirectView extends AbstractUrlBasedView implements SmartView {
    private boolean contextRelative = false;
    private boolean http10Compatible = true;
    private boolean exposeModelAttributes = true;
    private String encodingScheme;
    private HttpStatus statusCode;
    private boolean expandUriTemplateVariables = true;
    private boolean propagateQueryParams = false;
    private String[] hosts;
}
```

RedirectView 实现了 View 接口，可以直接作为控制器方法的返回类型。看到 RedirectView，DispatcherServlet 不会使用通常的视图解析方案。它调用 HttpServletResponse.sendRedirect( ) 发送 HTTP 重定向码到浏览器。

示例代码如下。

```
@RequestMapping("/user/logout")
public View logout(HttpServletRequest request) {
    request.getSession().invalidate();
    String path = request.getContextPath();
    String basePath = request.getScheme() + "://" +
        request.getServerName() + ":" + request.getServerPort() + path + "/";
    String target = basePath + "main.do";
    return new RedirectView(target);
}
```

重定向的路径有相对路径和绝对路径。上面的示例使用的是绝对路径，相对路径容易出错，要小心。

关于相对路径的示例如下。

```
@RequestMapping("/user/logout")
public View logout(HttpServletRequest request) {
    request.getSession().invalidate();
    return new RedirectView("main.do ");
}
```

希望用户退出后，转向主页 http://localhost:8081/HelloRedirect/main.do。然而，如果发送请求 http://localhost:8081/HelloRedirect/user/logout.do，则会转向 http://localhost:8081/HelloRedirect/user/main.do，报 404 错误，因为 new RedirectView("main.do")是相对路径。

关于的相对路径的另外一个示例如下。

```
@RequestMapping("/user/logout")
    public View logout(HttpServletRequest request) {
        request.getSession().invalidate();
        return new RedirectView("/main.do ");
    }
```

如果发送请求 http://localhost:8081/HelloRedirect/user/logout.do，会转向 http://localhost:8081/main.do，报 404 错误。

3. 传输数据到重定向目标

重定向的本质是客户端向浏览器重新发送了两次 HTTP 请求，因此基于 HTTP 请求的 request 对象携带的数据无法用重定向方式传递到后面的页面。如何通过重定向传递信息？Spring MVC 提供了多种解决方案。

下面给出一个示例。

（1）在 Model 属性中携带数据，测试目标地址是否能接收数据。

```
@RequestMapping("/user/logout")
public View logout(HttpServletRequest request,Model model) {
    request.getSession().invalidate();
    String path = request.getContextPath();
    String basePath = request.getScheme() + "://" +
        request.getServerName() + ":" + request.getServerPort() + path + "/";
    String target = basePath + "main.do";
    model.addAttribute("data", " xiaohp");
    return new RedirectView(target);
}
@RequestMapping("/main")
public String main(@ModelAttribute String data) {
    System.out.println("data=" + data);
    return "/main/main.jsp";
}
```

由测试结果可知，在重定向模式下，在 Model 属性中携带数据，目标地址无法接收到数据。

（2）在重定向目标地址的 URL 中携带参数。

```
@RequestMapping("/user/logout")
public View logout2(HttpServletRequest request,RedirectAttributes ra) {
    request.getSession().invalidate();
    String path = request.getContextPath();
    String basePath = request.getScheme() + "://" +
                    request.getServerName() + ":" + request.getServerPort() + path + "/";
   String target = basePath + "main.do";
    ra.addAttribute("data", "xiaohp");
    return new RedirectView(target);
}
```

在上面的代码，在 RedirectAttributes 中携带参数信息，其本质是在目标地址后面携带参数，参见如下代码。

```
String target = basePath + "main.do?data=xiaohp";
```

对于通过 HTTP 参数传递的数据，目标地址采用如下方式直接接收即可。

```
@RequestMapping("/main")
public String main(String data) {
    System.out.println("data=" + data);
    return "/main/main.jsp";
}
```

（3）通过临时对象刷新传递的数据。

```
@RequestMapping("/user/logout")
public View logout2(HttpServletRequest request,RedirectAttributes ra) {
    request.getSession().invalidate();
```

```
    String path = request.getContextPath();
    String basePath = request.getScheme() + "://" +
                request.getServerName() + ":" + request.getServerPort() + path + "/";
    String target = basePath + "main.do";
    ra.addFlashAttribute("token", "admesgs");
    return new RedirectView(target);
}
```

ra.addFlashAttribute("token", "admesgs")会把数据暂时存放于会话中，数据传递后再删除这个临时的会话键值。刷新模式有一定的局限性，它可以在 main.jsp 中接收这个数据，但是无法在 main.do 的方法中接收。

```
<html>
    <body>
        welcome you ${token}
    </body>
</html>
```

4. “redirect:” 前缀

使用 RedirectView 传递数据不是最佳方案，它会产生过多的耦合。控制器方法一般只需要返回视图名字。使用“redirect:”前缀可以更好地解决问题。

如果返回的视图名字带有“redirect:”前缀，UrlBasedViewResolver 会知道这是一个特殊的指示，即重定向到“redirect:”后面的 URL 中。

“redirect:”前缀后面的 URL 可以使用相对地址，也可以使用绝对地址。

示例如下。

- 相对地址：redirect:/myapp/some/resource。
- 绝对地址：redirect:http://myhost.com/some/arbitrary/path。

示例代码如下。

```
@RequestMapping("/user/logout")
public View logout(HttpServletRequest request,RedirectAttributes ra) {
    request.getSession().invalidate();
    String path = request.getContextPath();
    String basePath = request.getScheme() + "://" +
                request.getServerName() + ":" + request.getServerPort() + path + "/";
    String target = basePath + "main.do";
    ra.addAttribute("data", "xiaohp");
    return "redirect:" + target;
}
```

5. “forward:” 前缀

带“forward:”前缀的视图最终被 UrlBasedViewResolver 解析。这个前缀对于 InternalResourceView Resolver 和 InternalResourceView 没有特殊作用，最终都调用 Servlet API 的 RequestDispatcher.forward( )

方法。

当从一个 Action 方法转到另一个 Action 方法时，可以使用“forward:”前缀。服务器资源之间的跳转应该优先使用相对地址。

示例代码如下。

```
@RequestMapping("/m1")
public String method1(Model model) {
    model.addAttribute("token", "xiaohp24678");
    return "forward:m2.do" ;
}
@RequestMapping("/m2")
public String method2(@ModelAttribute String token,HttpServletRequest request) {
    System.out.println("token=" + token);
    System.out.println("req:" + request.getAttribute("token"));
    return "/main/main.jsp" ;
}
```

注意：在 m1 的方法中，用 model.addAttribute("token", "xiaohp24678")存储数据；在 m2 的方法中，@ModelAttribute String token 无法取得数据。request.getAttribute("token")可以读取 m1 中传递的数据，因为@ModelAttribute 是从 HTTP 请求参数中查找填充数据的。

7.9 使用 Flash 属性

Flash 属性提供方法，在一个 HTTP 请求中存储数据，在另一个请求中提取数据。这在重定向的请求模式下很常见。

在重定向到新 URL 之前，临时保存数据（如放在会话中）。重定向之后，立即移除该数据。

Spring MVC 有两个主要抽象，分别是 FlashMap 和 FlashMapManager。

FlashMap 用于管理 Flash 属性数据。

```
public final class FlashMap
                extends HashMap<String, Object>
                implements Comparable<FlashMap> {
}
```

在 Web 控制器中通常无须直接使用 FlashMap，可以通过注入 RedirectAttributes 调用 FlashMap。

FlashMapManager 用于存储、读取 Flash 属性，并管理 FlashMap 实例。

```
public interface FlashMapManager {
    FlashMap retrieveAndUpdate(HttpServletRequest request,
                               HttpServletResponse response);
    void saveOutputFlashMap(FlashMap flashMap,
                               HttpServletRequest request,
                               HttpServletResponse response);
}
```

Flash 属性数据的输入和输出使用 RequestContextUtils 的相关静态方法实现。

```
public abstract class RequestContextUtils {
    public static FlashMap getOutputFlashMap(
                          HttpServletRequest request) {}
    public static FlashMapManager getFlashMapManager(
                         HttpServletRequest request) {}
}
```

7.10 使用 Locale

7.10.1 Locale 对象

Locale 对象代表某个区域的具体地理信息、政治信息、文化信息，如国家名称、语言、时区等。需要通过 Locale 执行任务的操作称为语言环境敏感的操作。例如，显示一个数字是一个区域设置敏感的操作，该数字应该根据用户的国家、地区或文化的习惯和惯例进行格式化。参见 JDK 8 中 Locale 的定义。

```
public final class Locale
              implements Cloneable, Serializable {
    static public final Locale US
                        = createConstAnt("en", "US");
    static public final Locale SIMPLIFIED_CHINESE
                        = createConstAnt("zh", "CN");
    static public final Locale CHINA = SIMPLIFIED_CHINESE;
    public Locale(String language, String country) {
    }
    public static Locale getDefault() {}
}
```

示例代码如下。

```
Locale locale = new Locale("zh", "CN");
Locale  locale = new Locale("en", "US");
```

7.10.2 Locale 解析器

Spring 架构支持国际化。DispatcherServlet 调用 LocaleResolver 自动解析消息，使用本地的信息。LocaleResolver 解析器的接口参见如下代码。

```
public interface LocaleResolver {
    Locale resolveLocale(HttpServletRequest request);
    void setLocale(HttpServletRequest request,
            HttpServletResponse response, Locale locale);
}
```

调用 LocaleResolver::setLocale 设置时区信息。DispatcherServlet 调用 RequestContext.getLocale( )

方法获取时区信息。

```
public class RequestContext {
    private Locale locale;
    private TimeZone timeZone;
    public final Locale getLocale() {
        return this.locale;
    }
    public TimeZone getTimeZone() {
        return this.timeZone;
    }
}
```

LocaleResolver 接口的常用实现类如下。

- AcceptHeaderLocaleResolver：其实没有任何具体实现，是通过浏览器头部的语言信息来进行多语言选择的。
- FixedLocaleResolver：设置固定的语言信息，这样整个系统的语言是一成不变的，用处不大。
- CookieLocaleResolver：将语言信息设置到 Cookie 中，这样整个系统通过 Cookie 就可以获得语言信息。
- SessionLocaleResolver：与 CookieLocaleResolver 类似，将语言信息放到会话中，这样整个系统就可以从会话中获得语言信息。

7.10.3 Locale 拦截器

使用 Locale 拦截器，通过变化参数，可以改变 Locale 的设置。在拦截器中，调用当前上下文中的 LocaleResolver 的 setLocale( )方法，可以改变当前 Locale 的设置。

使用 LocaleChangeInterceptor 拦截器，可以在每一次 HTTP 请求中改变 Locale 的设置，这可以应用到国际化项目，按照用户的需求，显示不同语言的资源。

```
public class LocaleChangeInterceptor extends HandlerInterceptorAdapter {
    public static final String DEFAULT_PARAM_NAME = "locale";
    private String paramName = DEFAULT_PARAM_NAME;
    public void setParamName(String paramName) {
        this.paramName = paramName;
    }
    public boolean preHandle(HttpServletRequest request,
                        HttpServletResponse response, Object handler){
        LocaleResolver localeResolver
                = RequestContextUtils.getLocaleResolver(request);
        String newLocale = request.getParameter(getParamName());
        localeResolver.setLocale(request, response, parseLocaleValue(newLocale));
    }
}
```

在以下示例代码中，设置 LocaleChangeInterceptor 的参数值，调用所有*.view 的资源。

```
<bean id="localeChangeInterceptor"
        class="org.springframework.web.servlet.i18n.LocaleChangeInterceptor">
    <property name="paramName" value="siteLanguage"/>
</bean>
<bean id="localeResolver"
        class="org.springframework.web.servlet.i18n.CookieLocaleResolver"/>
<bean id="urlMapping"
     class="org.springframework.web.servlet.handler.SimpleUrlHandlerMapping">
    <property name="interceptors">
        <list>
            <ref bean="localeChangeInterceptor"/>
        </list>
    </property>
    <property name="mappings">
        <value>/**/*.view=someController</value>
    </property>
</bean>
```

7.10.4 项目案例：国际化应用

在本案例中，需求使用国际化设置，使用户信息页既可以用中文展示，也可以用英文显示。

操作步骤如下。

（1）配置 LocaleResolver，此处使用 SessionLocaleResolver。SessionLocaleResolver 类通过一个预定义会话名将区域化信息存储在会话中。

```
<bean id="localeResolver"
    class="org.springframework.web.servlet.i18n.SessionLocaleResolver" />
```

（2）配置国际化资源。在项目的 src 目录下，在资源文件中新建 messages_en.properties 和 messages_zh.properties，分别使用中文和英文进行配置。

messages_en.properties 的资源配置如下。

```
userManage=userManagement
userName=username
age=age
photoName=photo name
photo=photo
addUser=add user
showUserInfo= display user information
```

messages_zh.properties 的资源配置如下(输入中文,在 properties 文件中自动转换成如下格式)。

```
userManage=\u7528\u6237\u7BA1\u7406
userName=\u59D3\u540D
age=\u5E74\u9F84
photoName=\u7167\u7247\u540D\u79F0
```

```
photo=\u7167\u7247
addUser=\u589E\u52A0\u7528\u6237
showUserInfo=\u8FD9\u91CC\u662F\u5C55\u73B0\u7528\u6237\u4FE1\u606F
```

（3）配置消息源，加载资源文件。

```
<bean id="messageSource" class="org.springframework.context
                        .support.ReloadableResourceBundleMessageSource">
  <property name="basename" value="classpath:messages" />
</bean>
```

（4）配置拦截器，改变 Locale。此处使用自定义拦截器，没有使用 LocaleChangeInterceptor，这样更加灵活。

```
<mvc:interceptors>
   <bean class="com.icss.intercept.LangInterceptor">
     <property name="paramName" value="lang" />
   </bean>
</mvc:interceptors>
```

（5）实现自定义拦截器。参考 LocaleChangeInterceptor 的源代码，自定义拦截器的实现原理与之完全一致。

```
public class LangInterceptor extends HandlerInterceptorAdapter{
   private String paramName;
   public void setParamName(String paramName) {
      this.paramName = paramName;
   }
   public String getParamName() {
      return this.paramName;
   }
   @Override
   public boolean preHandle(HttpServletRequest request,
           HttpServletResponse response, Object handler) throws Exception {
        Locale locale = new Locale("zh", "CN");
        String language = request.getParameter(this.paramName);
            if(language!=null && language.equals("en")){
                locale = new Locale("en", "US");
            }
      LocaleResolver localeResolver =
                  RequestContextUtils.getLocaleResolver(request);
      if (localeResolver == null) {
         throw new IllegalStateException("No LocaleResolver found");
      }
      localeResolver.setLocale(request, response, locale);
      return true;
   }
}
```

（6）在控制器的代码中提取资源文件中的 key。此处的 showUserInfo 是资源文件中的某个 key 值。

```
@RequestMapping("/user")
public ModelAndView getUsers(HttpServletRequest request){
```

```
        List<User> userList = new ArrayList<>();
        userList.add(new User("tom",18,"aa"));
        userList.add(new User("jack",28,"bb"));
        userList.add(new User("rose",38,"cc"));
        ModelAndView mv =  new ModelAndView();
        RequestContext requestContext = new RequestContext(request);
        String value = requestContext.getMessage("showUserInfo");
        mv.addObject("showUserInfo", value);
        mv.addObject("userList", userList);
        mv.setViewName("/main/user.jsp");
        return mv;
    }
```

（7）在视图层使用标签调用资源。例如，使用<spring:message code="userName"/>。此处需要引用 Spring 的标签库　<%@ taglib prefix="spring" uri="http://www.springframework. org/tags"%>。核心代码如下。

```
<c:forEach var="user" items="${userList}">
    <div class="form-group">
    <label for="inputEmail3" class="col-sm-2 control-label">
                <spring:message code="userName"/>
</label>
     <div class="col-sm-10">
         <input class="form-control" placeholder="Email" value="${user.name }">
     </div>
 </div>
  <div class="form-group">
     <label for="inputPassword3" class="col-sm-2 control-label">
            <spring:message code="age"/>
</label>
<div class="col-sm-10">
         <input class="form-control" placeholder="Password" value="${user.age }">
     </div>
  </div>
</c:forEach>
```

（8）使用 http://localhost:8081/HelloLocale/user.do?lang=zh 和 http://localhost:8081/HelloLocale/user.do（默认语言为中文），测试浏览器（见图 7-39）。

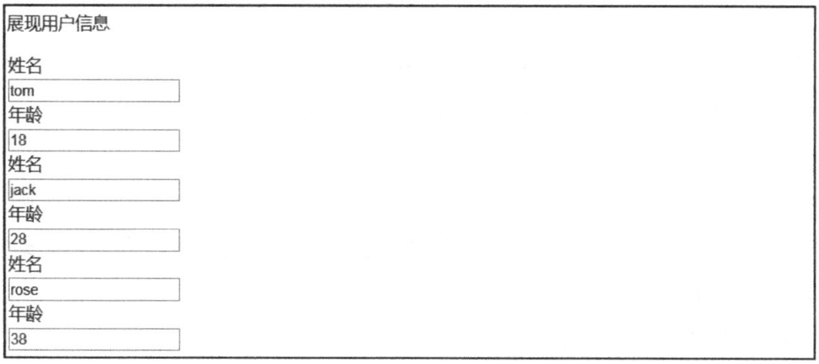

展现用户信息

姓名

tom

年龄

18

姓名

jack

年龄

28

姓名

rose

年龄

38

图 7-39　使用中文界面测试浏览器

当请求参数为 lang=en（使用 http://localhost:8081/HelloLocale/user.do?lang=en）时，拦截器动态设置 Locale 为英文环境，测试结果见图 7-40。

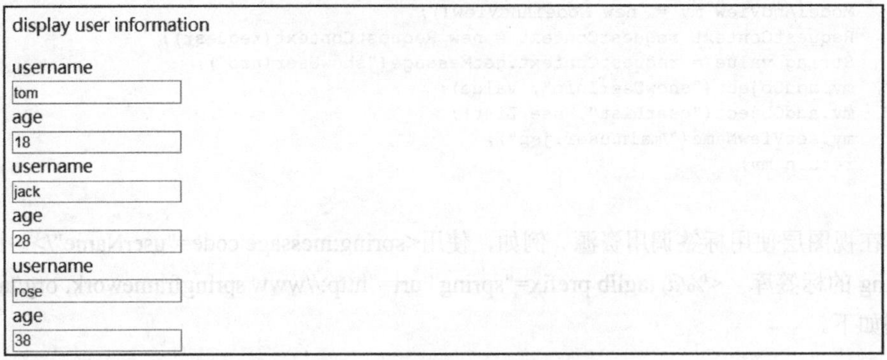

display user information

username

tom

age

18

username

jack

age

28

username

rose

age

38

图 7-40　使用英文界面测试浏览器

7.11　主题

7.11.1　主题介绍

在 Spring MVC 中，可以使用主题动态改变页面的显示效果。

示例如下。

（1）在 properties 文件配置以下属性。

```
styleSheet=/themes/cool/style.css
background=/themes/cool/img/coolBg.jpg
```

（2）在 JSP 文件中，通过<spring:theme>标签库应用主题。

```
<%@ taglib prefix="spring" uri="http://www.springframework.org/tags"%>
<html>
    <head>
        <link rel="stylesheet" href="<spring:theme code='styleSheet'/>" type="text/css"/>
    </head>
    <body style="background=<spring:theme code='background'/>">
        ...
    </body>
</html>
```

（3）使用 ResourceBundleThemeSource，设置从哪个位置加载主题配置文件，即需要说明前面定义的 properties 文件所在路径。

（4）使用 ThemeResolver 接口解析主题。

```
public interface ThemeResolver {
    String resolveThemeName(HttpServletRequest request);
```

```
    void setThemeName(HttpServletRequest request,
                      HttpServletResponse response, String themeName);
}
```

ThemeResolver 的实现类见表 7-4。

表 7-4　　ThemeResolver 的实现类

ThemeResolver 的实现类	描述
FixedThemeResolver	使用 defaultThemeName 属性，设置一个固定的主题
SessionThemeResolver	基于 HTTP 会话存储的主题，不能在会话间共享
CookieThemeResolver	主题存储于客户端的 Cookie 中

7.11.2　项目案例：主题的应用

本节通过完整的案例演示如何应用主题改变页面的显示效果。具体步骤如下。

（1）定义两套简单的 CSS 文件，在 JSP 页面中动态选择使用哪套主题。分别定义 blue.css 和 yellow.css 文件，并放置在 Web 项目的 css 文件夹下。

```
blue.css:
body{
     color: blue;
     font-size:20px;
}
a{
    font-size:16px;
}
yellow.css:
body{
    color: yellow;
    font-size:28px;
}
a{
    font-size:20px;
}
```

（2）在 properties 文件中设置主题。在 src 下新建 themes 文件夹，然后新建 3 个配置文件。

```
blue.properties:
    css.link=http://localhost:8081/HelloTheme/css/blue.css
default.properties:
    css.link=
yellow.properties:
    css.link=http://localhost:8081/HelloTheme/css/yellow.css
```

（3）在 spring-mvc.xml 中配置资源绑定。

```
<bean id="themeSource" class="org.springframework.ui
                        .context.support.ResourceBundleThemeSource">
```

```
    <property name="basenamePrefix" value="themes/" />
</bean>
```

（4）在 spring-mvc.xml 中配置主题解析器，此处使用 SessionThemeResolver。

```
<bean id="themeResolver" class="org.springframework
                        .web.servlet.theme.SessionThemeResolver">
    <!--主题文件的默认名字是 default-->
    <property name="defaultThemeName" value="default" />
</bean>
```

（5）在控制器代码中注入 SessionThemeResolver。

```
@Controller
public class OrderAction {
    @Autowired
    private ThemeResolver themeResolver;
    @GetMapping("/order")
    public String orderIndex(Model model) {
        model.addAttribute("order", new Order("12", "预输入", "email", 188));
        return "/main/orders.jsp";
    }
    @GetMapping("/themes/{css}")
    public String changeCss(@PathVariable("css") String css,
                HttpServletRequest req, HttpServletResponse resp) {
        themeResolver.setThemeName(req, resp, css);
        return "redirect:/order.do";
    }
}
```

（6）显示页面，设置引用主题。在 orders.jsp 中，通过*.properties 文件中的 key 引用主题<spring:theme>中的 code 值，指向 properties 文件中的 key。

```
<%@ taglib prefix="spring" uri="http://www.springframework.org/tags" %>
<html>
<head>
    <link rel="stylesheet" href="<spring:theme code="css.link"/>"/>
</head>
<body>
     <h1>主题测试</h1>
     <div align="center">
         <p>
         <a href="<%=basePath%>themes/yellow.do">黄色主题</a>  
          <a href="<%=basePath%>themes/blue.do">蓝色主题</a>
    </p>
    <p>${order.id}</p>
    <p>${order.name}</p>
    <p>${order.email}</p>
    <p>${order.price}</p>
   </div>
</body>
</html>
```

（7）测试主题。分别发送如下请求，显示不同的主题样式。

- http://localhost:8080/HelloTheme/themes/blue.do。
- http://localhost:8080/HelloTheme/themes/yellow.do。

7.12 multipart 文件的上传

multipart 文件为<form>的数据格式，表示非文本信息，里面包含字节流内容。

把实体文件（如 ZIP 包、图片等）从客户端浏览器上传到服务器与普通的文本信息上传是不同的。首先需要使用<input type="file"> 标签，然后需要使用<form>提交，不能使用 HTTP 参数形式。

<form>有一个 enctype 属性，它表示表单内容的 MIME 类型。表单的 enctype 属性默认值为 application/x-www-form-urlencoded，它表示上传的信息为文本信息。当使用表单上传文件时，必须要配置<form enctype="multipart/form-data">。

由于通过<form>上传的数据格式的变化，在服务器接收数据时，使用传统的 ServletRequest. getParameter( )将无法接收到数据。传统的解决方案是使用 smartUpload 控件解析数据。

7.12.1 MultipartResolver

Spring MVC 使用 MultipartResolver 支持文件上传，这个方案需要 Apache 的 commons-fileupload.jar 的支持。

示例代码如下。

```
<bean id="multipartResolver"
              class="org.springframework.web.multipart
                    .commons.CommonsMultipartResolver">
    <!-- 设置上传文件的最大 size 值 -->
    <property name="maxUploadSize" value="100000"/>
</bean>
```

在 Servlet 3.0 之后，Java EE 6 可以使用内置的 Part 来接收数据，可以不需要 commons-fileupload.jar 的支持。

```
public interface Part {
       public InputStream getInputStream() throws IOException;
       public String getContentType();
       public long getSize();
       public Collection<String> getHeaders(String name);
}
```

基于 Servlet 3.0 的解析器使用 StandardServletMultipartResolver。

```
<bean id="multipartResolver"
     class="org.springframework.web.multipart.support.StandardServletMultipartResolver">
</bean>
```

CommonsMultipartResolver 和 StandardServletMultipartResolver 都实现了 MultipartResolver 接口，使用哪种解析方案都可以。

```
public class CommonsMultipartResolver
                  extends CommonsFileUploadSupport
                  implements MultipartResolver, ServletContextAware {}

public class StandardServletMultipartResolver
                implements MultipartResolver {}
```

7.12.2 项目案例：上传图片

本节演示如何使用传统的 CommonsMultipartResolver 提取上传的图片数据。具体步骤如下。

（1）导入 Spring 的核心包和 Web 包，导入依赖包 commons-fileupload.jar 和 commons-io.jar。

（2）配置 CommonsMultipartResolver，配置文件最大为 100KB。

```
<bean id="multipartResolver"
    class="org.springframework.web.multipart.commons.CommonsMultipartResolver">
    <property name="maxUploadSize" value="102400" />
</bean>
```

（3）配置 upload.jsp 中 form 的 enctype="multipart/form-data"。

```
<form method="post" action="<%=basePath%>form.do" enctype="multipart/form-data">
          文件名：<input type="text" name="name" id="fname"/>
                  <input type="file" name="file" onchange="show(this)" />
                  <br>
         <input type="submit" value="提交" />
         ${msg}
</form>
```

（4）使用 JavaScript 接收选定的文件名。

```
function show(source) {
    var arrs = $(source).val().split('\\');
    var filename=arrs[arrs.length-1];
    $('#fname').val(filename);
}
```

（5）控制器接收上传的图片，保存在 Web 项目根目录下的 pic 文件夹下，使用 Apache 的 FileUtils 将字节流存储到文件中。

```
@PostMapping("/form")
public String handleFormUpload( String name,
              @RequestParam("file") MultipartFile file,Model model) {
    if (!file.isEmpty()) {
        try {
            byte[] bytes = file.getBytes();
```

```
            File pic = new File( context.getRealPath("/") + "pic/" + name);
            FileUtils.writeByteArrayToFile(pic, bytes);
            model.addAttribute("msg", name+"上传成功");
        } catch (Exception e) {
            e.printStackTrace();
            model.addAttribute("msg", name+"失败");
        }
    }
    return "/main/uploadFile.jsp";
}
```

7.13 异常处理

7.13.1 HandlerExceptionResolver

Spring 的 HandlerExceptionResolver 用于处理控制器执行过程中未知的异常。对于预期的异常，应该选择性捕获并进行处理。

```
public interface HandlerExceptionResolver {
    ModelAndView resolveException(HttpServletRequest request,
                        HttpServletResponse response,
                        Object handler, Exception ex);
}
```

HandlerExceptionResolver 只有一个方法，其功能是处理异常，然后根据异常类型，转到不同的错误页。

7.13.2 SimpleMappingExceptionResolver

SimpleMappingExceptionResolver 实现了 HandlerExceptionResolver 接口。SimpleMappingException Resolver 是经常使用的异常解析器。

```
public class SimpleMappingExceptionResolver
                extends AbstractHandlerExceptionResolver {
}
public abstract class AbstractHandlerExceptionResolver
                implements HandlerExceptionResolver, Ordered {}
```

示例代码如下。

在 beans.xml 中配置 SimpleMappingExceptionResolver，把所有控制层统一处理的已知错误抛到对应的错误处理页中，把未知错误统一抛到 error.jsp 中。

```
<bean
class="org.springframework.web.servlet.handler.SimpleMappingExceptionResolver">
    <property name="exceptionMappings">
        <props>
            <prop key="java.lang.Throwable">/error/error.jsp</prop>
            <prop key="org.springframework
```

```
                    .web.multipart.MaxUploadSizeExceededException">
                /error/OverMaxUploadSize.jsp</prop>
            <prop key="com.icss.exception.InputEmptyException">
                /error/InputEmpty.jsp</prop>
        </props>
    </property>
</bean>
```

7.13.3 @ExceptionHandler

前面讲了如何同时使用@ControllerAdvice 与@ExceptionHandler。示例代码如下。

```
@ControllerAdvice
public class GlobalExceptionHandler {
    @ExceptionHandler(Exception.class)
    public ModelAndView otherException(Exception e) {
        ModelAndView mv = new ModelAndView();
        mv.addObject("msg", e.getMessage());
        mv.setViewName("/error/error.jsp");
        return mv;
    }
}
```

HandlerExceptionResolver 和实现类 SimpleMappingExceptionResolver 都允许配置指定的异常到对应的视图页面，显示异常提示。但是，对于@ResponseBody 注释的方法，没有返回视图。这时出现异常，因此在 response 信息中携带状态码更合适。示例代码如下。

```
@Controller
public class SimpleController {
    @ExceptionHandler
    public ResponseEntity<String> handle(IOException ex) {
        HttpHeaders responseHeaders = new HttpHeaders();
        responseHeaders.set("code", "059");
        responseHeaders.set("msg", ex.getMessage());
        return new ResponseEntity<String>(
               "-1",responseHeaders,HttpStatus.FAILED_DEPENDENCY);
    }
    @RequestMapping("/check")
    @ResponseBody
    public int checkName(String name) throws IOException{
        throw new IOException("io exeception test");
    }
}
```

直接使用浏览器发出请求 http://localhost:8081/HelloException/check.do，通过抓包分析控制器的响应，对于@ResponseBody 异常，通过 ResponseEntity 携带的头和体信息返回错误（见图 7-41（a））。

如图 7-41（b）所示，如果客户端收到响应码-1，就表示出现了异常。从 SimpleMapping ExceptionResolver 头中提取错误码和错误消息，就可以把@ResponseBody 注解的异常信息顺利传递到客户端。

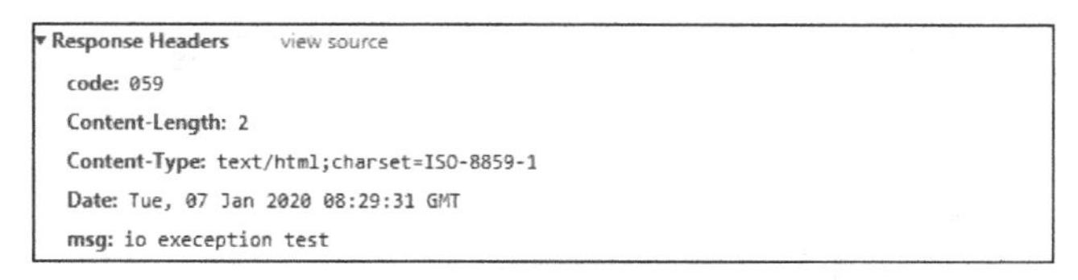

（a）错误

（b）异常

图 7-41　错误和异常

7.13.4　标准异常解析

对于客户端的 HTTP 请求，当出现异常时，用错误码提示客户端，如客户端错误（4xx）、服务器错误（5xx）。

最常见的 404 错误表示客户端访问的服务器目标资源没找到。

Spring MVC 使用 DefaultHandlerExceptionResolver 把指定的异常（见表 7-5）转换为状态码，参见 API 的定义（只显示了部分代码）。

```
public class DefaultHandlerExceptionResolver
          extends AbstractHandlerExceptionResolver {
    protected ModelAndView handleHttpMediaTypeNotSupported(...);
    protected ModelAndView handleHttpMediaTypeNotAcceptable(...);
    protected ModelAndView handleMissingPathVariable(...);
    protected ModelAndView handleTypeMismatch(...);
}
```

表 7-5　异常

异常	HTTP 状态码
BindException 异常	400（Bad Request）
ConversionNotSupportedException 异常	500（Internal Server Error）
HttpMediaTypeNotAcceptableException 异常	406（Not Acceptable）
HttpMediaTypeNotSupportedException 异常	415（Unsupported Media Type）
HttpMessageNotReadableException 异常	400（Bad Request）
HttpMessageNotWritableException 异常	500（Internal Server Error）
HttpRequestMethodNotSupportedException 异常	405（Method Not Allowed）
MethodArgumentNotValidException 异常	400（Bad Request）
MissingPathVariableException 异常	500（Internal Server Error）
MissingServletRequestParameterException 异常	400（Bad Request）
MissingServletRequestPartException 异常	400（Bad Request）
NoHandlerFoundException 异常	404（Not Found）
NoSuchRequestHandlingMethodException 异常	404（Not Found）
TypeMismatchException 异常	400（Bad Request）

在以下代码中，可以抛出不同的异常类型，显示相应的状态码提示（见图 7-42～图 7-44）。

```
@Controller
public class DefaultExceptionAction {
    @GetMapping("/m1")
    public String method1() throws Exception{
        throw new BindException(new Object(),"test");
    }
    @GetMapping("/m2")
    public void method2() throws HttpRequestMethodNotSupportedException {
        throw new HttpRequestMethodNotSupportedException("test");
    }
    @GetMapping("/m3")
    public void method3() throws Exception{
        throw new HttpMediaTypeNotAcceptableException("test");
    }
}
```

localhost:8081/HelloException/m1.do

应用

HTTP Status 400 -

type Status report

message

description The request sent by the client was syntactically incorrect.

图 7-42　BindException 异常

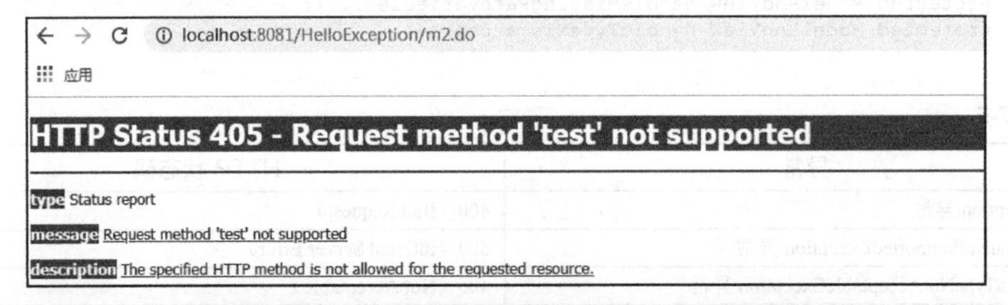

图 7-43　HttpRequestMethodNotSupportedException 异常

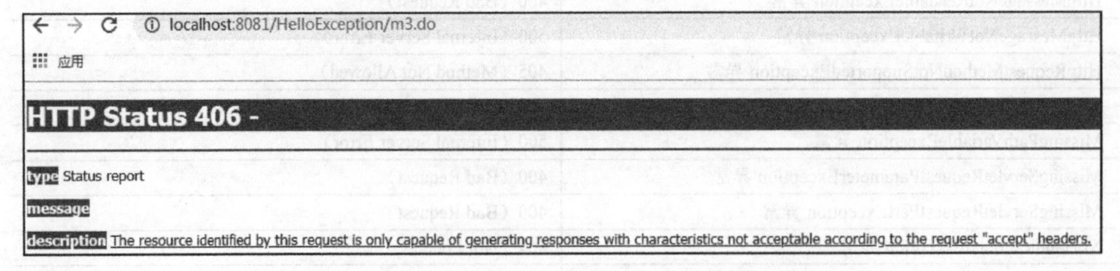

图 7-44　HttpMediaTypeNotAcceptableException 异常

注意：DefaultHandlerExceptionResolver 是 DispatcherServlet 默认激活的，无须配置。如果已经配置了 SimpleMappingExceptionResolver 或 ExceptionHandler，则异常将会转到错误页，客户端无法收到错误码提示。

7.14 使用 JSP 与 JSTL

7.14.1 JSP 与 JSTL

要解析 JSP，常用的解析器是 InternalResourceViewResolver 和 ResourceBundleViewResolver。两者都定义在 WebApplicationContext 中。示例代码如下。

```
<bean id="viewResolver"
     class="org.springframework.web.servlet.view.ResourceBundleViewResolver">
     <property name="basename" value="views"/>
</bean>
<bean id="viewResolver"
     class="org.springframework.web.servlet.view.InternalResourceViewResolver">
     <property name="viewClass" value="org.springframework.web.servlet.view.JstlView"/>
     <property name="prefix" value="/WEB-INF/jsp/"/>
     <property name="suffix" value=".jsp"/>
</bean>
```

为了使用 Java 标准标签库（Java Standard Tag Library，JSTL），需要用专门的 JstlView。为了便于 JSP 的开发，使请求参数与命令对象方便地进行绑定，可以使用 Spring 标签库，它定义在 spring-webmvc.jar 中。使用 Eclipse 开发 Spring MVC 项目，需要导入 jstl-impl.jar 和 javax.servlet.jsp.jstl.jar。

7.14.2 Spring 的基本标签

在前面几节中使用了如下标签库。

```
<%@ taglib prefix="spring" uri="http://www.springframework
.org/tags" %>
```

从国际化资源文件中，根据键值，使用 Spring 标签提取数据。

```
<spring:message code="userName"/>
<spring:message code="age"/>
```

参见标签库描述文件 spring.tld，文件位于 Spring-webmvc-4.3.19-RELEASE.jar--META-INF 文件夹下。message 标签的定义如下。

```
<taglib xmlns="http://java.sun.com/xml/ns/j2ee"
        http://java.sun.com/xml/ns/j2ee/web-jsptaglibrary_2_0.xsd"
    <short-name>spring</short-name>
    <uri>http://www.springframework.org/tags</uri>
```

```
    <tag>
        <name>message</name>
        <tag-class>org.springframework.web
                  .servlet.tags.MessageTag</tag-class>
        <body-content>JSP</body-content>
        <attribute>
            <name>code</name>
            <required>false</required>
        </attribute>
    <tag>
</taglib>
public class MessageTag
             extends HtmlEscapingAwareTag implements ArgumentAware {
    private String code;
    protected String resolveMessage()
                   throws JspException, NoSuchMessageException {}
}
```

从主题文件中，根据键值提取数据，如<spring:theme code="css.link"/>。主题标签的定义如下。

```
<taglib xmlns="http://java.sun.com/xml/ns/j2ee"
        http://java.sun.com/xml/ns/j2ee/web-jsptaglibrary_2_0.xsd"
    <short-name>spring</short-name>
    <uri>http://www.springframework.org/tags</uri>
    <tag>
    <name>theme</name>
        <tag-class>org.springframework.web.servlet.tags.ThemeTag</tag-class>
        <body-content>JSP</body-content>
        <attribute>
            <name>code</name>
            <required>false</required>
        </attribute>
</taglib>
public class ThemeTag extends MessageTag {
    protected MessageSource getMessageSource() {}
}
```

7.14.3 Spring 的 form 标签库

Spring 的 form 标签库定义在 Spring-webmvc-4.3.19-RELEASE.jar--META-INF 文件夹下的 spring-form.tld 文件中。

```
<taglib xmlns="http://java.sun.com/xml/ns/j2ee">
    <tlib-version>4.3</tlib-version>
    <short-name>form</short-name>
    <uri>http://www.springframework.org/tags/form</uri>
</taglib>
```

可以使用的 form 标签非常多，简单的示例如下。

```
<form:input id="" path="" />
<form:form  id="" name="" />
<form:password id="" path="" />
```

```
<form:hidden id="" path="" />
<form:select id="" path="" />
```

案例 7-7：使用 from 标签库提交用户数据。

（1）编写 JSP 文件 form.jsp，这里必须在 JSP 的头部引入 form 标签库。

```
<%@ taglib prefix="form"
                uri="http://www.springframework.org/tags/form"%>
<form:form  modelAttribute="user"
            action=" ${pageContext.request.contextPath}/form.do">
    <table>
        <tr>
            <td>First Name:</td>
            <td><form:input path="firstName" /></td>
        </tr>
        <tr>
            <td>Last Name:</td>
            <td><form:input path="lastName" /></td>
        </tr>
        <tr>
            <td colspan="2"><input type="submit" value="Save Changes" /></td>
        </tr>
    </table>
</form:form>
```

（2）定义 User 实体。

```
public class User {
    private String firstName;
    private String lastName;
}
```

（3）编写控制器代码。

```
@Controller
public class UserAction {
    @GetMapping("/form")
    public String form(@ModelAttribute User user) {
        return "/main/form.jsp";
    }
    @PostMapping(value = "/form")
    public String formPost(@ModelAttribute User user) {
        UserBiz biz = new UserBiz();
        boolean bRet = biz.formPost();
        if(bRet) {
            return "/main/main.jsp";
        }else {
            return "/main/form.jsp";
        }
    }
}
```

如上所示，在 get 请求中，必须有@ModelAttribute User user 这个参数，否则会报错——Neither

Binding result nor plain target object for bean name 'command' available as request attribute。

（4）定义 form2.jsp。与 form.jsp 比较，form2.jsp 为传统模式的 JSP 写法。注意，此处使用 value="${...}接收控制器返回的 Model 属性值。

```
<form method="POST"
        action=" ${pageContext.request.contextPath}/form.do">
    <table>
        <tr>
            <td>First Name:</td>
            <td>
              <input name="firstName" type="text" value="${user.firstName} "/>
            </td>
        </tr>
        <tr>
            <td>Last Name:</td>
            <td>
                <input name="lastName" type="text" value="${user.lastName}"/>
            </td>
        </tr>
        <tr>
            <td colspan="2"><input type="submit" value="Save Changes" /></td>
        </tr>
    </table>
</form>
@Controller
public class UserAction {
    @GetMapping("/form2")
    public String form2() {
        return "/main/form2.jsp";
    }
    @PostMapping(value = "/form2")
    public String formPost2(@ModelAttribute User user) {
        UserBiz biz = new UserBiz();
        boolean bRet = biz.formPost();
        if(bRet) {
            return "/main/main.jsp";
        }else {
            return "/main/form2.jsp";
        }
    }
}
```

form.jsp 与 form2.jsp 的显示效果一样。

第 8 章
基于 Spring MVC 的书城项目实战

8.1 项目结构与用户权限

书城项目分为前台和后台两个部分，前台用于普通用户和游客访问，后台用于管理员访问。用户的角色分为游客、注册用户、会员、管理员（会员为预留用户）。游客为非注册用户，可以浏览主页和图书详情。注册用户可以添加商品到购物车，并提交、结算。会员也可称为 VIP 用户，可以进入会员专区。管理员进入后台，可以浏览用户订单、上传图书信息。

8.2 开发环境

项目开发环境是以 Java EE 7 为基础的，具体为 Java 8、Tomcat 8、Servlet 3.1、Spring 4.3.19 和 MyBatis 3.5.3。数据库采用 Oracle Database 11g 和 MySQL 8。IDE 为 Eclipse，选择 eclipse-jee-photon 或 eclipse-jee-oxygen 等均可。数据库的设计工具使用 PowerDesigner，逻辑设计工具使用 Rational Rose。

8.3 表的结构设计

网上书城项目同时支持两套数据库，即 MySQL 和 Oracle，还可以扩展为支持其他数据库。

书城项目的主要目的是在实际项目中强化前面的知识点，应该尽量覆盖更多知识点，而不强调功能完善，因此设计上不能过于复杂，重复功能都尽量省略了，这与大型的实际企业级项目是有一定区别的。

书城项目中 Oracle 表的结构如图 8-1 所示。

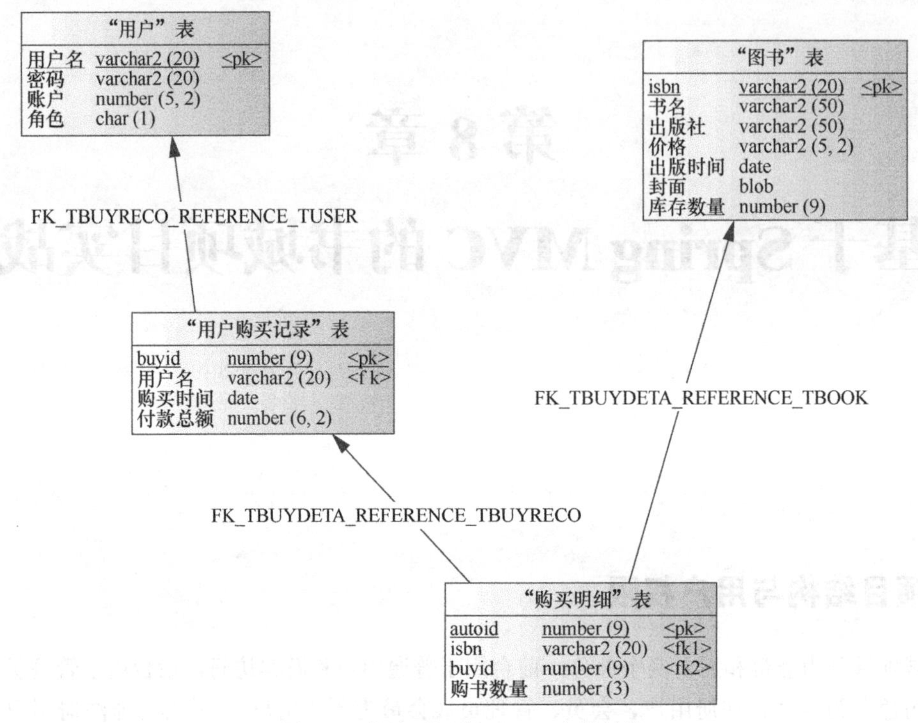

图 8-1 书城项目中 Oracle 表的结构

用户购买记录也可称为订单表。一个用户可以有多个订单，每个订单有多个订单明细。“图书”表用 ISBN 作为主键，即每个 ISBN 只存储一条记录，不是每本书对应一条记录。这与大型设备的管理模式不同，对于大型设备（如汽车），必须满足一辆车对应一条记录。

书城项目中 MySQL 的表结构如图 8-2 所示。

注意以下几点区别。

- Oracle 中使用 varchar2，MySQL 中使用 varchar。
- 对于整型和浮点型，Oracle 中推荐使用 number，MySQL 中使用 int 和 double。
- 两个数据库的所有字段名称和数量相同，唯一的区别是“用户购买记录”表的主键 buyid。Oracle 使用的是 number，而 MySQL 使用更有意义的 varchar。其实 Oracle 和 MySQL 中的 buyid 统一使用字符型更好，这里故意设计成不同的，用于对比事务操作。

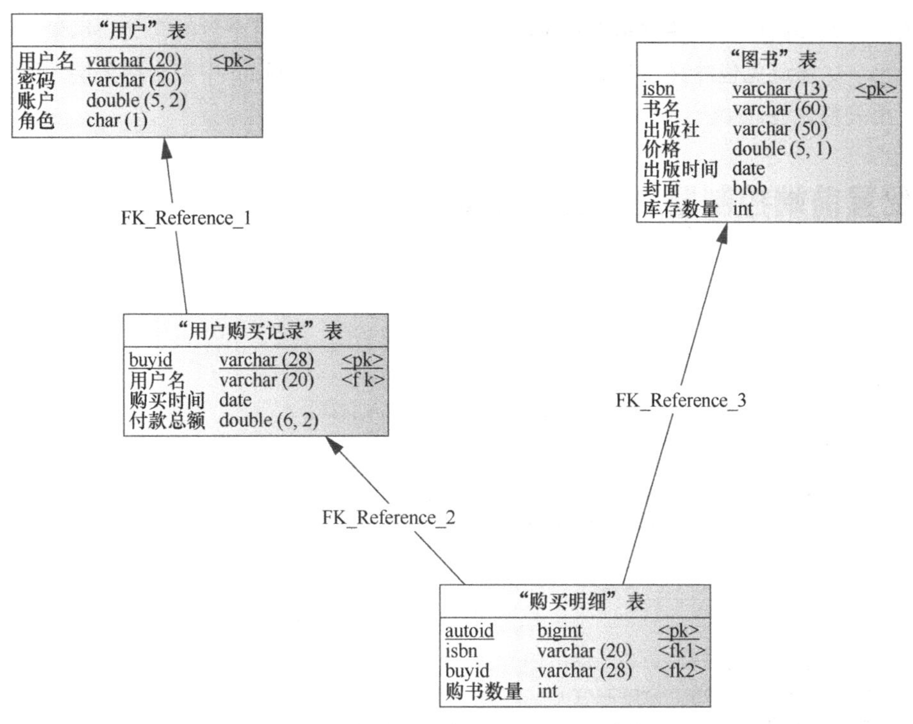

图 8-2　书城项目中 MySQL 表的结构

- “购买明细”表的主键为自增长的整型，为了提高代码的可移植性，项目都采用的是最大值+1 模式，没有采用数据库的专有设计 sequence 或 auto_increment。在非高并发的环境下，最大值+1 模式的效果与 sequence 的效果相同，但在高并发环境下则必须使用 sequence 或 auto_increment。

8.4 项目所需 JAR 包

首先，导入 Spring 核心包，包括 Spring-core.jar、Spring-context.jar、Spring-beans.jar、Spring-expression.jar。

然后，导入 Spring AOP 包，包括 Spring-AOP.jar 和 Spring-aspects.jar。

接下来，导入 Spring Web 包，包括 Spring-web.jar 和 Spring-webmvc.jar。

再接下来，导入依赖包，包括 commons-logging.jar、log4j-1.2.8.jar、jstl-impl.jar、javax.servlet.jsp.jstl、jackson-annotations-2.6.3.jar、jackson-core-2.6.3.jar、jackson-databind-2.6.3.jar、commons-fileupload.jar、commons-io.jar。

最后，导入数据库驱动包，包括 mysql-connector-java-8.0.11.jar 和 ojdbc6.jar。

在项目开发实践中，包冲突是很常见的现象。因此应该知道每个包的含义，尽量不要导入无用的冗余包，尽量避免包冲突。在书城项目的前期，不支持在 Spring 中整合持久层，所以没有导入持久层相关整合包。

8.5 配置前端控制器 DispatcherServlet

在 web.xml 中配置前端控制器。

```
<servlet>
    <servlet-name>aa</servlet-name>
    <servlet-class>
        org.springframework.web.servlet.DispatcherServlet
    </servlet-class>
    <init-param>
        <param-name>contextConfigLocation</param-name>
        <param-value>/WEB-INF/spring-mvc.xml</param-value>
    </init-param>
    <load-on-startup>1</load-on-startup>
</servlet>
<servlet-mapping>
    <servlet-name>aa</servlet-name>
    <url-pattern>*.do</url-pattern>
</servlet-mapping>
```

这里，<servlet-name>可以任意填写。此处采用的是传统的配置模式，没有采用 Servlet 3 的配置模式。

在书城项目中，没有设计服务器异步场景，因此没有使用如下配置。

```
<async-supported>true</async-supported>
```

8.6 配置 spring-mvc.xml

在项目的 WEB-INF 目录下新建 spring-mvc.xml 文件。配置 spring-mvc.xml 文件的步骤如下。

（1）配置 schema，注意，schema 的版本是 4.3。

```
<beans xmlns="http://www.springframework.org/schema/beans"
    xmlns:xsi="http://www.w3.org/2001/XMLSchema-instance"
    xmlns:mvc="http://www.springframework.org/schema/mvc"
    xmlns:context="http://www.springframework.org/schema/context"
    xsi:schemaLocation="
    http://www.springframework.org/schema/beans
    http://www.springframework.org/schema/beans/spring-beans-4.3.xsd
    http://www.springframework.org/schema/context
    http://www.springframework.org/schema/context/spring-context-4.3.xsd
    http://www.springframework.org/schema/mvc
    http://www.springframework.org/schema/mvc/spring-mvc-4.3.xsd">
```

（2）配置组件扫描位置。

```
<context:component-scan base-package="com.icss.action" />
```

（3）配置 MVC 支持。

```
<mvc:annotation-driven />
```

（4）配置 JSP 视图的解析器和默认前缀。

```
<bean
    class="org.springframework.web.servlet.view.InternalResourceViewResolver">
    <property name="prefix" value="/WEB-INF/views/" />
</bean>
```

（5）配置异常解析器。把所有未捕获的异常统一发给 error.jsp 并显示，不能直接抛给客户端浏览器。

```
<bean  class="org.springframework.web.servlet
                .handler.SimpleMappingExceptionResolver">
    <property name="exceptionMappings">
        <props>
            <prop key="java.lang.Throwable">/error/error.jsp</prop>
            <prop key="org.springframework.web
                        .multipart.MaxUploadSizeExceededException">
                /error/OverMaxUploadSize.jsp</prop>
            </props>
    </property>
</bean>
```

（6）配置 multiPartResolver。

```
<bean id="multipartResolver" class="org.springframework.web
                            .multipart.commons.CommonsMultipartResolver">
    <property name="maxUploadSize" value="102400" />
</bean>
```

（7）配置消息转换器。

```
<mvc:annotation-driven>
    <mvc:message-converters register-defaults="true">
        <bean class="org.springframework
                    .http.converter.StringHttpMessageConverter">
            <constructor-arg value="utf-8" />
            <property name = "supportedMediaTypes">
            <list>
                <value>application/json;charset=utf-8</value>
                <value>text/html;charset=utf-8</value>
              </list>
             </property>
        </bean>
```

```
        <bean class="org.springframework.http
                    .converter.ByteArrayHttpMessageConverter"/>
        <bean class="org.springframework.http
                .converter.json.MappingJackson2HttpMessageConverter" />
    </mvc:message-converters>
</mvc:annotation-driven>
```

注意：暂时没有使用拦截器，可以在需要时再配置。

8.7 配置 log4j 日志

log4j 日志对 Spring MVC 的调试非常有帮助。在 src 下新建 log4j.properties。

```
log4j.rootLogger=INFO,BB,AA
log4j.appender.AA=org.apache.log4j.ConsoleAppender
log4j.appender.AA.layout=org.apache.log4j.SimpleLayout
log4j.appender.BB=org.apache.log4j.FileAppender
log4j.appender.BB.File=book.log
log4j.appender.BB.layout=org.apache.log4j.PatternLayout
log4j.appender.BB.layout.ConversionPattern
                =%d{yyyy-MM-dd HH:mm:ss} %l %F %p %m%n
```

系统启动后，观察控制器中的 mapping 配置是否成功。

```
INFO - Mapped "{[/back/bookadd],methods=[POST]}"
INFO - Mapped "{[/back/buyinfo]}" onto public java.lang.String
                com.icss.action.back.OrderAction.readUserBuyRecord
INFO - Mapped "{[/main]}" onto public java.lang.String
        com.icss.action.BookAction.main(org.springframework.ui.Model)
INFO - Mapped "{[/user/pay],methods=[POST]}"
INFO - Mapped "{[/user/logout]}"
```

实际开发过程中经常遇到 404 错误。检查 log4j 的输出与请求的 URL 是否一致，对于调试非常有帮助。

8.8 配置数据库连接

在 src 下新建 db.properties，配置 MySQL 8 的数据库连接（对于其他版本的 MySQL，如 MySQL 5.5，配置信息可能不同）。

```
driver=com.mysql.cj.jdbc.Driver
url=jdbc:mysql://localhost:3306/book2?useSSL=false
      &serverTimezone=UTC&allowPublicKeyRetrieval=true
username=root
password=123456
```

Oracle 与 MySQL 使用同一个配置文件，即 key 值相同，而 value 不同。当根据业务需求要切换数据库时，修改相关配置信息即可。

8.9 实现权限校验

所有游客可以访问书城主页和图书明细页，无须权限校验。对于加入购物车、商品结算、退出等，需要注册用户登录后才能操作。只有管理员才能在后台查看图书上传、用户购买记录等。

对于权限校验，统一使用过滤器处理。对于用户注册模块，统一使用的 URL 为 http://localhost:8081/BookShopSpringMvc/user/xxx.do。对于后台管理员模块，统一使用的 URL 为 http://localhost:8081/BookShopSpringMvc/back/xxx.do。/user/*表示注册用户访问的资源，/back/*表示管理员访问的资源。注册用户使用 UserFilter 校验权限。拦截所有"/user/*"的请求，只有登录成功的用户才能访问/user/* 下的资源。

```
@WebFilter(filterName="/UserFilter",urlPatterns="/user/*")
public class UserFilter implements Filter {
    public void doFilter(ServletRequest request,
                         ServletResponse response, FilterChain chain)
                        throws IOException, ServletException {
        if(request instanceof HttpServletRequest){
            HttpServletRequest req = (HttpServletRequest)request;
            Object object = req.getSession().getAttribute("user");
            if(object != null){
                    chain.doFilter(request, response);//向下访问资源
            }else{
                request.setAttribute("msg","请先登录");
                req.getRequestDispatcher("/WEB-INF/views/main/login.jsp")
                                            .forward(request, response);
            }
        }
    }
}
```

后台用户使用 AuthFilter 校验权限。拦截所有/back/*的请求，只有登录成功且身份为 ADMIN 的用户才能访问后台资源。

```
@WebFilter(filterName="/AuthFilter",urlPatterns="/back/*")
public class AuthFilter implements Filter{
    public void doFilter(ServletRequest request, ServletResponse response,
                         FilterChain chain) throws IOException, ServletException {
        if(request instanceof HttpServletRequest){
            HttpServletRequest req = (HttpServletRequest)request;
            Object object = req.getSession().getAttribute("user");
            if(object != null){
                TUser user = (TUser)object;
                if(user.getRole() == IRole.ADMIN){
                    chain.doFilter(request, response);
                }else{
                    request.setAttribute("msg","管理员才能访问");
                    req.getRequestDispatcher("/WEB-INF/views/main/login.jsp")
                                            .forward(request, response);
                }
```

```
            }else{
                request.setAttribute("msg","请先登录");
                req.getRequestDispatcher("/WEB-INF/views/main/login.jsp")
                                   .forward(request, response);
            }
        }
    }
}
```

8.10 显示主页图书列表

图 8-3 为网上书城的主页显示样式。主页的功能是从数据库中动态提取已上架的图书，用列表显示（此处暂无翻页功能）。进入主页 http://localhost:8080/BookShopSpringMvc/。

welcome you admin 购物车 退出 后台

	三国演义	
	商品价格	78.5
	出版社	[illegible]出版社
	水浒传	
	商品价格	56.0
	出版社	[illegible]出版社
	红楼梦	
	商品价格	58.0
	出版社	[illegible]出版社
	西游记	
	商品价格	89.0
	出版社	[illegible]出版社

图 8-3　网上书城的主页显示样式

实现控制器的代码如下。

```
@Controller
public class BookAction {
    @RequestMapping("/main")
    public String main(Model model) throws Exception {
    BookBiz biz = new BookBiz();
        List<TBook> books = biz.getAllBooks();
        model.addAttribute("books",books);
        return "/main/main.jsp";
    }
}
```

注意：本章只讲解控制层的代码，关于逻辑层和持久层的代码，参见项目源代码，这里不再说明。第 9 章会讲解通过 Spring 整合书城的持久层代码。

8.11 实现图书明细页

在主页中，单击图书列表中的任意图书，跳转到 http://localhost:8080/BookShopSpringMvc/detail.do?

isbn=is003，出现的图书详情页如图 8-4 所示。

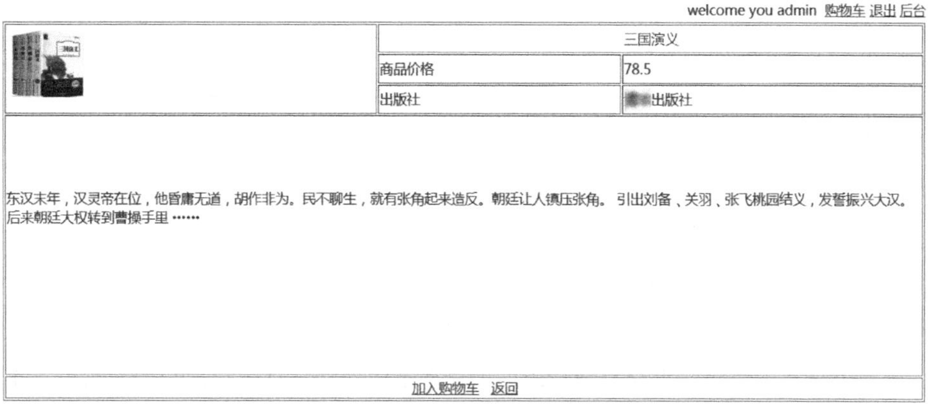

图 8-4　图书详情页

实现控制器的代码如下。

```
@Controller
public class BookAction {
    @RequestMapping("/detail")
    public String getBookDetail(String isbn,Model model) throws Exception {
        if(isbn == null || isbn.trim().equals("")) {
            model.addAttribute("msg","isbn 参数为空，请检查");
            return "/error/error.jsp";
        }
        BookBiz biz = new BookBiz();
        TBook book = biz.getBookDetail(isbn);
        model.addAttribute("book", book);
        return "/main/BookDetail.jsp";
    }
}
```

8.12　用户管理

8.12.1　用户登录

如果用户登录成功，则创建一个购物车，存放于会话中，用户对象也存于会话中，然后跳转到主页；若登录失败，则返回登录页。

实现用户登录的步骤如下。

（1）访问 http://localhost:8080/BookShopSpringMvc/login.do，进入登录页（见图 8-5）。

```
@Controller
public class UserAction {
```

```
    @GetMapping("/login")
    public String login() {
        return "/main/login.jsp";
    }
}
```

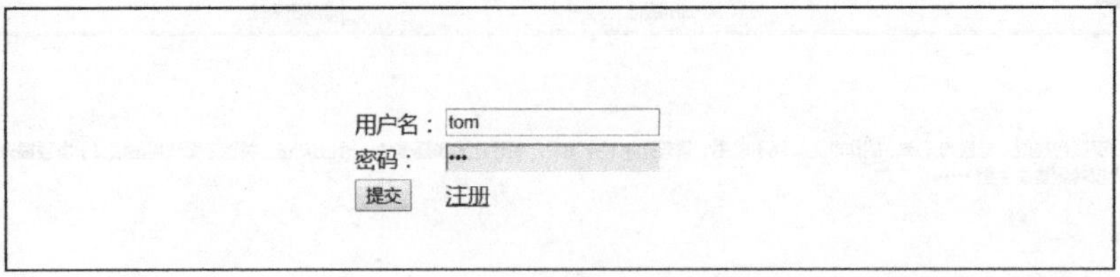

用户名：tom
密码：•••
提交 注册

图 8-5 登录页

（2）填写用户名与密码，提交登录请求。用户登录成功后，马上创建一个购物车，在后面的操作中无须判断购物车是否存在，这样会节省很多代码。如果用户登录成功，跳转到书城主页；如果用户名或密码错误，则返回登录页；如果用户名或密码输入为空，则返回登录页；如果系统错误，则跳转到错误页。

```
@PostMapping("/login")
public String login(String uname, String pwd,
                Model model, HttpSession session) throws Exception{
    if (uname == null || pwd == null || uname.equals("") || pwd.equals("")) {
        model.addAttribute("msg", "用户名或密码为空");
        return "/main/login.jsp";
    }
    UserBiz biz = new UserBiz();
    try {
        TUser user = biz.login(uname, pwd);
        if (user != null) {
            //登录成功
            session.setAttribute("user", user);
            Map<String, Integer> shopCar = new HashMap<String, Integer>();
            session.setAttribute("shopCar", shopCar);  //给用户一个购物车
            return "forward:main.do";
        } else {
            model.addAttribute("msg", "用户名或密码错误，请检查");
            return "/main/login.jsp";
        }
    } catch (InputNullExcepiton e) {
        model.addAttribute("msg", "用户名或密码为空");
        return "/main/login.jsp";
    }
}
```

8.12.2 用户退出

单击主页右上角的“退出”按钮，用户退出，清空会话，返回书城主页。具体代码如下。

```
@Controller
public class UserAction {
    @RequestMapping("/user/logout")
    public String logout(HttpServletRequest request){
        request.getSession().invalidate();
        String path = request.getContextPath();
        String basePath = request.getScheme() + "://" + request.getServerName()
                    + ":" + request.getServerPort() + path     + "/";
        return "redirect:" + basePath + "main.do";
    }
}
```

在退出时，需要注销会话，然后重定向到书城主页。使用重定向，可以改变浏览器地址栏的URL，防止重复提交。

注意：http://localhost:8080/BookShopSpringMvc/user/logout.do 这个请求路径中的/user 为所有注册用户使用的连接头，用于权限校验。

8.12.3 用户注册

游客通过注册，可以成为注册用户。注册用户可以在线购买图书。在用户注册时，要注意校验用户名的唯一性。

访问 http://localhost:8080/BookShopSpringMvc/regist.do，进入用户注册页（见图 8-6）。

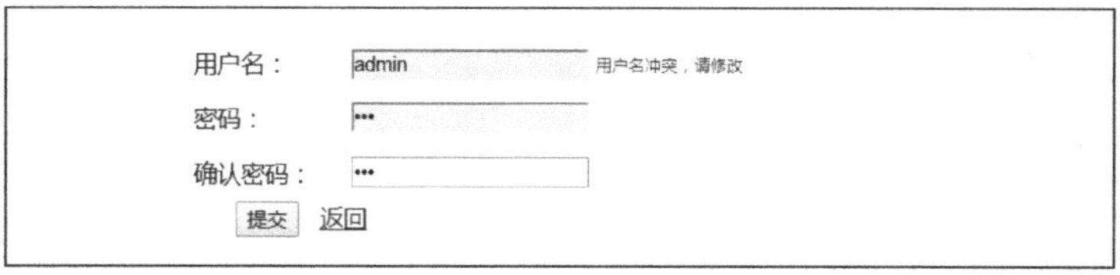

图 8-6 用户注册页

UserAction 中的用户注册方式如下。

```
@RequestMapping(value = "/regist", method = RequestMethod.GET)
public String regist() {
    return "/main/regist.jsp";
}
@RequestMapping(value = "/regist", method = RequestMethod.POST)
public String regist(String uname, String pwd, Model model) throws Exception{
    UserBiz biz = new UserBiz();
    TUser user = new TUser();
    user.setUname(uname);
    user.setPwd(pwd);
    user.setAccount(0);
    user.setRole(IRole.CUSER);
    try {
        biz.addUser(user);
```

```
        model.addAttribute("msg", "注册成功，请重新登录");
        return "/main/login.jsp";
    } catch (java.sql.SQLIntegrityConstraintViolationException e) {
        model.addAttribute("msg", "用户名冲突，请修改");
        return "/main/regist.jsp";
    }
}
```

注意：即使用 AJAX 做了用户名唯一性校验，也会有用户执意提交，这会导致用户名冲突，因此再次使用 SQLIntegrityConstraintViolationException 进行了异常校验。用户名冲突的异常不能用 error.jsp 页处理，应该在注册页友好地提示。

8.12.4 用户名校验

通过用户名文本框的 onblur 或 onkeyup 事件，触发 AJAX 请求，检查输入的用户名是否与库中的用户名冲突，然后在<span>中友好提示。

下面的代码使用 jQuery 发送 AJAX 请求。

（1）在脚本中提交用户名校验请求。

```
function unameValid(){
    var uname = $('#uname').val();
    if(uname != ""){
        var destUrl = "<%=basePath%>checkUname.do?uname=" + uname;
          $.ajax({
               type : "GET",
               url : destUrl,
               dataType : "text",
               timeout : 3000,
               success : function(msg) {
                    if (msg == 0) {
                         document.getElementById("unameAlert")
                                        .innerHTML = "用户名可用";
                    } else if (msg == 1) {
                         document.getElementById("unameAlert")
                                        .innerHTML = "用户名冲突，请修改";
                    } else if (msg == -1) {
                         document.getElementById("unameAlert")
                                        .innerHTML = "用户名为空";
                    }else if (msg == -2) {
                         document.getElementById("unameAlert")
                                        .innerHTML = "系统异常，请检查";
                    }
               },
               error : function() {
                    alert("连接超时，请重新连接");
               }
          });
    }else{
```

```
            document.getElementById("unameAlert").innerHTML = "用户名为空";
        }
    }
```

（2）在输入框的 onblur="..."中触发 unameValid( )。

```
<form action="<%=basePath%>regist.do"
    id="myform" onsubmit="return checkUserInfo()" method="post">
<table border="0" cellpadding="0" cellspacing="0" align="center">
    <tr>
        <td width="107" height="36">用户名：</td>
        <td width="524">
            <INPUT name="uname" id="uname" type="text"
                                maxlength="16" onblur="unameValid()">
            <span id="unameAlert" style="color: red; font-size: 8px"></span>
        </td>
    </tr>
</table>
</form>
```

（3）控制器使用@ResponseBody 响应 AJAX 请求。

```
@RequestMapping("/checkUname")
@ResponseBody
public int checkUname(String uname) {
    int ret;
    if(uname==null || uname.trim().equals("")) {
        return -1;
    }
    UserBiz biz = new UserBiz();
    try {
        boolean bRet = biz.isHaveUserName(uname);
        if (bRet) {
            ret = 1; //用户名冲突
        } else {
            ret = 0; //用户名不冲突
        }
    } catch (Exception e) {
        ret = -2;
    }
    return ret;
}
```

8.13 购物车实现

8.13.1 购物车设计

书城系统的购物车采用会话存储，数据结构为 Map<String, Integer>，键为 isbn，Integer 表示购买数量。

购物车中没有直接存储 TBook 实体，因为当用户数量庞大时，服务器会话占用的空间非常大，所以不能在会话中存储大数据对象。

当需要显示购物车中的图书信息时，需要通过 ISBN 直接到数据库中提取图书详细信息。

8.13.2 我的购物车

在用户浏览图书时，可以把选中的图书先添加到购物车中。已经登录的用户随时可以查看购物车（http://localhost:8080/BookShopSpringMvc/user/shopcar.do）中的数据（见图 8-7）。

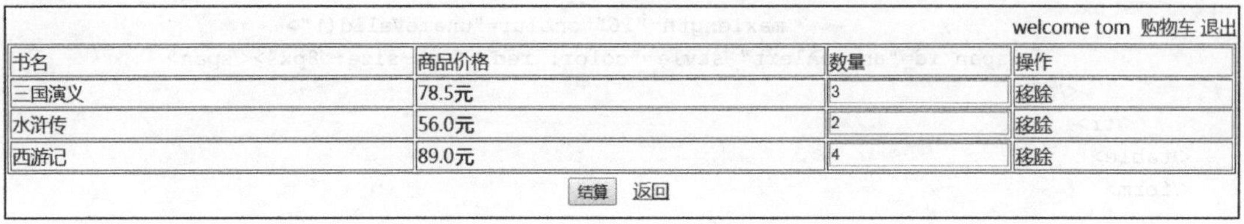

welcome tom 购物车 退出

书名	商品价格	数量	操作
三国演义	78.5元	3	移除
水浒传	56.0元	2	移除
西游记	89.0元	4	移除

结算 返回

图 8-7 购物车中的数据

因为用户登录成功时已经创建了一个购物车，所以这里直接通过@SessionAttribute Map<String, Integer> shopCar 提取购物车中的数据。

```
@Controller
@RequestMapping("/user")
public class ShopCarAction {
    @RequestMapping("/shopcar")
    public String getShopCar(@SessionAttribute Map<String, Integer> shopCar,
                        Model model) throws Exception {
        if(shopCar.size() == 0) {
            return "/main/ShopCar.jsp";
        }
        BookBiz biz = new BookBiz();
        List<TBook> books = biz.getBooks(shopCar.keySet());
        model.addAttribute("books", books);
        return "/main/ShopCar.jsp";
    }
}
```

8.13.3 加入购物车

在图书详情页，可以把图书添加到购物车中（对于未登录的用户，转向登录页，即 http://localhost:8080/BookShopSpringMvc/user/addShopcar.do?isbn=is002）。

具体代码如下。

```
@RequestMapping("/addShopcar")
public String addShopCar(String isbn,
```

```
        @SessionAttribute Map<String, Integer> shopCar) throws Exception {
    if (isbn == null) {
        throw new RuntimeException("isbn 为空");
    }
    shopCar.put(isbn, 1);
    return "forward:/user/shopcar.do";
}
```

8.13.4 移除购物车

单击购物车中的“移除”按钮，可以移除购物车中的某个商品。具体代码如下。

```
@RequestMapping("/removeShopcar")
public String removeShopCar(String isbn,
                            @SessionAttribute Map<String, Integer>
                            shopCar) throws Exception {
    if (isbn == null) {
        throw new RuntimeException("isbn 为空");
    }
    shopCar.remove(isbn);
    return "forward:/user/shopcar.do";
}
```

8.14 用户付款

8.14.1 结算

在购物车中，填写图书购买数量，然后单击“结算”按钮（见图 8-8）。

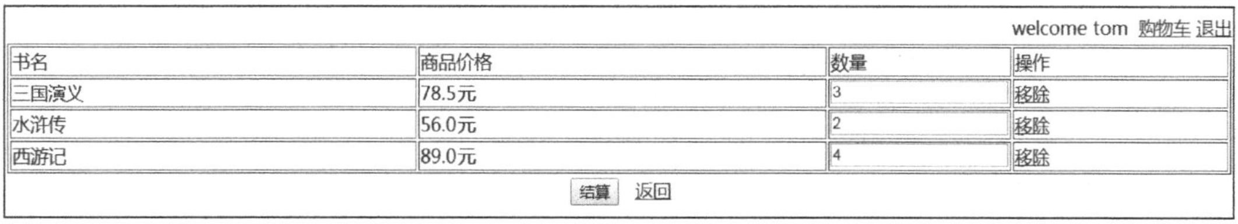

welcome tom 购物车 退出

书名	商品价格	数量	操作
三国演义	78.5元	3	移除
水浒传	56.0元	2	移除
西游记	89.0元	4	移除

结算 返回

图 8-8 结算

为了在控制器接收到每本书的数量，必须要知道每个输入框的 name。注意，每个文本框的 name 是按照 ISBN 动态设置的，如 name="${bk.isbn}"。具体代码如下。

```
<table border="1" width=100%>
    <tr>
        <td>书名</td>
        <td>商品价格</td><td width="5%">数量</td><td>操作</td>
    </tr>
    <c:forEach var="bk" items="${books}">
        <tr><td>${bk.bname}</td>
```

```
        <td>${bk.price}</td>
        <td ><input type="text"  name="${bk.isbn}" value="1" /></td>
        <td>
            <a href="<%=basePath%>user/removeShopcar.do?isbn=${bk.isbn}">
            移除</a>
        </td>
      </tr>
  </c:forEach>
</table>
```

会话中存储的购物车数据与页面中提交的购物车数据必须要保持一致。因此，遍历会话中的购物车数据，并用 **name** 对应的 **ISBN** 即可获取所有用户上传的图书购买数量。提取图书购买数量后，把购买数量存储于会话中。具体代码如下。

```
@RequestMapping("/checkout")
public String checkout(@SessionAttribute Map<String, Integer> shopCar,
                       HttpServletRequest request) throws Exception {
    BookBiz biz = new BookBiz();
    List<TBook> books = biz.getBooks(shopCar.keySet());
    for (TBook book : books) {
        //从购物车中获取 ISBN，然后通过 ISBN 在请求参数中接收用户输入的购买数量
        String value = request.getParameter(book.getIsbn());
        //默认值为 1，若输入错误，认为输入的值是 1
        int bookCount = 1;
        book.setBuyCount(1);
        try {
            if (value != null && !value.trim().equals("")) {
                bookCount = Integer.parseInt(value);
                book.setBuyCount(bookCount);   //存储图书购买数量
                //把书的购买数量存储到会话中
                shopCar.put(book.getIsbn(), bookCount);
            }
        } catch (Exception e) {
            Log.logger.error("购买图书的数量应该为整数:" + e.getMessage());
        }
    }
    double allMoney = 0;
    for (TBook bk : books) {
        allMoney = allMoney + bk.getPrice() * bk.getBuyCount();
    }
    request.setAttribute("books", books);
    request.setAttribute("allMoney", allMoney);
    return "/main/Checkout.jsp";
}
```

8.14.2 付款

如果选定图书的总价大于账户余额，则不能显示付款按钮（见图 8-9）。

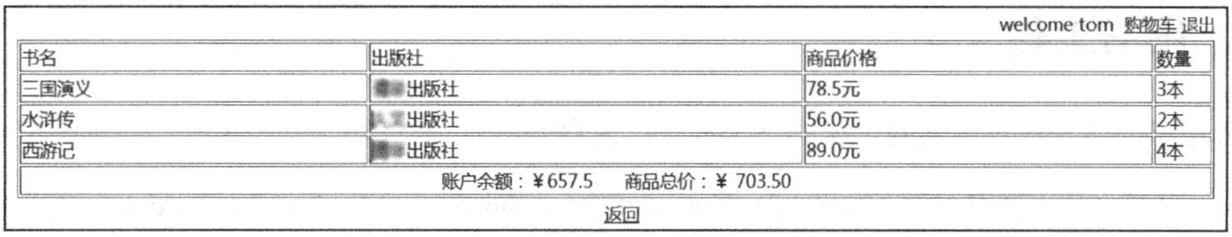

welcome tom 购物车 退出

书名	出版社	商品价格	数量
三国演义	出版社	78.5元	3本
水浒传	出版社	56.0元	2本
西游记	出版社	89.0元	4本
账户余额：¥657.5　商品总价：¥ 703.50			

返回

图 8-9　账户余额不足

如果余额充足，则可以付款（见图 8-10）。

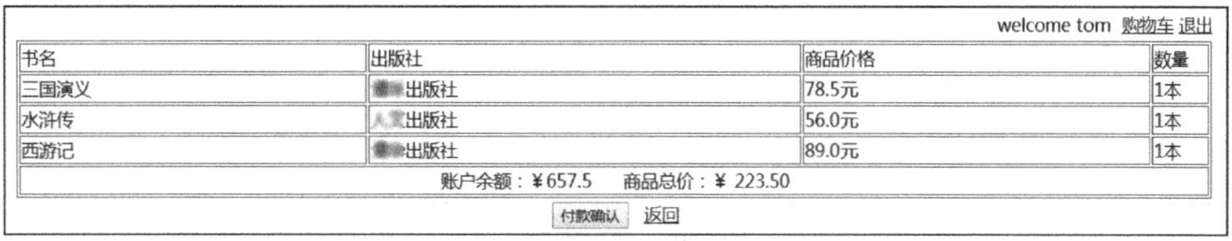

welcome tom 购物车 退出

书名	出版社	商品价格	数量
三国演义	出版社	78.5元	1本
水浒传	出版社	56.0元	1本
西游记	出版社	89.0元	1本
账户余额：¥657.5　商品总价：¥ 223.50			

付款确认　返回

图 8-10　付款

具体代码如下。

```
@PostMapping("/pay")
public String pay(@SessionAttribute TUser user,
                @SessionAttribute Map<String, Integer> shopCar,
                double allMoney,HttpServletRequest request) throws Exception {
    if(user.getAccount()>=allMoney) {
        UserBiz biz = new UserBiz();
        biz.buyBooks(user.getUname(), allMoney,shopCar);
        shopCar.clear();              //付款结束，清空购物车
        user.setAccount(user.getAccount()-allMoney);   //更新会话中的账户余额
        String path = request.getContextPath();
        return "redirect:" + path + "user/payok.do?allMoney=" + allMoney ;
    }else {
        request.setAttribute("msg", "余额不足，请先充值");
        return "/error/error.jsp";
    }
}
```

如果付款成功（见图 8-11），则需要清空购物车，并更新账户余额。

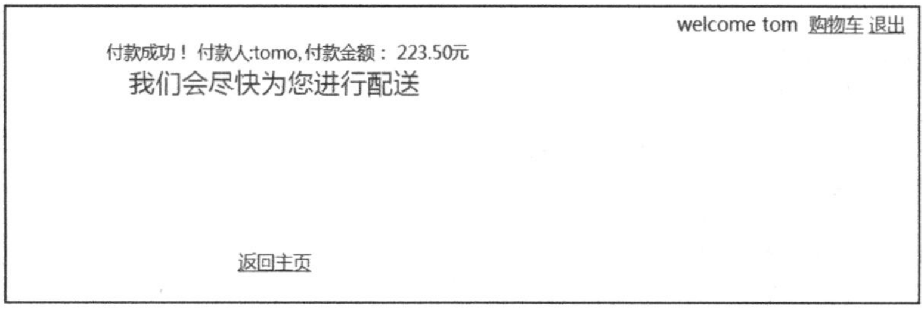

welcome tom 购物车 退出

付款成功！付款人:tomo,付款金额：223.50元

我们会尽快为您进行配送

返回主页

图 8-11　付款成功

8.15 图书上传

管理员在后台管理图书的上架和下架操作。具体操作如下。

（1）从后台进入图书上传页 http://localhost:8080/BookShopSpringMvc/back/bookadd.do（见图 8-12）。

图 8-12　图书上传页

（2）选择上传图片。当图片大于 100KB 时，异常解析器把用户带入指定错误页。注意，明确的异常类型不能在通用的 error.jsp 中显示。

```
<bean id="multipartResolver"
      class="org.springframework.web.multipart.commons.CommonsMultipartResolver">
          <property name="maxUploadSize" value="102400" />
</bean>
<bean
    class="org.springframework.web.servlet.handler.SimpleMappingExceptionResolver">
    <property name="exceptionMappings">
        <props>
            <prop key="java.lang.Throwable">/error/error.jsp</prop>
            <prop key="org.springframework
                .web.multipart.MaxUploadSizeExceededException">
                /error/OverMaxUploadSize.jsp</prop>
        </props>
    </property>
</bean>
```

（3）在绑定属性前，设置日期格式，禁止直接绑定 pic 图片。对于价格输入框，在视图层已经约束了浮点格式，但是如果为空，则禁止属性绑定。属性绑定的原理在后面讲解。

```
@Controller
@RequestMapping("/back")
public class BookAddAction {
    @InitBinder
    public void initBinder(WebDataBinder binder,String price) {
       binder.registerCustomEditor(String.class,
            new StringTrimmerEditor(true));
       binder.registerCustomEditor(Date.class,
            new CustomDateEditor(new SimpleDateFormat("MM/dd/yyyy"), true));
       if(price==null || price.equals("")) {
           binder.setDisallowedFields("price");
```

```
        }
        binder.setDisallowedFields("pic");
    }
}
```

（4）接收图片。

```
@RequestMapping(value = "/bookadd", method = RequestMethod.GET)
public String addBook() {
    return "/back/BookAdd.jsp";
}
@RequestMapping(value = "/bookadd", method = RequestMethod.POST)
public String addBook(TBook book,
      @RequestParam("pic") MultipartFile file, Model model) throws Exception {
    byte[] bytes = file.getBytes();
    book.setPic(bytes);
    BookBiz biz = new BookBiz();
    biz.addBook(book);
    model.addAttribute("msg", book.getBname() + "--录入成功");
    return "/back/BookAdd.jsp";
}
```

（5）编写视图层的代码。对于表单的属性，必须设置 enctype="multipart/form-data"。

```
<form action="<%=basePath%>back/bookadd.do" method="post" enctype="multipart/form-data">
    <table border="0" width=60% align="center">
        <tr><td>书号 ISBN</td><td><input type="text" name="isbn" value="${isbn}"/></td></tr>
         <tr><td>书名</td><td><input type="text" name="bname" value="${bname}" />
                             <span id="NameNull"></span></td></tr>
         <tr><td>出版社</td><td><input type="text" name="press" value="${press}"/></td></tr>
         <tr><td>出版日期</td><td><input type="text" name="pdate" class="easyui-datebox"/>
             </td></tr>
         <tr><td>价格</td><td><input type="text" name="price"
                      class="easyui-numberbox" value=0 precision="2"/></td></tr>
         <tr><td>图片上传</td><td><input type="file" name="pic"/></td></tr>
         <tr><td colspan=2 align=center><input type=submit value=提交 />${msg}</td></tr>
    </table>
</form>
```

8.16 查询用户购买记录

管理员有权在后台查看所有用户的购买记录，即订单及其详情。具体步骤如下。

（1）访问 http://localhost:8080/BookShopSpringMvc/back/buyinfo.do，查询用户购买记录（见图 8-13），第一次进入该页面时，URL 中没有带参数。在翻页查询时，URL 中需要指定查询条件，如 http://localhost:8080/ BookShopSpringMvc/back/buyinfo.do?page=2&uname=tom&beginDate=11/12/2019&endDate=11/28/2019。

用户名	书名--购买数量(本)	书单价	出版社	作者	购买日期	总付款
tom	红楼梦--2	58.0	[illegible]出版社	托尔斯泰	2019-11-25 08:00:00.0	194.5元
tom	三国演义--1	78.5	[illegible]出版社	托尔斯泰	2019-11-25 08:00:00.0	194.5元
tom	红楼梦--1	58.0	[illegible]出版社	托尔斯泰	2019-11-25 08:00:00.0	147.0元
tom	西游记--1	89.0	[illegible]出版社	托尔斯泰	2019-11-25 08:00:00.0	147.0元
tom	西游记--1	89.0	[illegible]出版社	托尔斯泰	2019-11-25 08:00:00.0	246.0元
tom	三国演义--2	78.5	[illegible]出版社	托尔斯泰	2019-11-25 08:00:00.0	246.0元
tom	水浒传--1	56.0	[illegible]出版社	托尔斯泰	2019-11-25 08:00:00.0	223.5元
tom	西游记--1	89.0	[illegible]出版社	托尔斯泰	2019-11-25 08:00:00.0	223.5元

图 8-13 购买记录

（2）定义属性绑定。定制日期绑定的格式，允许日期为空。

```
@Controller
@RequestMapping("/back")
public class OrderAction {
    @InitBinder
    public void initBinder(WebDataBinder binder) {
        binder.registerCustomEditor(String.class,
                new StringTrimmerEditor(true));
        binder.registerCustomEditor(Date.class,
            new CustomDateEditor(new SimpleDateFormat("MM/dd/yyyy"), true));
    }
}
```

（3）进行查询。注意，输入参数 page 在第一次进入订单详情页时可能为空，因此默认值设置为 1。而开始时间和结束时间不能使用 String 类型，必须要使用 java.util.Date 类型。

```
@RequestMapping("/buyinfo")
public String readUserBuyRecord(@RequestParam(name = "page", defaultValue = "1")
        int page,String uname,Date beginDate, Date endDate, Model model)
        throws Exception {
    UserBiz biz = new UserBiz();
    TurnPage tp = new TurnPage();
    tp.rowsOnePage = 8;
    if (page < 1)
        page = 1;
    tp.currentPage = page;
    List<BuyRecord> records = biz.getUserBuyRecord(uname, beginDate, endDate, tp);
    model.addAttribute("records", records);
    model.addAttribute("uname", uname);
    SimpleDateFormat sd = new SimpleDateFormat("MM/dd/yyyy");
    if (beginDate != null) {
        model.addAttribute("beginDate", sd.format(beginDate));
    }
    if (endDate != null) {
```

```
            model.addAttribute("endDate", sd.format(endDate));
        }
        model.addAttribute("CurrentPageNo", tp.currentPage);
        model.addAttribute("maxPageNo", tp.allPages);
        model.addAttribute("RecordAllCount", tp.allRows);
        return "/back/BuyRecord.jsp";
    }
```

（4）编写视图层的代码。注意，对于每个翻页按钮，必须要有页码以及 3 个查询条件（用户名、开始时间、结束时间）等参数。对于跳转按钮，此处未实现。

```
    <tr>
    <td>总记录数：${RecordAllCount}</td>
    <td>总页数：${maxPageNo}</td>
    <td>当前页：${CurrentPageNo}</td>
    <td>
        <a href="<%=basePath%>back/buyinfo.do?page=1&uname=${uname}
                &beginDate=${beginDate}&endDate=${endDate}">首页|</a>
        <a href="<%=basePath%>back/buyinfo.do?page=${CurrentPageNo-1}
            &uname=${uname}&beginDate=${beginDate}&endDate=${endDate}">《前页|
        </a>
        <a href="<%=basePath%>back/buyinfo.do?page=${CurrentPageNo+1}
            &uname=${uname}&beginDate=${beginDate}&endDate=${endDate}">后页》|
        </a>
        <a href="<%=basePath%>back/buyinfo.do?page=${maxPageNo}
                &uname=${uname}&beginDate=${beginDate}&endDate=${endDate}">末页|
        </a>
    </td>
    </tr>
```

第 9 章

通过 Spring 整合书城项目

使用 Spring 整合书城项目，即用 Spring 整合 JDBC 项目，实现声明性事务管理。

9.1 配置整合环境

配置整合环境的步骤如下。

（1）导入 Spring 持久层整合包 spring-jdbc.jar 和 spring-tx.jar。

（2）配置 Schema，支持 tx 命名空间。

```
<beans xmlns="http://www.springframework.org/schema/beans"
    xmlns:xsi="http://www.w3.org/2001/XMLSchema-instance"
    xmlns:mvc="http://www.springframework.org/schema/mvc"
    xmlns:tx="http://www.springframework.org/schema/tx"
    xmlns:context="http://www.springframework.org/schema/context"
    xsi:schemaLocation="
    http://www.springframework.org/schema/beans
    http://www.springframework.org/schema/beans/spring-beans-4.3.xsd
    http://www.springframework.org/schema/context
    http://www.springframework.org/schema/context/spring-context-4.3.xsd
    http://www.springframework.org/schema/tx
    http://www.springframework.org/schema/tx/spring-tx-4.3.xsd
    http://www.springframework.org/schema/mvc
    http://www.springframework.org/schema/mvc/spring-mvc-4.3.xsd">
</beans>
```

（3）配置数据源，删除 db.properties 文件。

```
<bean id="dataSource"
        class="org.springframework.jdbc.datasource.DriverManagerDataSource">
    <property name="driverClassName" value="com.mysql.cj.jdbc.Driver" />
    <property name="url"
            value="jdbc:mysql://localhost:3306/book2?useSSL=false
              &amp;serverTimezone=UTC&amp;allowPublicKeyRetrieval=true" />
    <property name="username" value="root" />
    <property name="password" value="123456" />
</bean>
```

（4）配置事务管理器。

```
<bean id="txManager"
        class="org.springframework.jdbc.datasource.DataSourceTransactionManager">
    <property name="dataSource" ref="dataSource" />
</bean>
<tx:annotation-driven transaction-manager="txManager" />
```

（5）配置组件扫描。

```
<context:component-scan base-package="com.icss.action" />
<context:component-scan base-package="com.icss.biz" />
<context:component-scan base-package="com.icss.dao" />
```

9.2 配置业务 Bean

常用的 Bean 对象是业务逻辑对象和持久层对象。Spring 倾向于把所有的业务逻辑对象和持久层对象都作为单例 Bean 进行统一管理，这样性能最高。

配置业务 Bean 的具体步骤如下。

（1）配置服务层和持久层的 Bean。

```
@Service("bookBiz")
public class BookBiz {}

@Service("userBiz")
public class UserBiz {}

@Repository("bookDao")
public class BookDaoMysql
            extends BaseDao implements IBookDao{}
@Repository("userDao")
public class UserDaoMysql
             extends BaseDao implements IUserDao{}
```

（2）所有持久层对象继承自 JdbcDaoSupport，把 JdbcDaoSupport 注入 dataSource。

```
public class BaseDao extends JdbcDaoSupport {
    public Connection openConnection() {
        return this.getConnection();
```

```
    }
    @Autowired
    public void setDataSource(org.springframework.jdbc
                .datasource.DriverManagerDataSource dataSource) {
        super.setDataSource(dataSource);
    }
}
```

9.3 配置依赖注入

优先使用依赖注入对象，尽量不要自己去容器中查找 Bean。使用依赖，可以解决 Bean 对象之间的协作与装配问题。

配置依赖注入的步骤如下。

（1）配置控制器依赖的服务层对象。此处只描述了两个控制器，关于其他控制器，参见项目源代码。

```
@Controller
public class BookAction {
    @Autowired
    private BookBiz bookBiz;
}
@Controller
@RequestMapping("/user")
public class PayAction {
    @Autowired
    private BookBiz bookBiz;
    @Autowired
    private UserBiz userBiz;
}
```

（2）配置服务层依赖的持久层对象。

```
@Service("bookBiz")
public class BookBiz {
    @Autowired
    private IBookDao bookDao;
}
@Service("userBiz")
public class UserBiz {
    @Autowired
    private IUserDao userDao;
}
```

9.4 配置声明性事务

优先使用注解的方式配置事务，这样更简单、更直观。注意，只有服务层能够使用 @Transactional 注解，持久层和控制层都不允许事务控制。

为了通过付款购买图书，需要配置事务。

```java
@Transactional(rollbackFor=Throwable.class)
public void buyBooks(String uname, double allMoney,
                     Map<String, Integer> shopCar) throws Exception {
    if(uname== null || uname.equals("")) {
        throw new Exception("用户名不能为空");
    }
    if(allMoney<=0) {
        throw new Exception("金额错误");
    }
    if(shopCar==null || shopCar.size()==0) {
        throw new Exception("购物车为空");
    }
    userDao.updateUserAccount(uname, -allMoney);
    userDao.addBuyRecord(uname, allMoney, shopCar);
}
```

对于其他只需要数据库、不需要事务操作的业务方法，配置只读事务（确保数据库连接的释放）。

```java
@Transactional(readOnly=true)
public List<TBook> getAllBooks() throws Exception{  }

@Transactional(readOnly=true)
public void addBook(TBook book) throws Exception{}

@Transactional(readOnly=true)
public byte[] getBookPic(String isbn) throws Exception{  }
```

9.5 处理异常

对比常用的两种异常设置，看看项目开发中使用哪种异常设置更好。

如果使用简单异常处理程序配置，则只能把指定的错误发送到约定的页面。以这种模式无法输出日志。

```xml
<bean class="org.springframework.web.servlet
          .handler.SimpleMappingExceptionResolver">
    <property name="exceptionMappings">
         <props>
              <prop key="java.lang.Throwable">/error/error.jsp</prop>
    </property>
</bean>
```

如果采用@ExceptionHandler模式的全局异常配置，则既可以转向指定的错误页，也可以在异常处理的代码中记录日志。

```java
@ControllerAdvice
public class GlobalExceptionHandler {
    @ExceptionHandler(Exception.class)
    public ModelAndView otherException(Exception e) {
        ModelAndView mv = new ModelAndView();
        mv.addObject("msg", "网络异常，请和管理员联系");
```

```
            mv.setViewName("/error/error.jsp");
            Log.logger.error(e.getMessage(),e);
            return mv;
        }
    }
```

把所有控制器未捕获的异常和 Exception 类型的异常统一抛给 GlobalExceptionHandler，这样控制器的代码就简单很多。同时增加日志信息，更加完善。

示例代码如下。

```
@RequestMapping("/shopcar")
public String getShopCar(@SessionAttribute Map<String, Integer> shopCar,
                        Model model) throws Exception {
    if (shopCar.size() == 0) {
        return "/main/ShopCar.jsp";
    }
    List<TBook> books = bookBiz.getBooks(shopCar.keySet());
    model.addAttribute("books", books);
    return "/main/ShopCar.jsp";
}
```

由于转向位置不同，自定义异常需要单独处理。如果用户名或密码为空，则不能在 error.jsp 中抛出异常，应该在 login.jsp 中显示。InputNullExcepiton 异常为自定义异常，应该在具体控制器方法中捕获，不能抛给 GlobalExceptionHandler。

示例代码如下。

```
@PostMapping("/login")
public String login(String uname, String pwd,
            Model model, HttpSession session) throws Exception {
    try {
        TUser user = userBiz.login(uname, pwd);
        if (user != null) {
            session.setAttribute("user", user);
            Map<String, Integer> shopCar = new HashMap<String, Integer>();
            session.setAttribute("shopCar", shopCar);
            return "forward:main.do";
        } else {
            model.addAttribute("msg", "用户名或密码错误，请检查");
            return "/main/login.jsp";
        }
    } catch (InputNullExcepiton e) {
        model.addAttribute("msg", "用户名或密码为空");
        return "/main/login.jsp";
    }
}
```

已知类型的异常需要单独处理。如用户注册时的主键冲突，应该在 regist.jsp 中显示。把 SQLIntegrityConstraintViolationException 异常放到 GlobalExceptionHandler 去处理显然是不合适的，因为在 GlobalExceptionHandler 中只能转到一个固定的异常提醒页，这无法在用户名冲突时友好地提示。

```
@RequestMapping(value = "/regist", method = RequestMethod.POST)
public String regist(String uname, String pwd, Model model) throws Exception {
    TUser user = new TUser();
    user.setUname(uname);
    user.setPwd(pwd);
    user.setAccount(0);
    user.setRole(IRole.CUSER);
    try {
        userBiz.addUser(user);
        model.addAttribute("msg", "注册成功，请重新登录");
        return "/main/login.jsp";
    } catch (java.sql.SQLIntegrityConstraintViolationException e) {
        model.addAttribute("msg", "用户名冲突，请修改");
        return "/main/regist.jsp";
    }
}
```

AJAX 的异常由服务器返回错误码，在客户端显示会更加友好。如果需要传递给客户端更多的异常信息，则返回类型可以使用 HttpEntity。

```
@RequestMapping("/checkUname")
@ResponseBody
public int checkUname(String uname){
    int ret;
    if(uname==null || uname.trim().equals("")) {
        return -1;
    }
    try {
        boolean bRet = userBiz.isHaveUserName(uname);
        if (bRet) {
            ret = 1;
        } else {
            ret = 0;
        }
    } catch (Exception e) {
        ret = -2;            //异常
        Log.logger.error(e.getMessage());
    }
    return ret;
}
```

9.6 常见错误

在使用 Spring 与 Spring MVC 时，会遇到一些常见错误。本节介绍这些常见错误。

1. 没有使用注入的依赖对象，而通过 new 重新创建了一个对象

例如，在下面的示例中，bookBiz = new BookBiz( )，因为 bookBiz 的引用指向的不是 Bean，

这会导致 bookBiz 依赖的 bookDao 对象为 null。

```
@Controller
public class BookAction {
    @Autowired
    private BookBiz bookBiz;
    @RequestMapping("/main")
    public String main(Model model) throws Exception {
        BookBiz bookBiz = new BookBiz();
        List<TBook> books = bookBiz.getAllBooks();
        model.addAttribute("books", books);
        return "/main/main.jsp";
    }
}
```

在运行时的错误提示如下。

```
ERROR - java.lang.NullPointerException
DEBUG - Method [com.icss.action.GlobalExceptionHandler.otherException] returned
    [ModelAndView: reference to view with name '/error/error.jsp'; model is {msg=网络异常，
    请和管理员联系}]
at com.icss.biz.BookBiz.getAllBooks(BookBiz.java:26)
at com.icss.action.BookAction.main(BookAction.java:26)
```

同理，如果在服务层通过 new 重新创建了 bookDao 对象，则也会出现错误提示。

```
@Service("bookBiz")
public class BookBiz {
    @Autowired
    private IBookDao bookDao;
    @Transactional(readOnly=true)
    public List<TBook> getAllBooks() throws Exception{
        IBookDao bookDao = new BookDaoMysql();
        return bookDao.getAllBooks();
    }
}
```

错误提示如下。

```
ERROR - java.lang.IllegalArgumentException: No DataSource specified
at org.springframework.util.Assert.notNull(Assert.java:134)
at org.springframework.jdbc.datasource.
        DataSourceUtils.doGetConnection(DataSourceUtils.java:97)
```

这是因为主动通过 new 创建的 bookDao，会导致 BaseDao 中的 JdbcDaoSupport 无法注入数据源。

2．服务层没有使用@Transactional(readOnly=true)

如果服务层没有使用@Transactional(readOnly=true)，将会导致数据库连接没有及时释放，造成严重的性能问题。注意，这个错误很隐蔽，只有在高并发的性能测试中才能发现。

```
//@Transactional(readOnly=true)
public List<TBook> getAllBooks() throws Exception{
    return bookDao.getAllBooks();
}
```

3. 从数据源中直接获取数据库连接

从数据源中直接获取数据库连接与从 JdbcDaoSupport 直接调用 getConnection( )含义完全不同。前者从数据源获取新的数据库连接，后者从 Spring 上下文获取数据库环境。

```
public Connection openConnection() throws Exception{
        //return this.getConnection();
        return this.getDataSource().getConnection();
}
```

第 10 章
通过 Spring 进行数据校验

10.1 数据校验的概念

在企业级项目中，数据校验是一个非常重要的工作。那么，在一个典型的 MVC 架构（见图 10-1）中，在哪里进行数据校验呢？

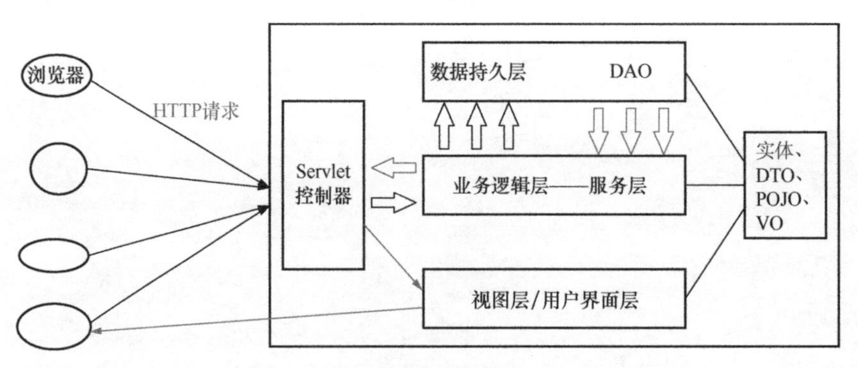

图 10-1 MVC 架构

最稳妥的方案当然是在每一层都进行数据校验，但是这样做工作量太大了。推荐的方案如下。

- 对于浏览器的视图层，在用户输入数据时必须要进行数据校验。

- 服务层代码不仅被控制器调用，还可能被其他服务对象调用，因此必须在服务层做数据校验。
- 把实体对象传给 Bean 对象，进行业务操作前要进行数据校验。

基于实体对象的数据校验是 JSR 中的重要规范，参见 JSR 的相关规范。

- JSR 330：Dependency Injection for Java 1.0。
- JSR 303：Bean Validation 1.0。
- JSR 349: Bean Validation 1.1。

数据校验的主要实现者是 Hibernate，不是 Spring。因此，在数据校验项目中必须要导入 hibernate-validator.jar。Spring 不负责数据校验的实现，它为支持 JSR 330 和 JSR 349 提供整合接口，实现由 Hibernate 来负责。必须要清楚 Java EE、Spring、Hibernate 三者之间的关系。Java EE 提供的是数据校验规范，Hibernate 提供的是数据校验实现，Spring 提供的是整合方案。

Java EE 容器（如 Tomcat 8）整合了 Hibernate-validator.jar 后，也可以实现数据校验工作。这种基于容器的数据校验方案是 Java EE 的官方推荐模式。

Spring 提供的数据校验工作，实际是把原来由 Java EE 容器实现的功能用 Spring 的容器代替了。Spring 整合数据校验工作后，代码更加简单了，这就是 Spring 的神奇之处。

10.2 在 Spring 中实现数据校验

10.2.1 Validator 接口

Spring 中的数据校验需要实现 Validator 接口。这个接口与任何基础架构或环境无关。Validator 接口与对象在 Web 层、数据访问层或其他层的校验并不耦合，可用于任何层。

Validator 接口与 Errors 协同工作，在 Spring 做校验的时候，它会将所有校验的错误汇总到 Errors 对象中。在 Spring Web MVC 中，可以使用<spring:bind/>标签来检查错误消息。当然，也可以自行处理错误。

```
package org.springframework.validation;
public interface Validator {
    boolean supports(Class<?> clazz);
    void validate(Object target, Errors errors);
}
```

要实现 org.springframework.validation.Validator 接口中的两个方法，需要为 Person 类添加校验行为。

- supports(Class)：表示这个 Validator 是否支持 Person 类的实例。
- validate(Object, org.springframework.validation.Errors)：对提供的对象进行校验，并将校验

的错误注册到传入的 Errors 对象中。

示例 10-1：实现 Validator 接口，校验人名是否为空，年龄是否介于 0～110 岁。

```
public class Person {
    private int age;
    private String name;
}
```

实现一个 Validator 接口比较简单，尤其是当学会了 Spring 所提供的 ValidationUtils 以后。我们一起来看一下如何才能创建一个 Validator 接口。具体代码如下。

```
public class PersonValidator implements Validator {
    public boolean supports(Class clazz) {
        return Person.class.equals(clazz);
    }
    public void validate(Object obj, Errors e) {
        ValidationUtils.rejectIfEmpty(e, "name", "名字不能为空");
        Person p = (Person) obj;
        if (p.getAge() < 0) {
            e.rejectValue("age", "年龄不能为负值");
        } else if (p.getAge() > 110) {
            e.rejectValue("age", "年龄不能超过 110");
        }
    }
}
```

测试代码如下。

```
public static void main(String[] args) {
    Person p = new Person();
    p.setAge(150);
    p.setName("");
    PersonValidator pv = new PersonValidator();
    Errors e = new MapBindingResult(new HashMap<>(),"person");
    pv.validate(p, e);
    List<ObjectError> allError = e.getAllErrors();
    for(ObjectError error :allError) {
        System.out.println(error.getObjectName() + "," + error.getCode());
    }
}
```

当输入的姓名或年龄不符合要求时，错误消息会存储到 Errors 对象中。

示例 10-2：实现 Validator 接口，完成用户名和密码校验。用户名与密码不能为 null，也不能为空字符串；密码的长度不能少于 6 位。

```
public class UserLoginValidator implements Validator {
    private static final int MINIMUM_PASSWORD_LENGTH = 6;
    public boolean supports(Class clazz) {
     return UserLogin.class.isAssignableFrom(clazz);
    }
```

```
    public void validate(Object target, Errors errors) {
      ValidationUtils.rejectIfEmptyOrWhitespace(errors,
                                "userName", "field.required");
      ValidationUtils.rejectIfEmptyOrWhitespace(errors,
                                "password", "field.required");
      UserLogin login = (UserLogin) target;
      if (login.getPassword() != null&& login.getPassword().trim().length()
                                < MINIMUM_PASSWORD_LENGTH) {
            errors.rejectValue("password", "field.min.length",
            new Object[]{Integer.valueOf(MINIMUM_PASSWORD_LENGTH)},
              "The password must be at least [" + MINIMUM_PASSWORD_LENGTH + "]");
      }
    }
  }
```

LocalValidatorFactoryBean是JSR 303中javax.validation的主要实现类，是org.springframework.validation.Validator 的重要实现类，它同时实现了 javax.validation.ValidatorFactory 接口，通过ValidatorFactory::getValidator( )方法得到 javax.validation.Validator 对象。

API 描述如下。

```
public class LocalValidatorFactoryBean extends SpringValidatorAdapter
                        implements ValidatorFactory, ApplicationContextAware,
                        InitializingBean, DisposableBean {
    private ValidationProviderResolver validationProviderResolver;
    private MessageInterpolator messageInterpolator;
    private ConstraintValidatorFactory constraintValidatorFactory;
    private Resource[] mappingLocations;
    private final Map<String, String> validationPropertyMap;
    private ApplicationContext applicationContext;
    private ValidatorFactory validatorFactory;
    public Validator getValidator() {}
    public MessageInterpolator getMessageInterpolator() {}
    public ConstraintValidatorFactory getConstraintValidatorFactory() {}
}
<bean id="validator"
      class="org.springframework.validation.beanvalidation.LocalValidatorFactoryBean"/>
```

注意，javax.validation.Validator 接口与 org.springframework.validation.Validator 完全不同，javax.validation.Validator 源于 Java EE 的 bean-validation 规范。下面给出了几个 Validator 所属的包，不要与 Spring 的 Validator 混淆。

```
package javax.xml.bind;
public interface Validator {}

package javax.xml.validation;
public abstract class Validator {}

Package javax.validation;
public Interface Validator{}
```

10.2.2 DataBinder 类

对于 DataBinder 类，当绑定 set 属性值到目标对象时，支持 validation 和绑定结果分析。绑定过程可以定制，如设置哪些属性允许绑定，哪些属性不允许绑定，哪些属性允许为空，哪些属性可以定制等。

DataBinder 类的描述如下。

```
public class DataBinder implements PropertyEditorRegistry, TypeConverter {
    private final Object  target;
    private final String  objectName;
    private AbstractPropertyBindingResult  bindingResult;
    private SimpleTypeConverter  typeConverter;
    private String[] allowedFields;
    private String[] disallowedFields;
    private String[] requiredFields;
    private MessageCodesResolver  messageCodesResolver;
    private BindingErrorProcessor  bindingErrorProcessor;
    private final List<Validator> validators = new ArrayList<Validator>();
    protected PropertyEditorRegistry  getPropertyEditorRegistry() {}
    public BindingResult getBindingResult() {}
}
public class WebDataBinder extends DataBinder {}
```

在 Spring MVC 中，使用 WebDataBinder 绑定数据，即把 HTTP 请求中的参数动态绑定到域对象上。WebDataBinder 是 DataBinder 的子类。调用 DataBinder.getBindingResult( )，获得数据绑定结果——BindingResult。对于 Validator 接口，校验方法的错误结果存放在 Errors 对象中，BindingResult 就是 Errors 的子接口。

```
public interface BindingResult extends Errors {
    Object getTarget();
    Map<String, Object> getModel();
    PropertyEditor findEditor(String field, Class<?> valueType);
    PropertyEditorRegistry getPropertyEditorRegistry();
    String[] resolveMessageCodes(String errorCode);
}
```

示例 10-3：基于书城项目中的图书上传，使用 WebDataBinder 进行属性绑定前的校验。

具体代码如下。

```
@Controller
@RequestMapping("/back")
public class BookAddAction {
    @InitBinder
    public void initBinder(WebDataBinder binder,String price) {
      binder.registerCustomEditor(String.class,
              new StringTrimmerEditor(true));
      binder.registerCustomEditor(Date.class,
              new CustomDateEditor(new SimpleDateFormat("MM/dd/yyyy"), true));
```

```
            if(price==null || price.equals("")) {
                binder.setDisallowedFields("price");
            }
            binder.setDisallowedFields("pic");
        }
    }
```

10.2.3 BeanWrapper 接口

当实体对象通过 get( )和 set( )进行属性操作时，DataBinder 会通过调用 BeanWrapper 来实现相关操作。

这是 Spring 对 Bean 操作的低级接口，很少由用户直接调用。一般通过 BeanFactory 和 DataBinder 隐式调用这个接口。

```
public interface BeanWrapper extends ConfigurablePropertyAccessor {
    Class<?>  getWrappedClass();
    Object   getWrappedInstance();
}
```

示例 10-4：通过 BeanWrapper 进行实体的属性设置与读取。

具体实现方式如下。

（1）编写 Company 实体类的代码。

```
public class Company {
    private String name;
    private Employee director;
    public String getName() {
        return this.name;
    }
    public void setName(String name) {
        this.name = name;
    }
    public Employee getDirector() {
        return director;
    }
    public void setDirector(Employee director) {
        this.director = director;
    }
}
```

（2）编写 Employee 实体类的代码。

```
public class Employee {
    private String name;
    private float salary;
    public String getName() {
        return this.name;
    }
```

```
    public void setName(String name) {
        this.name = name;
    }
    public float getSalary() {
        return salary;
    }
    public void setSalary(float salary) {
        this.salary = salary;
    }
}
```

（3）使用 BeanWrapper 进行实体的属性设置与读取。

```
public static void main(String[] args) {
    BeanWrapper company = new BeanWrapperImpl(new Company());
    PropertyValue value = new PropertyValue("name", "中软国际北京ETC");
    company.setPropertyValue(value); //设置公司名字

    BeanWrapper jim = new BeanWrapperImpl(new Employee());
    jim.setPropertyValue("name", "Elon Musk");
    jim.setPropertyValue("salary", "8800");

    company.setPropertyValue("director", jim.getWrappedInstance());

    Float salary = (Float) company.getPropertyValue("director.salary");
    System.out.println(salary);
    String ename = (String)company.getPropertyValue("director.name");
    System.out.println(ename);
    String cname = (String)company.getPropertyValue("name");
    System.out.println(cname);
}
```

输出结果如下。

```
8800.0
Elon Musk
中软国际北京ETC
```

10.2.4 属性编辑器

当输入的信息与实体对象的属性进行绑定时，使用属性编辑器进行数据类型之间的转换。所有输入的信息都是字符串类型，而接收实体的属性则可以为整数、双精度浮点数、日期等类型。

Spring 在 org.springframework.beans.propertyeditors 包下定义了一些属性编辑器（见表 10-1）。如果这些属性编辑器还不能满足要求，则可以定义自己的属性编辑器。

表 10-1　属性编辑器

属性编辑器	说明
ByteArrayPropertyEditor	把字符串转换为字节数组。使用 BeanWrapperImpl 注册
ClassEditor	分析字符串中的类和实际的类

续表

属性编辑器	说明
CustomBooleanEditor	把字符串转换为布尔类型
CustomCollectionEditor	把字符串转换为 Collection 类型
CustomDateEditor	把字符串转换为 java.util.Date 类型
CustomNumberEditor	把字符串转换为 java.lang.Number 及其子类型
FileEditor	解析 String 为 java.io.File
InputStreamEditor	把字符串转换为输入流
LocaleEditor	把字符串转换为 Locale 对象
PatternEditor	把字符串转换为 java.util.regex.Pattern 对象
PropertiesEditor	把字符串转换为 java.util.Properties 对象
StringTrimmerEditor	删除字符串中的空格
URLEditor	把字符串转换为 java.net.URL 对象

示例 10-5：自定义属性编辑器，把字符串转换为 ExoticType 对象。

（1）新建一个 Bean。

```
public class DependsOnExoticType {
    private ExoticType type;
    public void setType(ExoticType type) {
        this.type = type;
    }
}
```

（2）新建一个实体类，业务 Bean 依赖这个实体类。

```
public class ExoticType {
    private String name;
    public ExoticType(String name) {
        this.name = name;
    }
}
```

（3）配置 Bean，传入字符串。

```
<bean id="sample" class="com.icss.example.DependsOnExoticType">
    <property name="type" value="myExotic" />
</bean>
```

（4）自定义属性编辑器，把字符串转换为 ExoticType 对象。

```
public class ExoticTypeEditor extends PropertyEditorSupport {
    public void setAsText(String text) {
        setValue(new ExoticType(text.toUpperCase()));
        System.out.println("字符串转对象，text=" + text);
    }
}
```

（5）注册自定义属性转换器。

```
<bean
    class="org.springframework.beans.factory.config.CustomEditorConfigurer">
    <property name="customEditors">
        <map>
            <entry key="com.icss.example.ExoticType"
                value="com.icss.example.ExoticTypeEditor" />
        </map>
    </property>
</bean>
```

（6）测试。

① 测试代码如下。

```
public static void main(String[] args) {
    ApplicationContext app = new ClassPathXmlApplicationContext("beans.xml");
    System.out.println("ok");
}
```

② 测试结果如下。

```
信息: Loading XML bean definitions from class path resource [beans.xml]
字符串转对象，text=myExotic
ok
```

其中，DependsOnExoticType 依赖 ExoticType 对象，但是在 beans.xml 中，依赖对象配置的是 <property name="type" value="myExotic" />，即依赖的是 myExotic 字符串，并没有真正的 ExoticType 对象。IoC 容器在装载 Bean 对象时，根据注册的自定义属性编辑器 ExoticTypeEditor，会把字符串 myExotic 自动转换成 ExoticType 对象。

10.3 项目案例：用户注册校验

在用户注册时，校验用户名是否为空，用户名的长度是否介于 3～10 个字符，用户年龄是否介于 10～100 岁。对于非法数据，抛出异常。操作步骤如下。

（1）新建 Java 项目，导入 hibernate-validator-5.4.1.jar 和依赖包，导入 Spring 核心包和 AOP 包。

（2）Spring 配置文件使用 LocalValidatorFactoryBean 作为校验器。

```
<context:component-scan base-package="com.icss.biz" />
<bean id="validator" class="org.springframework.validation
                .beanvalidation.LocalValidatorFactoryBean" />

<bean class="org.springframework.validation
            .beanvalidation.MethodValidationPostProcessor">
    <property name="validator" ref="validator" />
</bean>
```

（3）使用 JSR 注解，校验 Person 类的属性。

```
public class Person {
    @Min(10) @Max(100)
```

```
    private int age;
    @NotEmpty
    @NotNull
    @Size(min = 3, max = 10, message = "用户名长度应该介于{min}和{max} ")
    private String name;
}
```

（4）调用业务类，进行用户注册。必须调用 Bean 方法，才能使校验器生效。

```
@Validated
@Service("userBiz")
public class UserBiz {
    public void regitst(@NotNull @Valid Person person) throws Exception {
        System.out.println("用户注册.........");
        System.out.println(person.getAge());
        System.out.println(person.getName());
    }
}
```

（5）测试。

① 测试代码如下。

```
public static void main(String[] args) {
    ApplicationContext ac = new ClassPathXmlApplicationContext("beans.xml");
    UserBiz biz = ac.getBean(UserBiz.class);
    Person person = new Person();
    person.setAge(15);
    person.setName("tom");
    try {
        biz.regitst(person);
    } catch (javax.validation.ConstraintViolationException e) {
        System.out.println("Person 的属性值不符合要求，请检查");
        e.printStackTrace();
    } catch (Exception e) {
        e.printStackTrace();
    }
}
```

② 进行测试后，正常的输出结果如下。

```
信息: Bean 'validator' of type [org.springframework.validation
.beanvalidation.LocalValidatorFactoryBean] is not  eligible for getting
processed by all BeanPostProcessors
用户注册.........
15
tom
```

（6）当 age 或 name 属性的值不符合要求时，抛出异常。

```
信息: Bean 'validator' of type [org.springframework.validation
.beanvalidation.LocalValidatorFactoryBean]
Person 的属性值不符合要求，请检查
javax.validation.ConstraintViolationException  at
org.springframework.validation.beanvalidation.MethodValidationInterceptor
.invok e(MethodValidationInterceptor.java:147)
```

第11章 MyBatis基础知识

MyBatis 是非常有名的持久层框架，它占据了 Hibernate 的很多市场。持久层使用 MyBatis 框架后，代码更加简单，而且 MyBatis 的性能优于 Hibernate。

对于同时支持多种数据库的需求，MyBatis 完全不同于 Hibernate 查询语言（Hibernate Query Language，HQL）的思路，采用了更加简洁、有效的解决方案。

MyBatis 也有不足之处，如复杂的 SQL 语句，这是对象关系映射（Object Relational Mapping，ORM）中最棘手的问题，MyBatis 的解决方案不如 Hibernate 优美。

11.1 下载 MyBatis 资源

在 MyBatis 官网，查看 MyBatis 3.5.3 的介绍（见图 11-1）。

MyBatis 的完整资源可以从 GitHub 上获取。

由于 MyBatis 3.5.3 支持 Java 8，这与我们的开发环境一致，因此选择下载 MyBatis 3.5.3（见图 11-2）。

MyBatis 3.5.3 released!

Dear Community,

We are pleased to announce the release of MyBatis 3.5.3.
The below is the list of user visible changes.

Enhancements:

- Support variable substitution in CDATA of included <sql />. #1615
- Support default method invocation on JDK 14+8 or later. #1626
- Avoid illegal reflective access warning when invoking default mapper method. #1636
- Ambiguous getter/setter now throws ReflectionException only when it is actually accessed. #1201

Bugs:

-
- Possible infinite loop when a SQL provider throws an exception. #1616
- Unable to iterate Cursor if the next element is null. #1653
- queryCursor() fails in streaming mode of MySQL Connector/J. #1654

图 11-1 MyBatis 官网上 MyBatis 3.5.3 的介绍

mybatis-3.5.3

harawata released this Oct 20, 2019 · 103 commits to master since this release

Enhancements:

- Support variable substitution in CDATA of included `<sql />`. #1615
- Support default method invocation on JDK 14+8 or later. #1626
- Avoid illegal reflective access warning when invoking default mapper method. #1636
- Ambiguous getter/setter now throws ReflectionException only when it is actually accessed. #1201

Bug fixes:

- Possible infinite loop when a SQL provider throws an exception. #1616
- Unable to iterate Cursor if the next element is null. #1653
- queryCursor() fails in streaming mode of MySQL Connector/J. #1654

There is no known backward incompatible change since 3.5.2.

Please see the 3.5.3 milestone page for the complete list of changes.

▾ Assets 3

mybatis-3.5.3.zip	7.09 MB
Source code (zip)	
Source code (tar.gz)	

图 11-2 下载 MyBatis 3.5.3

11.2 快速入门示例

使用原来的 StaffUser 项目，实现一个基于 MyBatis 的快速入门示例。该示例的功能是读取全部 User 信息，并在视图层显示。

11.2.1 创建 SqlSessionFactory

SqlSessionFactory 的主要作用是读取系统配置信息，生成 SqlSession 对象等。BaseDao 是所有持久层类的父类，在 BaseDao 中创建一个静态的 SqlSessionFactory，即整个项目只需要一个 SqlSessionFactory 实例即可。

在静态代码块中创建 SqlSessionFactory 对象，读取 mybatis.xml 配置文件。

调用 sqlSessionFactory.openSession( )，返回一个用于数据库操作的 SqlSession 对象。

```
public class BaseDao {
    private static SqlSessionFactory sqlSessionFactory;
    public SqlSession getSession() {
        return  sqlSessionFactory.openSession();
    }
    static {
        try {
            String resource = "mybatis.xml";
            InputStream inputStream = Resources.getResourceAsStream(resource);
            sqlSessionFactory =
                                  new SqlSessionFactoryBuilder().build(inputStream);
        } catch (Exception e) {
            e.printStackTrace();
        }
    }
}
```

在项目的 src 目录下新建 mybatis.xml，配置如下。

```
<?xml version="1.0" encoding="UTF-8" ?>
    <!DOCTYPE configuration
    PUBLIC "-//mybatis.org//DTD Config 3.0//EN"
    "http://mybatis.org/dtd/mybatis-3-config.dtd">
<configuration>
    <environments default="development">
    <environment id="development">
        <transactionManager type="JDBC" />
        <dataSource type="POOLED">
            <property name="driver" value="com.mysql.cj.jdbc.Driver" />
            <property name="url"
            value="jdbc:mysql://localhost:3306/staff?useSSL=false
                &amp;serverTimezone=UTC&amp;allowPublicKeyRetrieval=true" />
            <property name="username" value="root" />
            <property name="password" value="123456" />
        </dataSource>
    </environment>
</environments>
<mappers>
```

```
    <mapper resource="com/icss/mapper/userMappper.xml" />
</mappers>
</configuration>
```

注意以下几点。

- schema 的版本要配置正确。
- environments 为数据库驱动配置。
- 读取 MyBatis 配置信息和映射信息并创建 SqlSessionFactory 对象是一个复杂的操作过程，我们后面再分析这个过程。

11.2.2 从 SqlSessionFactory 获得 SqlSession

SqlSession 是 MyBatis 对数据库连接的封装，可以使用 SqlSession 操作数据库。我们可以先简单地把一个 SqlSession 看成一个 Connection 对象。通过以下代码，从 SqlSessionFactory 获得 SqlSession。

```
public class BaseDao {
    private static SqlSessionFactory sqlSessionFactory;
    public SqlSession getSession() {
          return  sqlSessionFactory.openSession();
    }
}
```

11.2.3 新建 Mapper 接口和映射文件

新建 Mapper 接口和映射文件，MyBatis 把所有 SQL 都放到了映射文件中，只需要接口与映射文件对应即可。所有执行 SQL 的过程都由 MyBatis 框架自动完成。

```
public interface IUserMapper {
    /**
     * 读取所有用户信息
    */
    public List<User> getAllUser() throws Exception;
}
```

映射文件的命名空间与接口对应。

```
<!DOCTYPE mapper
    PUBLIC "-//mybatis.org//DTD Mapper 3.0//EN"
    "http://mybatis.org/dtd/mybatis-3-mapper.dtd">
<mapper namespace="com.icss.mapper.IUserMapper">
    <select id="getAllUser" resultType="com.icss.entity.User">
        select * from tuser
    </select>
</mapper>
```

11.2.4 配置映射文件的指向

在 mybatis.xml 中配置映射文件的指向。在创建 SqlSessionFactory 时，需要加载 mybatis.xml，并且会读取映射文件。

```
<mappers>
    <mapper resource="com/icss/mapper/userMappper.xml" />
</mappers>
```

11.2.5 调用 Mapper 接口

在持久层代码中，通过 SqlSession 调用 Mapper 接口。

```
public class UserDaoMysql extends BaseDao implements IUserDao {
    public List<User> getAllUser() throws Exception {
    SqlSession session = this.getSession();
    IUserMapper mapper = session.getMapper(IUserMapper.class);
    return mapper.getAllUser();
    }
}
```

11.2.6 测试

编写测试代码。

```
public static void main(String[] args) {
    IUser iu = new UserBiz();
    try {
        List<User> users = iu.getAllUser();
        for(User u : users) {
        System.out.println(u.getUname());
        }
    } catch (Exception e) {
        System.out.println(e.getMessage());
    }
}
```

11.2.7 通过 log4j 跟踪 MyBatis

在 log4j.properties 中配置 MyBatis 后，可以输出 SQL，这对项目调试非常有帮助。配置示例如下。

```
log4j.logger.com.ibatis=DEBUG
log4j.logger.com.ibatis.common.jdbc.SimpleDataSource=DEBUG
log4j.logger.com.ibatis.common.jdbc.ScriptRunner=DEBUG
log4j.logger.com.ibatis.sqlmap.engine.impl.SqlMapClientDelegate=DEBUG
log4j.logger.java.sql.Connection=DEBUG
log4j.logger.java.sql.Statement=DEBUG
log4j.logger.java.sql.PreparedStatement=DEBUG
```

测试的输出如下。

```
DEBUG - PooledDataSource forcefully closed/removed all connections.
DEBUG - Opening JDBC Connection
DEBUG - Created connection 21563224.
DEBUG - Setting autocommit to false on JDBC Connection
        [com.mysql.cj.jdbc.ConnectionImpl@1490758]
DEBUG - ==>  Preparing: select * from tuser
DEBUG - ==> Parameters:
DEBUG - <==      Total: 7
```

11.3 MyBatis 的原理

11.3.1 SqlSession 与连接

SqlSession 接口是对 java.sql.Connection 接口的封装。通过 SqlSession 接口中的 getMapper( ) 方法返回 Mapper 接口，是我们后面用得最多的行为。通过调用 getConfiguration( )可以获取映射文件的配置信息。

接口定义如下。

```
public interface SqlSession extends Closeable {
  <T> T  selectOne(String statement);
  <E> List<E>  selectList(String statement);
  int  insert(String statement);
  int  update(String statement);
  int  delete(String statement);
  void  commit();
  void  rollback();
  void  close();
  <T> T  getMapper(Class<T> type);
  Connection  getConnection();
  Configuration  getConfiguration();
}
```

DefaultSqlSession 是 SqlSession 接口的实现类。Configuration 对象包含所有映射文件和核心配置文件的信息；Executor 接口封装了数据库的增删改查操作和事务控制操作；autoCommit 指定数据库的提交模式是自动提交，还是手动提交。

```
public class DefaultSqlSession implements SqlSession {
    private final Configuration configuration;
    private final Executor executor;
    private final boolean autoCommit;
    private boolean dirty;
    private List<Cursor<?>> cursorList;
    public DefaultSqlSession(Configuration configuration, Executor executor) {
    }
}
```

BaseExcecutor 是 Executor 接口的实现类，封装了对数据库中 Connection 的管理。在 DefaultSqlSession 的构造函数中传入 Executor 对象。

```
public abstract class BaseExecutor implements Executor {
    protected Transaction transaction;
    protected Executor wrapper;
    protected Configuration configuration;
    protected BaseExecutor(Configuration configuration, Transaction transaction) {
    }
}
```

数据库的 Connection 从 dataSource 中获取，并用 Transaction 包装。

```
public class JdbcTransaction implements Transaction {
     protected Connection connection;
     protected DataSource dataSource;
     protected TransactionIsolationLevel level;
     protected boolean autoCommit;
     public JdbcTransaction(Connection connection) { }
}
```

总之，SqlSession 的操作需要 Executor 对象，而 Executor 的操作需要 Configuration 和 Transaction。Transaction 的底层是连接。因此，SqlSession 是依赖 Connection 对象的，一个 SqlSession=数据库连接+配置信息。

11.3.2 SqlSession 的 getMapper

MyBatis 的代码非常简洁，通过 SqlSession::getMapper( )即可返回动态创建的接口实现对象，然后调用接口中的方法。如此简单、高效的代码是如何实现的呢？下面我们通过断点跟踪的方式观察一下 getMapper 的底层实现机制。这个跟踪过程对于深入了解 MyBatis 非常重要。

操作步骤如下。

（1）调用 SqlSession::getMapper( )，输入 Mapper 接口，返回 Mapper 接口的实现对象。例如：

```
IUserMapper mapper = session.getMapper(IUserMapper.class);
```

（2）根据输入的接口类型，返回接口对象，然后调用接口对象中的方法。这种操作模式非常简单、高效。其底层调用 configuration.getMapper( )来实现。

```
public class DefaultSqlSession implements SqlSession {
    private final Configuration configuration;
    private final Executor executor;
    public <T> T getMapper(Class<T> type) {
        return configuration.getMapper(type, this);
    }
}
```

（3）观察 Configuration::getMapper( )方法。Configuration 是 MyBatis 底层的重要实现类，它

存储了从配置文件 mybatis.xml 和映射文件（如 userMapper.xml）读取的大量信息。通过断点观察 SqlSession→Configuration→mappedStatements，可以看到映射文件中配置的 SQL 都存放在 Configuration 的 mappedStatements 属性（见图 11-3（a））中。mappedStatements 也存储了所有 Mapper 接口，如图 11-3（b）所示。

```
public class Configuration {
    protected final Map<String, MappedStatement> mappedStatements;
    protected final MapperRegistry mapperRegistry;
    public <T> T getMapper(Class<T> type, SqlSession sqlSession) {
       return mapperRegistry.getMapper(type, sqlSession);
    }
}
```

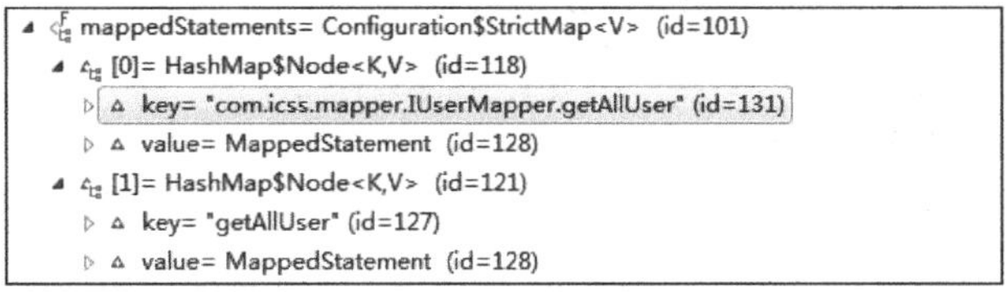

```
mappedStatements= Configuration$StrictMap<V>  (id=101)
  [0]= HashMap$Node<K,V>  (id=118)
    key= "com.icss.mapper.IUserMapper.getAllUser" (id=131)
    value= MappedStatement  (id=128)
  [1]= HashMap$Node<K,V>  (id=121)
    key= "getAllUser" (id=127)
    value= MappedStatement  (id=128)
```

（a）映射文件中配置的 SQL

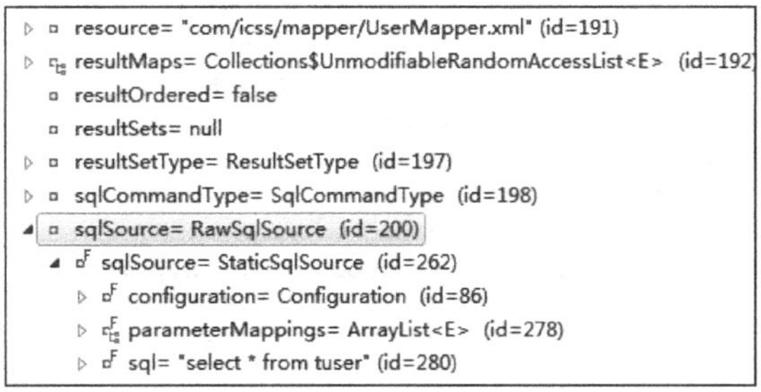

```
resource= "com/icss/mapper/UserMapper.xml" (id=191)
resultMaps= Collections$UnmodifiableRandomAccessList<E>  (id=192)
resultOrdered= false
resultSets= null
resultSetType= ResultSetType  (id=197)
sqlCommandType= SqlCommandType  (id=198)
sqlSource= RawSqlSource  (id=200)
  sqlSource= StaticSqlSource  (id=262)
    configuration= Configuration  (id=86)
    parameterMappings= ArrayList<E>  (id=278)
    sql= "select * from tuser" (id=280)
```

（b）所有 Mapper 接口

图 11-3　mappedStatements 中存放的内容

（4）MapperRegistry 的 knownMappers 存储了所有 Mapper 接口与映射代理工厂（见图 11-4）。

```
public class MapperRegistry {
   private final Configuration config;
   private final Map<Class<?>, MapperProxyFactory<?>> knownMappers;
   public <T> T getMapper(Class<T> type, SqlSession sqlSession) {
       return mapperProxyFactory.newInstance(sqlSession);
   }
}
```

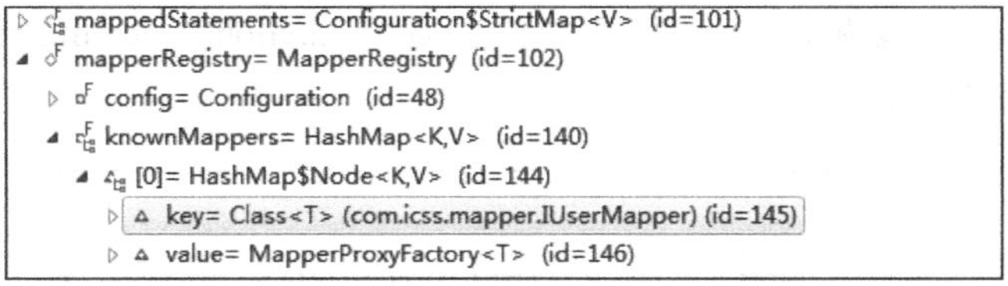

```
mappedStatements= Configuration$StrictMap<V>  (id=101)
mapperRegistry= MapperRegistry  (id=102)
  config= Configuration  (id=48)
  knownMappers= HashMap<K,V>  (id=140)
    [0]= HashMap$Node<K,V>  (id=144)
      key= Class<T> (com.icss.mapper.IUserMapper) (id=145)
      value= MapperProxyFactory<T>  (id=146)
```

图 11-4　knownMappers 中存储的内容

（5）使用 JDK 的动态代理类 Proxy，根据 Mapper 接口，创建动态代理对象。

```
public class MapperProxyFactory<T> {
    public MapperProxyFactory(Class<T> mapperInterface) {
        this.mapperInterface = mapperInterface;
    }
    public Class<T> getMapperInterface() {
        return mapperInterface;
    }
    protected T newInstance(MapperProxy<T> mapperProxy) {
        return (T) Proxy.newProxyInstance(mapperInterface.getClassLoader(),
                new Class[] { mapperInterface }, mapperProxy);
    }
    public T newInstance(SqlSession sqlSession) {
        final MapperProxy<T> mapperProxy =
             new MapperProxy<>(sqlSession, mapperInterface, methodCache);
        return newInstance(mapperProxy);
    }
}
```

（6）在 invoke( )中动态激活被代理的方法。

```
public class MapperProxy<T> implements InvocationHandler, Serializable {
    public Object invoke(Object proxy, Method method,
                        Object[] args) throws Throwable {
          if (Object.class.equals(method.getDeclaringClass())) {
                return method.invoke(this, args);
            } else if (method.isDefault()) {
                if (privateLookupInMethod == null) {
                return invokeDefaultMethodJava8(proxy, method, args);
            } else {
                return invokeDefaultMethodJava9(proxy, method, args);
          }
    }
}
```

总之，在创建 SqlSessionFactory 时读取映射文件，把映射文件中所有配置的 SQL 都读取到 MapperRegistry 中并存储成 Map。mapperRegistry 存储的是所有接口信息，mappedStatements 存储的是 Mapper 接口与 SQL。当调用接口中的方法时，通过 JDK 的动态代理类 Proxy 动态调用被代理对象的方法。

11.4 配置 MyBatis

查看 mybatis.xml，可以发现所有元素都配置在<configuration> </configuration>根下。注意元素的配置顺序（见图 11-5）。

<configuration>
The content of element type "configuration" must match "(properties?,settings?,typeAliases?,typeHandlers?,objectFactory?,objectWrapperFactory?,reflectorFactory?,plugins?,environments?,databaseIdProvider?,mappers?)".

图 11-5　元素的配置顺序

XML 是一个树状结构，很少要求元素顺序，但是在 MyBatis 配置文件中必须严格按照上面提示的顺序配置元素。在这里，很多元素可以省略，即使用默认配置。但是如果使用这些元素，则顺序不能错误。

配置示例如下。

```
<?xml version="1.0" encoding="UTF-8" ?>
<!DOCTYPE configuration
    PUBLIC "-//mybatis.org//DTD Config 3.0//EN"
    "http://mybatis.org/dtd/mybatis-3-config.dtd">
<configuration>
    <properties> </properties>
    <settings> </settings>
    <typeAliases > </typeAliases >
    <typeHandlers > </typeHandlers >
    <objectFactory > </objectFactory >
    <plugins> </plugins >
    <databaseIdProvider > </databaseIdProvider >
    <environments default="development">
        <environment id="development">
        </environment>
    </environments>
    <mappers>
        <mapper resource="com/icss/mapper/userMappper.xml" />
    </mappers>
</configuration>
```

11.4.1 配置属性文件

在 mybatis.xml 中可以读取外部的属性文件，这样做的好处是可以在属性文件中配置大量备用信息，而 Mybatis 根据键、值选择使用部分内容。

配置示例如下。

（1）在 src 下新建 db.properties 配置文件。

```
driver=com.mysql.cj.jdbc.Driver
url=jdbc:mysql://localhost:3306/staff?useSSL=false
username=root
password=123456
```

（2）在 mybatis.xml 中使用<properties>引用外部属性文件后，使用${key}调用配置文件中的键、值。

```
<configuration>
<properties resource="db.properties">
</properties>
<environments default="development">
    <environment id="development">
        <transactionManager type="JDBC" />
        <dataSource type="POOLED">
            <property name="driver" value="${driver}" />
```

```
            <property name="url" value="${url}" />
            <property name="username" value="${username}" />
            <property name="password" value="${password}" />
        </dataSource>
    </environment>
  </environments>
</configuration>
```

（3）在 mybatis.xml 中，也可以直接设置对象的属性值，这样就可以省略属性文件。注意，db.properties 中的连接符号&在 XML 文件中用“&”表示。

```
<dataSource type="POOLED">
    <property name="driver" value="com.mysql.cj.jdbc.Driver " />
    <property name="url"
            value="jdbc:mysql://localhost:3306/staff?useSSL=false
                    &amp serverTimezone=UTC " />
    <property name="username" value="root" />
    <property name="password" value="123456" />
</dataSource>
```

注意，此处的<dataSource>为下面的类，后面会详细讲解。

```
public class UnpooledDataSource implements DataSource {
    private String driver;
    private String url;
    private String username;
    private String password;
}
```

（4）属性文件还可以用于 SqlSessionFactory 对象的构建。

```
SqlSessionFactory factory =
          sqlSessionFactoryBuilder.build(reader, props);
```

或者

```
SqlSessionFactory factory =
          new SqlSessionFactoryBuilder
                    .build(reader, environment, props);
```

11.4.2 配置 setting 项

1．setting 项

使用 setting 项，可以在运行时对 MyBatis 的行为进行调整。如果对属性特征不熟悉，则尽量使用默认值。可用的 setting 项很多（见表 11-1），描述信息参见 MyBatis 3.5.3 官方文档。

表 11-1　　可用的 setting 项

可用的 setting 项	有效值	默认值
cacheEnabled	true、false	true
lazyLoadingEnabled	true、false	false

续表

可用的 setting 项	有效值	默认值
AggressiveLazyLoading	true、false	false（在 MyBatis 3.4.1 版本之前，为 true）
multipleResultSetsEnabled	true、false	true
useColumnLabel	true、false	true
useGeneratedKeys	true、false	false
autoMappingBehavior	NONE、PARTIAL、FULL	PARTIAL
autoMappingUnknownColumnBehavior	NONE、WARNING、FAILING	NONE
defaultExecutorType	SIMPLE、REUSE、BATCH	SIMPLE
defaultStatementTimeout	任何正整数	未设置（null）
defaultFetchSize	任何正整数	未设置（null）
defaultResultSetType	FORWARD_ONLY、SCROLL_SENSITIVE、SCROLL_INSENSITIVE、DEFAULT（与未设置的行为相同）	未设置（null）
safeRowBoundsEnabled	true、false	false
safeResultHandlerEnabled	true、false	true
mapUnderscoreToCamelCase	true、false	false
localCacheScope	会话语句	会话
jdbcTypeForNull	JDBC 类型的枚举，常见的包括 NULL、VARCHAR 和 OTHER	OTHER
lazyLoadTriggerMethods	用逗号分隔的方法名列表	equals、clone、hashCode、toString
defaultScriptingLanguage	一个类型的别名或者完全限定的类名	org.apache.ibatis.scripting.xmltags.XMLLanguage
defaultEnumTypeHandler	一个类型的别名或者完全限定的类名	org.apache.ibatis.type.EnumTypeHandler
callSettersOnNulls	true、false	false
returninstanceforemptyrow	true、false	false
logPrefix	任何字符串	未设置
logImpl	SLF4J、LOG4J、LOG4J2、JDK_LOGGING、COMMONS_LOGGING、STDOUT_LOGGING、NO_LOGGING	未设置
proxyFactory	CGLIB、JAVASSIST	JAVASSIST（用于 MyBatis 3.3 版本或更高版本）
vfsImpl	完全限定类名或者用逗号分隔的 VFS 实现方式	未设置
useActualParamName	true、false	true
configurationFactory	一个类型的别名或者完全限定的类名	未设置

使用 setting 项的示例如下。

```
<settings>
    <setting name="cacheEnabled" value="true" />
    <setting name="lazyLoadingEnabled" value="true" />
```

```
    <setting name="multipleResultSetsEnabled" value="true" />
    <setting name="useColumnLabel" value="true" />
    <setting name="useGeneratedKeys" value="false" />
    <setting name="autoMappingBehavior" value="PARTIAL" />
    <setting name="defaultExecutorType" value="SIMPLE" />
    <setting name="defaultStatementTimeout" value="25" />
    <setting name="defaultFetchSize" value="100" />
    <setting name="safeRowBoundsEnabled" value="false" />
    <setting name="mapUnderscoreToCamelCase" value="false" />
    <setting name="localCacheScope" value="SESSION" />
    <setting name="jdbcTypeForNull" value="OTHER" />
    <setting name="lazyLoadTriggerMethods"
            value="equals,clone,hashCode,toString" />
</settings>
```

2. 常用 setting 项

setting 项很多，对于绝大多数选项，我们尽量选择默认值。注意，对于 jdbcTypeForNull，推荐使用 NULL。在项目实践中，jdbcTypeForNull 使用默认值 OTHER，在书城项目的主页中，该项对性能有很大影响（对于不同版本，影响的大小可能不同）。

```
<settings>
    <setting name="jdbcTypeForNull" value="NULL" />
</settings>
```

11.4.3 配置 typeAliases

typeAliases 是 Java 类型的简写，它仅在 XML 配置文件中用于简化合格的类名。

```
<typeAliases>
    <typeAlias alias="Author" type="domain.blog.Author" />
    <typeAlias alias="Blog" type="domain.blog.Blog" />
    <typeAlias alias="Comment" type="domain.blog.Comment" />
    <typeAlias alias="Post" type="domain.blog.Post" />
    <typeAlias alias="Section" type="domain.blog.Section" />
    <typeAlias alias="Tag" type="domain.blog.Tag" />
</typeAliases>
```

对于常用的 Java 类型，MyBatis 内置了与之对应的别名，见表 11-2。这些别名都很简单。如果在映射文件中使用映射类型，需要包名和类型的全称，如 java.lang.Integer。对于别名，可以直接使用（如用 int 替代 java.lang.Integer）。

表 11-2　Java 类型的别名

Java 类型的别名	映射类型
_byte	byte
_long	long
_short	short

续表

Java 类型的别名	映射类型
_int	int
_integer	int
_double	double
_float	float
_boolean	boolean
string	String
byte	Byte
long	Long
short	Short
int	Integer
integer	Integer
double	Double
float	Float
boolean	Boolean
date	Date
decimal	BigDecimal
bigdecimal	BigDecimal
object	Object
map	Map
hashmap	HashMap
list	List
arraylist	ArrayList
collection	Collection
iterator	Iterator

下面介绍一下别名机制。

通过断点跟踪观察 session→configuration→typeAliasRegistry。如图 11-6 所示，这里注册了系统中的所有别名。前面的是系统定义的默认别名，后面的是开发人员自定义的别名。

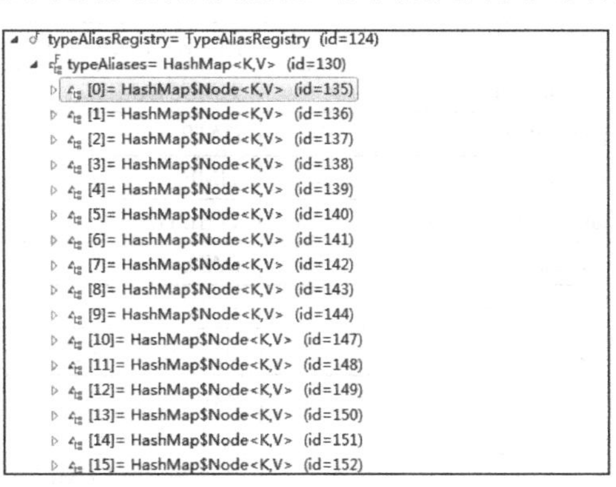

图 11-6　类型别名的注册

11.4.4 配置 typeHandlers

无论是在 MyBatis 中给 PreparedStatement 设置一个参数，还是从 PreparedStatement 中返回 ResultSet，都要面临一个问题——Java 类型与 JDBC 类型之间的转换。

Java 类型面向的是 JVM，JDBC 类型面向的是各种关系型数据库，因此二者有很大的不同。

Java 类型与 JDBC 类型之间如何相互转换，是 ORM 框架中最关键的问题。

1. 默认类型映射器

表 11-3 展示了 MyBatis 的默认类型映射器。

表 11-3　　MyBatis 的默认类型映射器

类型处理程序	Java 类型	JDBC 类型
BooleanTypeHandler	java.lang.Boolean、boolean	任何兼容的布尔类型
ByteTypeHandler	java.lang.Byte、byte	任何兼容的 NUMERIC 或者 BYTE
ShortTypeHandler	java.lang.Short、short	任何兼容的 NUMERIC 或者 SMALLINT
IntegerTypeHandler	java.lang.Integer、int	任何兼容的 NUMERIC 或者 INTEGER
LongTypeHandler	java.lang.Long、long	任何兼容的 NUMERIC 或者 BIGINT
FloatTypeHandler	java.lang.Float、float	任何兼容的 NUMERIC 或者 FLOAT
DoubleTypeHandler	java.lang.Double、double	任何兼容的 NUMERIC 或者 DOUBLE
BigDecimalTypeHandler	java.math.BigDecimal	任何兼容的 NUMERIC 或者 DECIMAL
StringTypeHandler	java.lang.String	CHAR、VARCHAR
ClobReaderTypeHandler	java.io.Reader	
ClobTypeHandler	java.lang.String	CLOB、LONGVARCHAR
NStringTypeHandler	java.lang.String	NVARCHAR、NCHAR
NClobTypeHandler	java.lang.String	NCLOB
BlobInputStreamTypeHandler	java.io.InputStream	
ByteArrayTypeHandler	byte[]	任何兼容的字节流类型
BlobTypeHandler	byte[]	BLOB、LONGVARBINARY
DateTypeHandler	java.util.Date	TIMESTAMP
DateOnlyTypeHandler	java.util.Date	DATE
TimeOnlyTypeHandler	java.util.Date	TIME
SqlTimestampTypeHandler	java.sql.Timestamp	TIMESTAMP
SqlDateTypeHandler	java.sql.Date	DATE
SqlTimeTypeHandler	java.sql.Time	TIME
ObjectTypeHandler	Any	OTHER 或者未指定
EnumTypeHandler	Enumeration Type	VARCHAR，任何兼容的字符串类型，因此已经存储了编码（不是索引）
EnumOrdinalTypeHandler	Enumeration Type	任何兼容的 NUMERIC 或者 DOUBLE，因为已经存储了位置（而不是编码）

续表

类型处理程序	Java 类型	JDBC 类型
SqlxmlTypeHandler	java.lang.String	SQLXML
InstAntTypeHandler	java.time.InstAnt	TIMESTAMP
LocalDateTimeTypeHandler	java.time.LocalDateTime	TIMESTAMP
LocalDateTypeHandler	java.time.LocalDate	DATE
LocalTimeTypeHandler	java.time.LocalTime	TIME
OffsetDateTimeTypeHandler	java.time.OffsetDateTime	TIMESTAMP
OffsetTimeTypeHandler	java.time.OffsetTime	TIME
ZonedDateTimeTypeHandler	java.time.ZonedDateTime	TIMESTAMP
YearTypeHandler	java.time.Year	INTEGER
MonthTypeHandler	java.time.Month	INTEGER
YearMonthTypeHandler	java.time.YearMonth	VARCHAR 或者 LONGVARCHAR
JapaneseDateTypeHandler	java.time.chrono.Japanese	DATE

注意：从 MyBatis 3.4.5 开始，默认支持 JSR 310。

通过断点跟踪（session→configuration→typeHandlerRegistry），观察内存中的类型映射器（见图 11-7）。

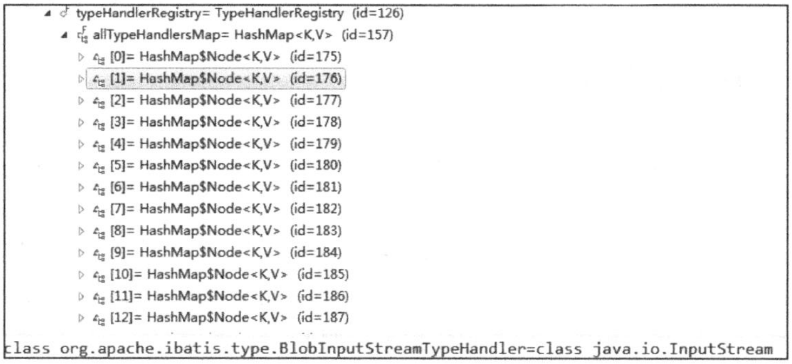

```
typeHandlerRegistry= TypeHandlerRegistry (id=126)
  allTypeHandlersMap= HashMap<K,V> (id=157)
    [0]= HashMap$Node<K,V> (id=175)
    [1]= HashMap$Node<K,V> (id=176)
    [2]= HashMap$Node<K,V> (id=177)
    [3]= HashMap$Node<K,V> (id=178)
    [4]= HashMap$Node<K,V> (id=179)
    [5]= HashMap$Node<K,V> (id=180)
    [6]= HashMap$Node<K,V> (id=181)
    [7]= HashMap$Node<K,V> (id=182)
    [8]= HashMap$Node<K,V> (id=183)
    [9]= HashMap$Node<K,V> (id=184)
    [10]= HashMap$Node<K,V> (id=185)
    [11]= HashMap$Node<K,V> (id=186)
    [12]= HashMap$Node<K,V> (id=187)
class org.apache.ibatis.type.BlobInputStreamTypeHandler=class java.io.InputStream
```

图 11-7　类型映射器

2. 自定义类型映射器

可以重写类型映射器，也可以创建自己的非标准类型映射器。要自定义类型映射器，需要实现 org.apache.ibatis.type.TypeHandler 接口，或继承自类 org.apache.ibatis.type.BaseTypeHandler。

下面给出一个示例。

（1）定义 String 与 JdbcType.VARCHAR 之间的类型映射器。

```
@MappedJdbcTypes(JdbcType.VARCHAR)
public class ExampleTypeHandler extends BaseTypeHandler<String> {
```

```
    @Override
    public void setNonNullParameter(PreparedStatement ps, int i,
            String parameter, JdbcType jdbcType)throws SQLException {
        ps.setString(i, parameter);
        Log.logger.info("ExampleTypeHandler-->>setNonNullParameter");
    }
    @Override
    public String getNullableResult(ResultSet rs,
                String columnName) throws SQLException {
        Log.logger.info("ExampleTypeHandler-->>getNullableResult");
        return rs.getString(columnName);
    }
    @Override
    public String getNullableResult(ResultSet rs,
                    int columnIndex) throws SQLException {
        return rs.getString(columnIndex);
    }
    @Override
    public String getNullableResult(CallableStatement cs,
                    int columnIndex) throws SQLException {
        return cs.getString(columnIndex);
    }
}
```

注意：@MappedJdbcTypes 注解的自定义类型映射器在 resultMap 中是默认无效的，需要显式设置。

（2）在 mybatis.xml 中注册这个类型映射器。

```
<typeHandlers>
    <typeHandler handler="com.icss.handler.ExampleTypeHandler"/>
</typeHandlers>
```

如上示例重写了 java.lang.String 与 JDBC 类型 VARCHAR 之间的映射器，默认映射器是 StringTypeHandler。@MappedTypes 表明了映射的 JDBC 类型；BaseTypeHandler<String>表明映射的 Java 类型是 String，因此 JDBC 的相关操作都是按照 String 处理的。

（3）StringTypeHandler 在 API 中的原始定义如下。

```
public class StringTypeHandler extends BaseTypeHandler<String> {
    @Override
    public void setNonNullParameter(PreparedStatement ps, int i,
            String parameter, JdbcType jdbcType)throws SQLException {
        ps.setString(i, parameter);
    }
    @Override
    public String getNullableResult(ResultSet rs,
                String columnName) throws SQLException {
        return rs.getString(columnName);
    }
    @Override
```

```
    public String getNullableResult(ResultSet rs,
                     int columnIndex) throws SQLException {
        return rs.getString(columnIndex);
    }
    @Override
    public String getNullableResult(CallableStatement cs,
                int columnIndex) throws SQLException {
        return cs.getString(columnIndex);
    }
}
```

（4）测试。测试代码如下（基于前面的入门示例测试自定义类型映射器的使用）。

```
<mapper namespace="com.icss.mapper.IUserMapper">
    <select id="getAllUser" resultType="User">
        select * from tuser
    </select>
</mapper>
```

（5）UI 调用的代码如下。

```
public static void main(String[] args) {
    IUser iu = new UserBiz();
    try {
        List<User> users = iu.getAllUser();
        for(User u : users) {
            System.out.println(u.getUname());
        }
    } catch (Exception e) {
        System.out.println(e.getMessage());
    }
}
```

（6）输出如下。

```
DEBUG - ==>  Preparing: select * from tuser
DEBUG - ==> Parameters:
INFO - ExampleTypeHandler-->>getNullableResult
INFO - ExampleTypeHandler-->>getNullableResult
INFO - ExampleTypeHandler-->>getNullableResult
INFO - ExampleTypeHandler-->>getNullableResult
INFO - ExampleTypeHandler-->>getNullableResult
INFO - ExampleTypeHandler-->>getNullableResult
DEBUG - <==      Total: 2
101000123
101000124
101000125
```

根据输出，发现 getNullableResult(ResultSet rs, String columnName)方法被调用了 6 次。为什么呢？tuser 表有 4 个字段，分别是 uname、sno、pwd、role。其中 uname、sno、pwd 这 3 个字段是 varchar 类型。现在库中有两条记录、3 个 varchar 字段，在读取每个 varchar 字段并把它转换为 User 实体的 String 属性时，都会调用一次类型映射器的 getNullableResult( )方法，因此输出 6（即 2×3）次。

11.4.5 配置 ObjectFactory

MyBatis 每次创建一个新的结果集对象时，都需要调用 ObjectFactory 实例。如果希望重写默认 ObjectFactory 的行为，可以自定义对象工厂，继承自 DefaultObjectFactory。

ObjectFactory 接口非常简单，它包含两个 create 方法，一个用于处理默认构造函数，另一个用于处理带参构造函数。

```
public interface ObjectFactory {
    <T> T  create(Class<T> type);
    <T> T create(Class<T> type,
            List<Class<?>> constructorArgTypes, List<Object> constructorArgs);
}
```

案例 11-1：创建自定义对象工厂

（1）自定义对象工厂继承自 DefaultObjectFactory。

```
public class ExampleObjectFactory extends DefaultObjectFactory {
    @Override
    public <T> T create(Class<T> type) {
        Log.logger.info("默认构造:" + type.toString() );
        return super.create(type);
    }
    @Override
    public <T> T create(Class<T> type, List<Class<?>> constructorArgTypes,
                List<Object> constructorArgs) {
        Log.logger.info("带参构造:" + type.toString() );
        return super.create(type, constructorArgTypes, constructorArgs);
    }
    @Override
    public void setProperties(Properties properties) {
        Log.logger.info("ExampleObjectFactory-->setProperties");
        super.setProperties(properties);
    }
    @Override
    public <T> boolean isCollection(Class<T> type) {
        Log.logger.info("ExampleObjectFactory-->isCollection");
        return Collection.class.isAssignableFrom(type);
    }
}
```

（2）在 mybatis.xml 中注册自定义的 objectFactory。

```
<objectFactory type="com.icss.factory.ExampleObjectFactory"></objectFactory>
```

（3）测试。仍然使用 11.4.4 节的用户查询代码。

```
public interface IUserMapper {
    public List<User> getAllUser() throws Exception;
}
<mapper namespace="com.icss.mapper.IUserMapper">
```

```
    <select id="getAllUser" resultType="User">
        select * from tuser
    </select>
</mapper>
```

测试结果如下。

```
DEBUG - ==>  Preparing: select * from tuser
DEBUG - ==> Parameters:
INFO - 默认构造:interface java.util.List
INFO - 带参构造: interface java.util.List
INFO - 默认构造:class com.icss.entity.User
INFO - 带参构造: class com.icss.entity.User
INFO - 类型转换-->>getNullableResult
INFO - 类型转换-->>getNullableResult
INFO - 类型转换-->>getNullableResult
INFO - 默认构造:class com.icss.entity.User
INFO - 带参构造: class com.icss.entity.User
INFO - 类型转换-->>getNullableResult
INFO - 类型转换-->>getNullableResult
INFO - 类型转换-->>getNullableResult
DEBUG - <==      Total: 2
101000123
101000124
```

为了查询所有用户，要创建一个集合对象，因此先调用一次 ExampleObjectFactory 的默认构造函数。对于两条用户记录，需要调用两次 ExampleObjectFactory 的默认构造函数。在 ExampleObjectFactory 的默认构造中又会调用父类 DefaultObjectFactory 的带参构造函数。因此，每次构造函数都会输出两次 create 方法。

```
public class DefaultObjectFactory
            implements ObjectFactory, Serializable {
    @Override
    public <T> T create(Class<T> type) {
        return create(type, null, null);
    }
}
```

11.4.6 配置 plugins 拦截器

默认情况下，MyBatis 允许 plug ins 拦截器拦截如下方法。

- Executor 包括（update、query、flushStatements、commit、rollback、getTransaction、close、isClosed）。
- ParameterHandler 包括（getParameterObject、setParameters）。
- ResultSetHandler 包括（handleResultSets、handleOutputParameters）。
- StatementHandler 包括（prepare、parameterize、batch、update、query）。

示例 11-1：自定义拦截器，过滤映射方法。

（1）实现 Interceptor 接口。

```
@Intercepts({ @Signature(type = Executor.class, method = "query",
        args = { MappedStatement.class,
        Object.class ,RowBounds.class,ResultHandler.class}) })
public class ExamplePlugin implements Interceptor {
    @Override
    public Object intercept(Invocation invocation) throws Throwable {
        Log.logger.info("自定义拦截器-预处理...");
        Object returnObject = invocation.proceed();
        Log.logger.info("自定义拦截器-后置处理...");
        return returnObject;
    }
}
```

（2）在 mybatis.xml 中配置自定义拦截器。

```
<plugins>
    <plugin interceptor="com.icss.plugin.ExamplePlugin"></plugin>
</plugins>
```

（3）测试。继续使用上一节的用户查询代码。

```
<mapper namespace="com.icss.mapper.IUserMapper">
    <select id="getAllUser" resultType="User">
        select * from tuser
    </select>
</mapper>
```

测试结果如下。

```
INFO - 自定义拦截器-预处理...
DEBUG - Opening JDBC Connection
DEBUG - Created connection 2081658.
DEBUG - Setting autocommit to false on JDBC Connection
DEBUG - ==>  Preparing: select * from tuser
DEBUG - ==> Parameters:
DEBUG - <==      Total: 2
INFO - 自定义拦截器-后置处理...
101000123
101000124
```

11.4.7　配置环境

1. 多环境配置

MyBatis 允许同时配置多个环境，如开发、测试、生产，还允许同时操作多个不同类型的数据库。

注意：虽然可以同时配置多个环境，但是对于一个 SqlSessionFactory 实例，只能选择一个环境。因此，如果要同时连接两个数据库，必须创建两个 SqlSessionFactory 实例。

示例 11-2： 同时连接员工库和书城库。

具体操作如下。

（1）配置 db.properties。

```
driver=com.mysql.cj.jdbc.Driver
url=jdbc:mysql://localhost:3306/staff?useSSL=false
url2=jdbc:mysql://localhost:3306/book2?useSSL=false
username=root
password=123456
```

（2）配置 mybatis.xml。

```
<environments default="development">
    <environment id="development">
        <transactionManager type="JDBC" />
        <dataSource type="UNPOOLED">
            <property name="driver" value="${driver}" />
            <property name="url" value="${url}" />
            <property name="username" value="${username}" />
            <property name="password" value="${password}" />
        </dataSource>
    </environment>
    <environment id="development2">
        <transactionManager type="JDBC" />
        <dataSource type="POOLED">
            <property name="driver" value="${driver}" />
            <property name="url" value="${url2}" />
            <property name="username" value="${username}" />
            <property name="password" value="${password}" />
            <property name="poolMaximumActiveConnections" value="50"/>
            <property name="poolMaximumIdleConnections" value="5"/>
        </dataSource>
    </environment>
</environments>
```

（3）传入不同环境，分别创建不同的 SqlSessionFactory。

```
public class BaseDao {
    private static SqlSessionFactory sqlSessionFactory;
    static {
        try {
            String resource = "mybatis.xml";
            InputStream inputStream = Resources.getResourceAsStream(resource);
            sqlSessionFactory = new
                    SqlSessionFactoryBuilder().build(inputStream,"development2");
        } catch (Exception e) {
            e.printStackTrace();
        }
    }
}
```

（4）对于相同的映射代码，根据传入环境的 id，可以从不同的数据库提取数据（书城库与员工库的 TUser 表的结构相同）

```
<mapper namespace="com.icss.mapper.IUserMapper">
    <select id="getAllUser" resultType="User">
        select * from tuser
    </select>
</mapper>
```

总之，通过环境设置，MyBatis 可以同时支持多个不同的数据库，如 Oracle 和 MySQL，这在项目实践中非常重要。

2．transactionManager

这里有两种事务管理类型可选，type=[JDBC | MANAGED]。

- JDBC：根据从 dataSource 返回的 Connection 对象，直接使用 JDBC 管理事务。
- MANAGED：由 Java EE 容器管理整个事务（如 CMT）。

注意： 如果使用 Spring 整合 MyBatis，则无须配置 transactionManager，因为 Spring 会用自己的事务环境重写 MyBatis 的配置。

示例代码如下。

```
<environment id="development">
    <transactionManager type="JDBC" />
    <dataSource type="UNPOOLED">
        <property name="driver" value="${driver}" />
        <property name="url" value="${url}" />
        <property name="username" value="${username}" />
        <property name="password" value="${password}" />
    </dataSource>
</environment>
```

3．dataSource

使用标准数据源接口 javax.sql.DataSource 配置数据库的 JDBC 连接。

```
<environment id="development">
    <transactionManager type="JDBC" />
    <dataSource type="UNPOOLED">
        <property name="driver" value="${driver}" />
        <property name="url" value="${url}" />
        <property name="username" value="${username}" />
        <property name="password" value="${password}" />
    </dataSource>
</environment>
```

对于此处的<dataSource >，底层会使用 javax.sql.DataSource 接口。

```java
public interface DataSource extends CommonDataSource, Wrapper {
      Connection getConnection() throws SQLException;
      Connection getConnection(String username, String password)
                                  throws SQLException;
}
```

这里有 3 种数据源类型 UNPOOLED、POOLED、JNDI 可选。

1）UNPOOLED 数据源

UNPOOLED 数据源默认使用 UnpooledDataSource 对象，采用非池化的方式访问数据库。

```java
public class UnpooledDataSource implements DataSource {
     private String driver;
     private String url;
     private String username;
     private String password;
     private Integer defaultTransactionIsolationLevel;
     private Integer defaultNetworkTimeout;
}
```

UnpooledDataSource 对象实现了 DataSource 接口，表示没有使用连接池的数据源。

采用 UNPOOLED 数据源，对于每次请求，都需要打开、关闭一次数据库连接。对于性能要求不高的系统，这个模式是一个不错的选择。

基本属性信息如下。

- driver：表示 JDBC 驱动。
- url: 表示连接数据库的 JDBC 地址。
- username: 表示数据库的用户名。
- password: 表示数据库校验密码。
- defaultTransactionIsolationLevel：表示默认事务隔离级别。
- defaultNetworkTimeout：表示等候数据库操作完成的倒计时，以毫秒（ms）计。

配置示例如下。

```xml
<environment id="development">
      <transactionManager type="JDBC" />
      <dataSource type="UNPOOLED">
          <property name="driver" value="${driver}" />
          <property name="url" value="${url}" />
          <property name="username" value="${username}" />
          <property name="password" value="${password}" />
      </dataSource>
  </environment>
```

2）POOLED 数据源

POOLED 数据源采用 MyBatis 数据库连接池配置，可以优化性能，是 MyBatis 推荐的 Web 项目配置。默认使用 PooledDataSource 对象访问数据库连接池。

```
public class PooledDataSource implements DataSource {
    protected int poolMaximumActiveConnections = 10;
    protected int poolMaximumIdleConnections = 5;
    protected int poolMaximumCheckoutTime = 20000;
    protected int poolTimeToWait = 20000;
    protected int poolMaximumLocalBadConnectionTolerance = 3;
}
```

除了 UNPOOLED 数据源的几个属性外，PooledDataSource 还增加了如下属性设置。

- poolMaximumActiveConnections：表示激活的最大数据库连接数。
- poolMaximumIdleConnections：表示最大的空闲连接数。
- poolMaximumCheckoutTime：默认为 20s，数据库连接被放回池之前需要检出。
- poolTimeToWait: 表示等待从池中获取数据库连接的时间，默认为 20s。
- poolMaximumLocalBadConnectionTolerance：当一个线程获得了一个无效连接时，有机会再次获取一个新的连接的次数。默认值为 3。
- poolPingQuery：发出 ping 请求，校验连接池是否可以返回有效连接。
- poolPingEnabled：打开或关闭 poolPingQuery 功能。默认值为 false。
- poolPingConnectionsNotUsedFor：为了避免不必要的 ping 请求的优化设置。默认值为 0，即当 poolPingEnabled 为 true 时，每次请求连接都需要 ping。

配置示例如下。

```
<environment id="development">
        <transactionManager type="JDBC" />
        <dataSource type="POOLED">
            <property name="driver" value="${driver}" />
            <property name="url" value="${url}" />
            <property name="username" value="${username}" />
            <property name="password" value="${password}" />
        </dataSource>
    </environment>
```

3）JNDI 数据源

JNDI 是 Java 平台访问资源的统一接口，可以用于访问文件系统、数据库、MOM、EJB 等。

从 JNDI 中提取数据库连接的方式与使用 JDBC 获取数据库连接的方式不同。在前一种方式下，需要创建 InitialContext 对象，然后调用 lookup( )找到数据源，示例如下。

```
InitialContext context = new InitialContext();
DataSource ds = (DataSource)context.lookup(strJNDIName) ;
Connection con = ds.getConnection();
```

使用容器管理的数据源（如部署 EJB 的应用服务器）只有两个属性配置。

- initial_context：从 InitialContext 中查找上下文。
- data_source：数据源实例的上下文路径。

配置 JNDI 数据源的示例如下。

（1）新建 Web 项目，部署在 Tomcat 8.0 上。Tomcat 8.0 内置了 DBCP 数据库连接池，此示例使用这个连接池。注意，把 MySQL 的驱动包复制到 Tomcat 的 lib 目录下，访问数据库并池化是 Tomcat 的职责，Web 站点只调用 Tomcat 建好的连接池。

（2）在 Web 项目的 META-INF 下，新建 context.xml。

```
<Context>
    <Resource name="jdbc/StaffJndi" auth="Container"
            type="javax.sql.DataSource" username="root"
            password="123456"  driverClassName="com.mysql.cj.jdbc.Driver"
            url="jdbc:mysql://localhost:3306/staff?useSSL=false"
            maxActive="100" maxIdle="30" maxWait="10000"/>
</Context>
```

（3）在 mybatis.xml 中配置 JNDI。

```
<environments default="development">
    <environment id="development">
        <transactionManager type="JDBC" />
        <dataSource type="JNDI">
            <property name="data_source" value="java:comp/env/jdbc/StaffJndi"/>
        </dataSource>
    </environment>
</environments>
```

（4）创建 SqlSessionFctory。此处创建的 SqlSessionFctory 是通过默认的环境 development 获取的。SqlSessionFctory 会使用配置信息，从 JNDI 中创建 InitialContext，并查找 DataSource。

```
String resource = "mybatis.xml";
InputStream inputStream = Resources.getResourceAsStream(resource);
SqlSessionFactory sqlSessionFactory =
                new SqlSessionFactoryBuilder( ).build(inputStream);
```

（5）默认使用 JndiDataSourceFactory 创建 DataSource。<dataSource type="JNDI">的实现类是 JndiDataSourceFactory。

```
public class JndiDataSourceFactory implements DataSourceFactory {
    public static final String INITIAL_CONTEXT = "initial_context";
    public static final String DATA_SOURCE = "data_source";
    public static final String ENV_PREFIX = "env.";
    private DataSource dataSource;
}
```

4）使用第三方数据源

通过实现 org.apache.ibatis.datasource.DataSourceFactory 接口，可以使用第三方数据源，参考 UnpooledDataSourceFactory、JndiDataSourceFactory 的实现。

```
public class UnpooledDataSourceFactory
                        implements DataSourceFactory {}
public class JndiDataSourceFactory
                        implements DataSourceFactory {}
```

c3p0 是用途最广的第三方数据库连接池。下面通过示例演示如何在 MyBatis 项目中使用 c3p0 搭建数据库连接池。具体操作如下。

（1）导入 c3p0-0.9.5.4.jar 和依赖包 mysql-connector-java-8.0.11.jar。

（2）自定义 C3P0DataSourceFactory 类，该类继承自 UnpooledDataSourceFactory，这间接实现了 DataSourceFactory 接口。

```
public class C3P0DataSourceFactory extends UnpooledDataSourceFactory {
    public C3P0DataSourceFactory() {
        this.dataSource = new ComboPooledDataSource();
    }
}
```

（3）在 mybatis.xml 中配置 C3P0DataSourceFactory。db.properties 的配置不变。

```
<environment id="development">
    <transactionManager type="JDBC" />
    <dataSource type="com.icss.factory.C3P0DataSourceFactory">
        <property name="driverClass" value="${driver}" />
        <property name="jdbcUrl" value="${url}" />
        <property name="user" value="${username}" />
        <property name="password" value="${password}" />
        <property name="initialPoolSize" value="5"/>
       <property name="maxPoolSize" value="20"/>
       <property name="minPoolSize" value="5"/>
    </dataSource>
</environment>
```

（4）调用 c3p0 配置的环境，创建 SqlSessionFactory。

```
String resource = "mybatis.xml";
InputStream inputStream = Resources.getResourceAsStream(resource);
SqlSessionFactory sqlSessionFactory = new
      SqlSessionFactoryBuilder().build(inputStream,"development");
```

11.4.8 配置 databaseIdProvider

databaseIdProvider 元素的主要作用是支持不同厂商的数据库，即同时支持多个数据库。

```
 <databaseIdProvider type="DB_VENDOR">
     <property name="MySQL" value="mysql"/>
     <property name="Oracle" value="oracle" />
</databaseIdProvider>
```

支持多数据库的传统做法是生成多套映射文件，在 mybatis.xml 中配置使用哪套映射文件。这个做法有很大的缺陷——在多套映射文件中，对于相同接口的实现，多数代码是相同的，这就产生了很多冗余配置。如果代码变动，则多套文件需要同时修改，这给开发额外增加了很大的工作量。

MyBatis 使用 databaseIdProvider 完美解决了这个问题。

示例 11-3：查询用户列表，同时支持 Oracle 和 MySQL。

具体操作如下。

（1）配置 mybatis.xml。

```
<environments default="development">
        <environment id="developmentMysql">
            <transactionManager type="JDBC" />
            <dataSource type="POOLED">
                <property name="driver" value="${driver}" />
                <property name="url" value="${url}" />
                <property name="username" value="${username}" />
                <property name="password" value="${password}" />
            </dataSource>
        </environment>
        <environment id="developmentOracle">
            <transactionManager type="JDBC" />
            <dataSource type="POOLED">
                <property name="driver" value="${driver2}" />
                <property name="url" value="${url2}" />
                <property name="username" value="${username2}" />
                <property name="password" value="${password2}" />
                <property name="poolMaximumActiveConnections" value="50" />
            </dataSource>
        </environment>
    </environments>
    <databaseIdProvider type="DB_VENDOR">
            <property name="MySQL" value="mysql"/>
        <property name="Oracle" value="oracle" />
    </databaseIdProvider>
```

（2）配置 db.properties。

```
driver=com.mysql.cj.jdbc.Driver
url=jdbc:mysql://localhost:3306/staff?useSSL=false
password=123456
```

```
driver2=oracle.jdbc.driver.OracleDriver
url2=jdbc:oracle:thin:@10.3.35.211:1521:orcl
username2=staff
password2=123456
```

（3）在创建 SqlSessionFactory 时，指明使用哪个数据库环境。

```
String resource = "mybatis.xml";
InputStream inputStream = Resources.getResourceAsStream(resource);
SqlSessionFactory sqlSessionFactory =
    new SqlSessionFactoryBuilder().build(inputStream,"developmentMysql");
```

（4）对于同一个 Mapper 接口，在不同数据库中的实现不同。在同一个映射文件中，同时配置一个接口的多个不同实现。

```
<mapper namespace="com.icss.mapper.IUserMapper">
    <select id="getAllUser" resultType="User" databaseId="mysql">
        select uname from tuser
    </select>
    <select id="getAllUser" resultType="User" databaseId="oracle">
        select * from tuser
    </select>
</mapper>
```

总之，MyBatis 的上述代码非常简洁，没有产生任何冗余。如果使用 Hibernate 框架，会采用同一个接口有两个实现类的方式，相同的代码通常会抽取到公用类中，可这又会造成实现类太多的问题。通过比较，还是 MyBatis 的这个实现方案更完美。

11.4.9 配置映射文件的路径

要配置 SQL 映射文件的路径，可以使用多种配置格式。

- 使用相对路径。

```
<mappers>
    <mapper resource="org/mybatis/builder/AuthorMapper.xml" />
    <mapper resource="org/mybatis/builder/BlogMapper.xml" />
    <mapper resource="org/mybatis/builder/PostMapper.xml" />
</mappers>
```

- 使用绝对路径。

```
<mappers>
    <mapper url="file:///var/mappers/AuthorMapper.xml" />
    <mapper url="file:///var/mappers/BlogMapper.xml" />
    <mapper url="file:///var/mappers/PostMapper.xml" />
</mappers>
```

- 注册 Mapper 接口。

```
<mappers>
    <mapper class="org.mybatis.builder.AuthorMapper"/>
```

```
    <mapper class="org.mybatis.builder.BlogMapper"/>
    <mapper class="org.mybatis.builder.PostMapper"/>
</mappers>
```

- 注册指定包内的所有接口。

```
<mappers>
    <package name="org.mybatis.builder"/>
</mappers>
```

11.5 配置映射文件

与传统的 JDBC 代码相比，使用 XML 映射文件，配置 SQL 语句，可以减少 50%以上的代码，极大地简化开发工作。

在映射文件中，可以使用如下元素。

- cache：对指定的命名空间配置缓存。
- cache-ref：从另外一个命名空间引用缓存。
- resultMap：描述如何从数据库中加载结果集。
- parameterMap：已废弃。
- sql：定义可重用的 SQL 代码块。
- insert：插入声明。
- update：更新声明。
- delete：删除声明。
- select：查询声明。

11.5.1 mapper 元素

在 mybatis.xml 中配置映射文件的位置。

```
<mappers>
    <mapper resource="com/icss/mapper/userMappper.xml" />
</mappers>
```

SQL 映射文件的根元素<mapper>的唯一属性 namespace 与 Mapper 接口对应。注意，schema 不能配置错误。

示例 11-4：配置映射文件 userMapper.xml 的根元素。

具体代码如下。

```xml
<?xml version="1.0" encoding="UTF-8"?>
<!DOCTYPE mapper
    PUBLIC "-//mybatis.org//DTD Mapper 3.0//EN"
    "http://mybatis.org/dtd/mybatis-3-mapper.dtd">
<mapper namespace="com.icss.mapper.IUserMapper">
</mapper>
```

11.5.2 select 元素

在 MyBatis 中，查询是应用最多的操作。一次插入、更新或者删除可能对应多次查询操作。

关于 select 元素的示例如下。

```xml
<select id="selectPerson" parameterType="int"
    parameterMap="deprecated" resultType="hashmap"
    resultMap="personResultMap" flushCache="false" useCache="true"
    timeout="10" fetchSize="256" statementType="PREPARED"
    resultSetType="FORWARD_ONLY">
</select>
```

1. select 元素的属性

select 元素可以使用的属性见表 11-4。

表 11-4　select 元素可以使用的属性

属性	描述
id	在当前命名空间中，必须是唯一识别符。它与 Mapper 接口中的某个方法名对应
parameterType	从接口中传入的类名或别名，此参数可选。因为 TypeHandler 可以根据实际的接口参数进行推测
parameterMap	已废弃
resultType	返回结果的类名或别名。如果返回集合，则此处只需要写集合中的元素类型，而不是集合自身
resultMap	对于外部 resultMap 的引用。返回值可以选择 resultType 或 resultMap，不能同时用两个
flushCache	若设置为 true，每次查询会引发本地缓存和二级缓存的刷新。默认值为 false
useCache	设置为 true，将引发结果集被缓存到二级缓存中。默认值为 true
timeout	等候数据库返回结果的最长时间，如果超时，抛出异常。默认值为 unset，即依赖数据库驱动的设置
fetchSize	暗示驱动批量返回结果数据。默认值是 unset
statementType	可选 STATEMENT、PREPARED 或 CALLABLE，指示 MyBatis 使用 Statement、PreparedStatement 或 CallableStatement。默认使用的是 PREPARED
resultSetType	可选 FORWARD_ONLY、SCROLL_SENSITIVE、SCROLL_INSENSITIVE、DEFAULT，默认值为 unset
databaseId	与 databaseIdProvider 中配置的数据库对应
resultOrdered	用于嵌套查询的结果集设置，默认值为 false
resultSets	用于多结果集查询

2. 为实现用户登录传递参数

使用 MyBatis，完成用户登录功能。具体步骤如下。

（1）定义 Mapper 接口。

```
public interface IUserMapper {
    public User login(@Param("sno")String sno,
                @Param("pwd") String pwd) throws Exception;
}
```

（2）映射 SQL。在映射文件中，使用“#{参数名}”形式接收参数。在 Mapper 接口中，必须使用@Param( )声明参数。

```
<select id="login" resultType="User" >
    select * from tuser where sno=#{sno} and pwd=#{pwd}
</select>
```

（3）测试，并观察 SQL 的输出。

```
public static void main(String[] args) {
    IUser iu = new UserBiz();
    try {
        User user = iu.login("101000123", "123456");
        if(user != null) {
            System.out.println(user.getSno() + "登录成功");
        }else {
            System.out.println("登录失败");
        }
    } catch (Exception e) {
        e.printStackTrace();
    }
}
```

输出结果如下。

```
DEBUG - Opening JDBC Connection
DEBUG - Checked out connection 8199481 from pool.
DEBUG - ==>  Preparing: select * from tuser where sno=? and pwd=?
DEBUG - ==> Parameters: 101000123(String), 123456(String)
DEBUG - <==      Total: 1
101000123 登录成功
```

11.5.3 插入、删除和更新元素

插入、删除、更新元素的属性有很多相似之处，本节介绍如何实现这些操作。

在 MyBatis 中，关于元素属性的插入、删除、更新的示例如下。

```
<insert id="insertAuthor"
        parameterType="domain.blog.Author"
        flushCache="true"
        statementType="PREPARED"
        keyProperty=""
        keyColumn=""
        useGeneratedKeys=""
```

```
        timeout="20">
<update id="updateAuthor"
        parameterType="domain.blog.Author"
        flushCache="true"
        statementType="PREPARED"
        timeout="20">
<delete id="deleteAuthor"
        parameterType="domain.blog.Author"
        flushCache="true"
        statementType="PREPARED"
        timeout="20">
```

插入、删除、更新元素属性的 SQL 示例如下。

```
<insert id="insertAuthor">
    insert into Author (id,username,password,email,bio)
    values (#{id},#{username},#{password},#{email},#{bio})
</insert>
<update id="updateAuthor">
    update Author set username = #{username},password = #{password},
    email = #{email},bio = #{bio} where id = #{id}
</update>
<delete id="deleteAuthor">
    delete from Author where id = #{id}
</delete>
```

插入、更新和删除元素的属性见表 11-5。

表 11-5　　插入、更新和删除元素的属性

属性	描述
id	在当前名称空间中，必须是唯一识别符。它与 Mapper 接口中的某个方法名对应
parameterType	从接口中传入的类名或别名，此参数可选。因为 TypeHandler 可以根据实际的接口参数进行推测
parameterMap	已废弃
flushCache	若设置为 true，每次查询会引发本地缓存和二级缓存的刷新。默认值为 true
timeout	等候数据库返回结果的最长时间，若超时，抛出异常。默认值为 unset，即依赖数据库驱动的设置
statementType	可选 STATEMENT、PREPARED 或 CALLABLE，指示 MyBatis 使用 Statement、PreparedStatement 或 CallableStatement。默认使用的是 PREPARED
databaseId	与 databaseIdProvider 中配置的数据库对应
useGeneratedKeys	仅对插入和更新有效。告诉 MyBatis 使用 JDBC getGeneratedKeys 方法返回数据库自增长的主键值。默认值为 false
keyProperty	仅对插入和更新有效。识别与数据库自增长主键对应的一个属性
keyColumn	设置表中某列的名字与自增长主键对应

注意：这里着重介绍 statementType，默认情况下使用的是 PreparedStatement，这是一个预编译的声明，会提高 SQL 的运行效率。在开发实践中，更新/删除/插入操作执行之后，系统没有任何错误提示，但是从数据库中确实找不到执行结果。这是由预编译产生的缓存造成的。这时修改 statementType 为 Statement，把参数当成字符串传入即可。

案例 11-2：修改用户密码

使用 update 元素，完成修改密码的功能。操作步骤如下。

（1）定义 Mapper 接口。

```
public interface IUserMapper {
    public void updatePassword(@Param("sno")String sno,
                    @Param("pwd")String newPwd) throws Exception;
}
```

（2）配置映射的 SQL。

```
<mapper namespace="com.icss.mapper.IUserMapper">
    <update id="updatePassword">
        update tuser set pwd=#{pwd} where sno=#{sno}
    </update>
</mapper>
```

（3）测试。

① 测试代码如下。

```
public static void main(String[] args) {
    IUser iu = new UserBiz();
    try {
        iu.updatePassword("101000123", "12345678");
        System.out.println("密码修改完毕");
    } catch (Exception e) {
        e.printStackTrace();
    }
}
```

② 测试结果如下。

```
DEBUG - Opening JDBC Connection
DEBUG - Checked out connection 8199481 from pool.
DEBUG - ==>  Preparing: select * from tuser where sno=? and pwd=?
DEBUG - ==> Parameters: 101000123(String), 123456(String)
DEBUG - <==      Total: 1
101000123 登录成功
```

（4）SqlSessionFactory 的 openSession( )应设置自动提交模式为 true，否则密码不会修改成功。

```
public class BaseDao {
    private static SqlSessionFactory sqlSessionFactory;
    public SqlSession getSession() {
          return  sqlSessionFactory.openSession(true);
    }
}
```

11.5.4 项目案例：新增员工

继续以员工系统为例，持久层采用 MyBatis。

新增员工的同时增加一个默认用户，在逻辑层使用事务保证员工和用户同时添加成功。

1. 事务封装

在 SqlSessionFactory 中，openSession(true)表示自动提交模式，openSession(false)表示编程式事务模式。事务封装模式与直接封装数据库连接非常相似。

```
public class BaseDao {
    private static SqlSessionFactory sqlSessionFactory;
    protected SqlSession session;
    public SqlSession getSession() {
        return session;
    }
    public void setSession(SqlSession session) {
        this.session = session;
    }
    static {
        try {
            String resource = "mybatis.xml";
            InputStream inputStream = Resources.getResourceAsStream(resource);
            sqlSessionFactory = new SqlSessionFactoryBuilder()
                                    .build(inputStream,"developmentMysql");
        } catch (Exception e) {
            e.printStackTrace();
        }
    }
    public void  openSession(){
        if(session == null ) {
            session = sqlSessionFactory.openSession(true);
        }
    }
    public void closeSession() {
        if(session != null) {
            session.close();
        }
    }
    public void beginTransaction() {
        session = sqlSessionFactory.openSession(false);
    }
    public void commit() {
        if(session != null) {
            session.commit();
        }
    }
    public void rollback() {
        if(session !=null) {
            session.rollback();
        }
    }
}
```

2. 设计 Mapper 接口

持久层接口与 Mapper 接口有时一致，有时并不一致。一个持久层的 DAO 方法可能会调用多个 Mapper 接口中的方法。

```
public interface IStaffDao extends IBaseDao{
    public void addStaff(Staff staff) throws Exception;
}
public interface IUserDao extends IBaseDao{
    public void addUser(User user) throws Exception ;
}
```

每个 Mapper 接口对应一条 SQL 语句，持久层的每个方法可能调用多个 Mapper 方法。有时持久层会对 Mapper 方法返回的内容进行处理，这也会导致两个接口不一致。

```
public interface IStaffMapper {
    public void addStaff(Staff staff) throws Exception;
}
public interface IUserMapper {
    public void addUser(User user) throws Exception ;
}
```

3. 映射 SQL

通过以下代码映射 SQL。

```
<mapper namespace="com.icss.mapper.IStaffMapper">
   <insert id="addStaff" parameterType="com.icss.entity.Staff">
        insert into tstaff
                values(#{sno},#{name},#{birthday},#{address},#{tel})
   </insert>
</mapper>
<mapper namespace="com.icss.mapper.IUserMapper">
    <insert id="addUser" parameterType="com.icss.entity.User">
        insert into tuser
                       values(#{uname},#{sno},#{pwd},#{role})
    </insert>
</mapper>
```

4. 逻辑层的事务控制

在逻辑层，事务控制的关键是多个持久层方法必须使用同一个 SqlSession 对象。

```
public void addStaffUser(Staff staff) throws Exception {
    Log.logger.info("StaffBiz-->>添加员工与用户");
    StaffDao staffDao = new StaffDao();
    UserDao userDao = new UserDao();
    User user = new User();
    user.setSno(staff.getSno());
    user.setUname(staff.getSno());
```

```
        user.setPwd("123456");
        user.setRole(IRole.COMMON_USER);
        try {
            staffDao.beginTransaction();
            staffDao.addStaff(staff);
            userDao.setSession(staffDao.getSession()); //session 重用
            userDao.addUser(user);
            staffDao.commit();
        } catch (Exception e) {
            staffDao.rollback();
            throw e;
        }finally {
            staffDao.closeSession();
        }
    }
```

5. 测试

事务代码必须要测试正常提交和异常回滚两种情况。尤其是异常测试，只有异常数据可以回滚，才能说明事务控制是正确的。

（1）通过以下代码，进行正常提交测试。

```
public static void main(String[] args) {
    Staff staff = new Staff();
    staff.setSno("101000135");
    staff.setName("rose");
    SimpleDateFormat sdf = new SimpleDateFormat("yyyy-MM-dd");
    try {
        staff.setBirthday(sdf.parse("1995-10-1"));
    } catch (Exception e) {
    }
    staff.setAddress("北京朝阳区建国门");
    staff.setTel("1352245466221");
    try {
        IStaff biz = new StaffBiz();
        biz.addStaffUser(staff);
        System.out.println(staff.getSno() + "创建成功....");
    } catch (Exception e) {
        e.printStackTrace();
    }
}
```

测试结果如下。

```
DEBUG - Setting autocommit to false on JDBC Connection
DEBUG - ==> Preparing: insert into tstaff(sno,name,birthday,address,tel)
            values(?,?,?,?,?)
DEBUG - ==> Parameters: 101000135(String), rose(String), 1995-10-01
            , 北京朝阳区建国门(String), 1352245466221(String)
DEBUG - <==   Updates: 1
DEBUG - ==> Preparing: insert into tuser values(?,?,?,?)
```

```
DEBUG - ==> Parameters: 101000135(String), 101000135(String)
                    , 123456(String), 2(Integer)
DEBUG - <==    Updates: 1
DEBUG - Committing JDBC Connection
DEBUG - Resetting autocommit to true on JDBC Connection
DEBUG - Closing JDBC Connection
DEBUG - Returned connection 4792741 to pool.
101000135创建成功....
```

（2）通过以下代码，进行异常回滚测试。

```
public void addUser(User user) throws Exception {
    this.openSession();
    IUserMapper mapper = this.session.getMapper(IUserMapper.class);
    mapper.addUser(user);
    throw new RuntimeException("异常测试 ....");
}
```

测试结果如下。

```
DEBUG - Setting autocommit to false on JDBC Connection
DEBUG - ==>  Preparing: insert into tstaff(sno,name,birthday
                    ,address,tel) values(?,?,?,?,?)
DEBUG - ==> Parameters: 101000136(String), rose(String), 1995-10-01
               , 北京朝阳区建国门(String), 1352245466221(String)
DEBUG - <==    Updates: 1
DEBUG - ==>  Preparing: insert into tuser values(?,?,?,?)
DEBUG - ==> Parameters: 101000136(String), 101000136(String)
                  , 123456(String), 2(Integer)
DEBUG - <==    Updates: 1
DEBUG - Rolling back JDBC Connection
DEBUG - Resetting autocommit to true on JDBC Connection
java.lang.RuntimeException: 异常测试 ....
    at com.icss.dao.impl.UserDao.addUser(UserDao.java:18)
    at com.icss.biz.impl.StaffBiz.addStaffUser(StaffBiz.java:28)
    at com.icss.ui.TestAddStaff.main(TestAddStaff.java:25)
DEBUG - Closing JDBC Connection
DEBUG - Returned connection 4792741 to pool.
```

11.5.5 项目案例：员工打卡

员工系统记录员工每天的打卡信息。对于“打卡记录”表，主键 aid 只用于唯一识别记录，没有业务含义，因此使用数据库的自增长模式。MySQL 使用 auto_increment 控制增长模式，Oracle 使用 sequence 控制增长模式。在高并发系统中，采用数据库自身的自增长主键非常有用。

1. 设计表结构

MySQL 表的结构如图 11-8 所示。

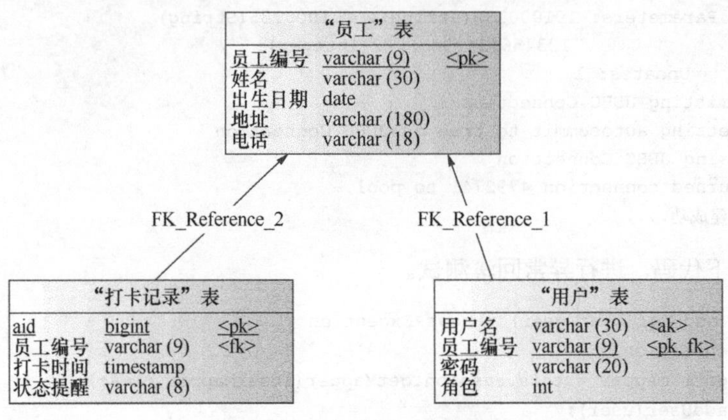

图 11-8　MySQL 表的结构

具体的实现方式如下。

```
create table TCard (
        aid bigint not null AUTO_INCREMENT,
        sno varchar(9),
        ctime timestamp not null,
        info varchar(8),
        primary key (aid)
);
alter table TCard add constraint FK_Reference_2 foreign key (sno)
        references TStaff (sno) ;
```

"打卡记录"表的主键 aid 采用 auto_increment 模式。打卡时间采用 timestamp 类型。

Oracle 表的结构如图 11-9 所示。

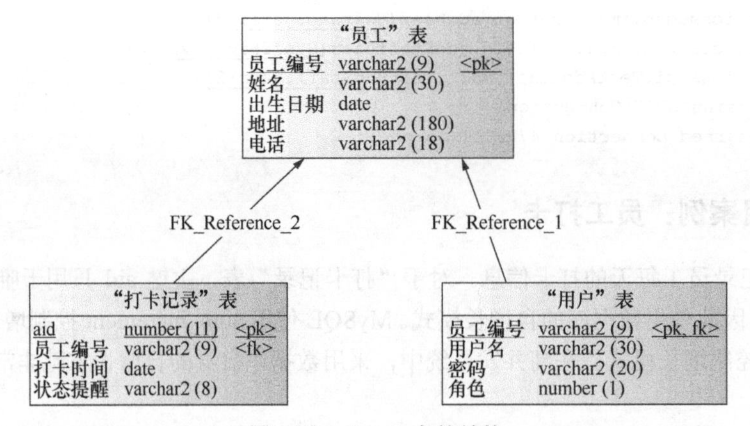

图 11-9　Oracle 表的结构

具体的实现方式如下。

```
create sequence card_aid;
     create table TCard (
         aid number(11) not null,
         sno varchar2(9),
         ctime date not null,
         info varchar2(8),
         primary key (aid)
);
alter table TCard add constraint FK_Reference_2 foreign key (sno)
            references TStaff (sno);
```

“打卡记录”表的主键 aid 采用 sequence 模式。Oracle 中的字符类型使用 varchar2，整型使用 number。在 Oracle 中，打卡时间使用 date 类型，虽然 Oracle 有 timestamp 类型，但是 timestamp 类型的精度太高，可以精确到秒后 6 位小数，这里采用 date 类型即可完全满足要求。

2. 设计 Mapper 接口

为了设计 Mapper 接口，添加打卡记录，传入 Card 实体对象。

```
public interface IStaff {
    /**
     * 添加打卡记录
    * @param card
    * @throws Exception
    */
    public void addCard(Card card) throws Exception;
}
public interface IStaffDao extends IBaseDao{
    public void addCard(Card card) throws Exception;
}
public interface IStaffMapper {
    public void addCard(Card card) throws Exception;
}
```

3. 配置映射

（1）配置 MySQL 映射。

```
<mapper namespace="com.icss.mapper.IStaffMapper">
    <insert id="addCard" parameterType="Card"  databaseId="mysql">
        insert into tcard(sno,ctime,info) values(#{sno},now(),#{info})
    </insert>
</mapper>
```

打卡时间采用 MySQL 的 now( )函数。主键 aid 采用 auto_increment 模式，aid 不用在 SQL 中出现。

（2）配置 Oracle 映射。

```
<mapper namespace="com.icss.mapper.IStaffMapper">
    <insert id="addCard" parameterType="Card" databaseId="oracle">
        insert into tcard(aid,sno,ctime,info)
                values(card_aid.nextval,#{sno},sysdate,#{info})
```

```
    </insert>
</mapper>
```

打卡时间采用 Oracle 的 sysdate( )函数。主键 aid 采用 sequence 模式，card_aid.nextval 获取自增长 id。

4. 测试

(1) 选择 MySQL 环境进行测试。

```
String resource = "mybatis.xml";
InputStream inputStream = Resources.getResourceAsStream(resource);
SqlSessionFactory sqlSessionFactory =
     new SqlSessionFactoryBuilder().build(inputStream,"developmentMysql");
```

① 测试代码如下。

```
public static void main(String[] args) {
    IStaff biz = new StaffBiz();
    Card card = new Card();
    card.setCtime(new Date());
    card.setSno("101000125");
    card.setInfo("正常");
    try {
        biz.addCard(card);
    } catch (Exception e) {
        e.printStackTrace();
    }
}
```

② 测试结果如下。

```
DEBUG - Opening JDBC Connection
DEBUG - Checked out connection 4792741 from pool.
DEBUG - ==>  Preparing: insert into tcard(sno,ctime,info) values(?,now(),?)
DEBUG - ==> Parameters: 101000125(String), 正常(String)
DEBUG - <==    Updates: 1
DEBUG - Closing JDBC Connection [com.mysql.cj.jdbc.ConnectionImpl]
DEBUG - Returned connection 4792741 to pool.
```

(2) 选择 Oracle 环境进行测试。

```
String resource = "mybatis.xml";
InputStream inputStream = Resources.getResourceAsStream(resource);
SqlSessionFactory sqlSessionFactory =
     new SqlSessionFactoryBuilder().build(inputStream,"developmentOracle");
```

测试代码不变，经过测试可以发现卡添加成功。

5. 返回自增长 ID

如果希望返回自增长 ID，如打卡后的自增长主键 aid，则需要使用 useGeneratedKeys 属性，操作如下。

（1）配置 SQL 映射。

```
<mapper namespace="com.icss.mapper.IStaffMapper">
    <insert id="addCard" parameterType="Card" databaseId="mysql"
            keyColumn="aid" keyProperty="aid" useGeneratedKeys="true">
        insert into tcard(sno,ctime,info) values(#{sno},now(),#{info})
    </insert>
    <insert id="addCard" parameterType="Card" databaseId="oracle"
            keyColumn="aid" keyProperty="aid" useGeneratedKeys="true">
        insert into tcard(aid,sno,ctime,info)
                    values(card_aid.nextval,#{sno},sysdate,#{info})
    </insert>
</mapper>
```

（2）接收返回的自增长主键。自增长主键会自动存放到输入参数的映射对象中，本例中的自增长主键会自动存储到 card 对象的 aid 属性中，通过 card.getAid( )即可取出。

```
public class StaffDao extends BaseDao implements IStaffDao{
    public void addCard(Card card) throws Exception {
        this.openSession();
        IStaffMapper mapper = this.session.getMapper(IStaffMapper.class);
        mapper.addCard(card);
        System.out.println("打卡成功，aid=" + card.getAid());
    }
}
```

11.5.6 配置参数

1. 传递基本类型参数

简单类型的参数在接口映射中需要使用@Param("参数名")声明。在 mapper 文件中使用#{变量名}调用时，无须指定类型，由 MyBatis 自动识别。

```
public interface IUserMapper {
    public User login(@Param("sno")String sno,
                @Param("pwd")String pwd) throws Exception;
}
<mapper namespace="com.icss.mapper.IUserMapper">
    <select id="login" resultType="com.icss.entity.User">
        select * from tuser where sno = #{sno} and pwd=#{pwd}
    </select>
</mapper>
```

2. 传递对象类型参数

对象类型的参数（如本例中的 User），无须使用@Param( )声明。在 SQL 映射中直接使用#{属性名}调用，非常简单。

```
public interface IUserMapper {
    public void addUser(User user) throws Exception ;
}
<mapper namespace="com.icss.mapper.IUserMapper">
    <insert id="addUser" parameterType="com.icss.entity.User">
     insert into tuser values(#{uname},#{sno},#{pwd},#{role})
    </insert>
</mapper>
```

3. 指定参数类型

使用如下格式，可以指定参数类型。

```
#{property,javaType=int,jdbcType=NUMERIC}
```

当 MyBatis 自动适配的 TypeHandler 转换的类型不符合预期时，可以手动对某个参数的类型进行转换，也可以单独使用 jdbcType。

```
<insert id="addCard" parameterType="Card" useGeneratedKeys="true"
              keyColumn="aid" keyProperty="aid" databaseId="oracle">
    insert into tcard(aid,sno,ctime,info)
                     values(card_aid.nextval,#{sno},
                    #{ctime, jdbcType=TIMESTAMP},#{info})
 </insert>
```

另外，还可以按如下格式，进一步指定类型映射器。

```
#{age,javaType=int,jdbcType=NUMERIC,typeHandler=MyTypeHandler}
```

示例代码如下。

```
<insert id="addCard" parameterType="Card" useGeneratedKeys="true"
              keyColumn="aid" keyProperty="aid" databaseId="oracle">
    insert into tcard(aid,sno,ctime,info)
         values(card_aid.nextval,#{sno},
        #{ctime,typeHandler=org.apache.ibatis.type.DateTypeHandler},#{info})
</insert>
```

还可以在参数中指定 number 类型的精度。

```
#{height,javaType=double,jdbcType=NUMERIC,numericScale=2}
```

对于值可能为 null 的列，常用 jdbcType 属性设置，以防止出错。

```
#{middleInitial,jdbcType=VARCHAR}
```

4. #{}与${}

默认情况下，使用#{}语法，MyBatis 会用传入的值代替 PreparedStatement 的参数“?”。有时，我们不希望使用传入的参数替换 PreparedStatement 的“?”，例如：

```
ORDER BY ${columnName}
```

其中，列的名字是变量，即按照传入的列名排序。

这种情况下，可以采用如下方案。

```
${...}
```

这表示用字符串替代参数“?”。

示例 11-5：查询所有用户，按照输入的列名排序。

具体实现方式如下。

```
public interface IUserMapper {
    public List<User> getAllUser(@Param("columnName")String columnName)
                            throws Exception;
}
<mapper namespace="com.icss.mapper.IUserMapper">
    <select id="getAllUser" resultType="com.icss.entity.User">
        select * from tuser order by ${columnName}
    </select>
</mapper>
```

测试代码如下。

```
public static void main(String[] args) {
    IUser iu = new UserBiz();
    List<User> users = iu.getAllUser();
}
```

测试结果如下。

```
DEBUG - Opening JDBC Connection
DEBUG - Created connection 25545510.
DEBUG - ==>  Preparing: select * from tuser order by uname
DEBUG - ==> Parameters:
DEBUG - <==      Total: 13
```

示例 11-6：修改密码。

方法 1：使用 PreparedStatement 模式。

如下代码{}使用#会替换 PreparedStatement 中的“?”，正常情况下会正确更新密码。但是由于 PreparedStatement 的预编译机制，少数情况下 SQL 语句不会执行。即使提示更新成功，数据库也并未发生变化。这时，可以采用 Statement 模式。

```
public interface IUserMapper {
    public void updatePassword(@Param("sno")String sno,
                        @Param("newPwd")String newPwd) throws Exception;
}
<mapper namespace="com.icss.mapper.IUserMapper">
    <update id="updatePassword"  >
        update tuser set pwd=#{newPwd} where sno=#{sno}
    </update>
</mapper>
```

测试代码如下。

```
public static void main(String[] args) {
    IUser biz = new UserBiz();
    try {
        biz.updatePassword("101000123", "12345");
        System.out.println("密码修改完毕...");
    } catch (Exception e) {
        e.printStackTrace();
    }
}
```

测试结果如下。

```
DEBUG - Opening JDBC Connection
DEBUG - Created connection 7526964.
DEBUG - ==>  Executing: update tuser set pwd = 12345 where sno='101000123'
密码修改完毕...
```

方法 2：使用 Statement 模式。

Statement 不是预编译模式，不会产生更新不成功的现象。

```
public interface IUserMapper {
    public void updatePassword2(@Param("strWhere")String strWhere)
                                throws Exception;
}
<mapper namespace="com.icss.mapper.IUserMapper">
    <update id="updatePassword2" statementType="STATEMENT" >
        update tuser set pwd = ${strWhere}
    </update>
</mapper>
```

持久层调用的 Mapper 接口如下。

```
public void updatePassword(String sno, String newPwd) throws Exception {
    SqlSession session = this.getSession();
    IUserMapper mapper = session.getMapper(IUserMapper.class);
    String strWhere = newPwd + " where sno='" + sno + "'";
    mapper.updatePassword2(strWhere);
}
```

11.5.7 resultMap

resultMap 元素是 MyBatis 映射文件中最重要、功能最强大的元素之一。使用 JDBC 查询，返回 ResultSet，是最常用的数据库操作之一。resultMap 的设计简化了在映射文件中编写复杂 SQL 的操作。

1. 列名与属性名之间的映射

通过衔接 ResultSet 与映射实体，resultMap 可以解决表中列的名字与实体中属性名不一致的问题。

下面给出一个示例。

（1）定义实体类的属性。

```
public class User {
    private int id;
    private String username;
    private String hashedPassword;
}
```

（2）指定表的列名，包括 user_id，user_name、hashed_password。

（3）定义 Mapper 接口。

```
public interface IUserMapper {
    public User selectUsers(@Param("id")String id);
}
```

（4）配置映射文件。传统上，通过转换别名，让列名与实体的属性名匹配。

```
<select id="selectUsers" resultType="User">
    select user_id as "id",
        user_name as "userName",
        hashed_password as "hashedPassword"
    from some_table where id = #{id}
</select>
```

也可使用 resultMap 解决列名与属性名之间的映射问题。

```
<mapper namespace="com.icss.mapper.IUserMapper">
    <resultMap id="userResultMap" type="User">
        <id property="id" column="user_id" />
        <result property="username" column="user_name"/>
        <result property="password" column="hashed_password"/>
    </resultMap>
    <select id="selectUsers" resultMap="userResultMap">
        select user_id, user_name, hashed_password
        from some_table where id = #{id}
    </select>
</mapper>
```

2. 数据类型适配

当 MyBatis 默认选择的 typeHandler 不能满足要求时，可以使用 resultMap 进行 Java 类型与 JDBC 类型之间的适配。

下面给出一个示例。

```
<resultMap id="userResultMap" type="User">
    <id property="id" column="user_id"
                javaType="int" jdbcType="INTEGER"/>
    <result property="username" column="user_name"
                javaType="String" jdbcType="VARCHAR"/>
```

```
    <result property="password" column="hashed_password"
                    javaType="String" jdbcType="VARCHAR"/>
</resultMap>
```

注意以下几点。

- resultMap 的子元素 id 表示唯一识别的属性或字段，如主键。
- result 子元素对应普通列和属性。
- 设置 id，以帮助 MyBatis 优化性能，如在缓存中用 id 识别。

3．resultMap 的高级配置

ORM 中的难点是复杂 SQL 的映射问题。Hibernate 在映射文件中配置实体与表的对应关系，HQL 查询结果按照映射配置文件自动填充到实体对象中，配合 Hibernate 懒加载机制，多数复杂 SQL 可以完成。MyBatis 没有表与实体之间的映射配置文件，它只有接口与 SQL 之间的映射配置文件。因此，要建立实体与表之间的映射关系，就需要使用额外的 resultMap 配置。

在博客系统中，一个博客有主题、id、内容等，每个博客有唯一的作者，以及多个 Post 对象。

下面给出一个示例。

（1）配置 resultMap。配置中使用的元素会后面逐一讲解。

```
<resultMap id="detailedBlogResultMap" type="Blog">
    <constructor>
        <idArg column="blog_id" javaType="int" />
    </constructor>
    <result property="title" column="blog_title" />
    <association property="author" javaType="Author">
        <id property="id" column="author_id" />
        <result property="username" column="author_username" />
        <result property="password" column="author_password" />
        <result property="email" column="author_email" />
        <result property="bio" column="author_bio" />
        <result property="favouriteSection"
            column="author_favourite_section" />
    </association>
    <collection property="posts" ofType="Post">
        <id property="id" column="post_id" />
        <result property="subject" column="post_subject" />
        <association property="author" javaType="Author" />
        <collection property="comments" ofType="Comment">
            <id property="id" column="comment_id" />
        </collection>
        <collection property="tags" ofType="Tag">
            <id property="id" column="tag_id" />
        </collection>
```

```
        <discriminator javaType="int" column="draft">
            <case value="1" resultType="DraftPost" />
        </discriminator>
    </collection>
</resultMap>
```

（2）配置 SQL 映射。

```
<select id="selectBlogDetails" resultMap="detailedBlogResultMap">
    select
    B.id as blog_id,
    B.title as blog_title,
    B.author_id as blog_author_id,
    A.id as author_id,
    A.username as author_username,
    A.password as author_password,
    A.email as
    author_email,
    A.bio as author_bio,
    A.favourite_section as
    author_favourite_section,
    P.id as post_id,
    P.blog_id as post_blog_id,
    P.author_id as post_author_id,
    P.created_on as post_created_on,
    P.section as post_section,
    P.subject as post_subject,
    P.draft as draft,
    P.body as post_body,
    C.id as comment_id,
    C.post_id as comment_post_id,
    C.name as comment_name,
    C.comment as comment_text,
    T.id as tag_id,
    T.name as tag_name
    from Blog B
    left outer join Author A on B.author_id =
    A.id
    left outer join Post P on B.id = P.blog_id
    left outer join Comment
    C on P.id = C.post_id
    left outer join Post_Tag PT on PT.post_id = P.id
    left outer join Tag T on PT.tag_id = T.id
    where B.id = #{id}
</select>
```

（3）定义实体类。

```
public class Blog {
    private int id;
    private String title;
```

```
    private Author author;
    private List<Post> posts;
    public Blog(int id, String title, Author author, List<Post> posts) {
    }
}
public class Author implements Serializable {
    protected int id;
    protected String username;
    protected String password;
    protected String email;
    protected String bio;
    protected Section favouriteSection;
}
public class Post {
    private int id;
    private Author author;
    private Blog blog;
    private Date createdOn;
    private Section section;
    private String subject;
    private String body;
    private List<Comment> comments;
    private List<Tag> tags;
}
```

其中，constructor 是 resultMap 的子元素。constructor 表示把结果集注入 Blog 的构造函数中。constructor 的参数如下。

- idArg：id 参数，标识结果集中的唯一列，注入构造函数中，用于性能优化。
- arg：普通参数。

下面给出一个示例。

```
public class User {
    // ...
    public User(Integer id, String username, int age) {
    // ...
    }
    // ...
}
<constructor>
    <idArg column="id" javaType="int" />
    <arg column="username" javaType="String" />
    <arg column="age" javaType="_int" />
</constructor>
```

association 是 resultMap 的子元素。association 通常用于表示 resultMap 映射实体与子对象是一对一的关系。

下面给出一个示例。

```
<association property="author" javaType="Author">
    <id property="id" column="author_id" />
    <result property="username" column="author_username" />
</association>
```

collection 是 resultMap 的子元素。collection 表示 resultMap 映射实体与子对象是一对多的关系。在博客系统中，一个 Blog 下有多个 Post 对象。

下面给出一个示例。

```
public class Blog {
    private int id;
    private String title;
    private Author author;
    private List<Post> posts;
}
<collection property="posts" ofType="Post">
        <id property="id" column="post_id" />
        <result property="subject" column="post_subject" />
        <association property="author" javaType="Author" />
</collection>
```

11.5.8 项目案例：查询员工打卡记录

员工打卡记录查询案例是一个多表联合查询的案例。下面我们采用两种方案分别实现这个案例。

该案例中表的结构如图 11-10 所示。

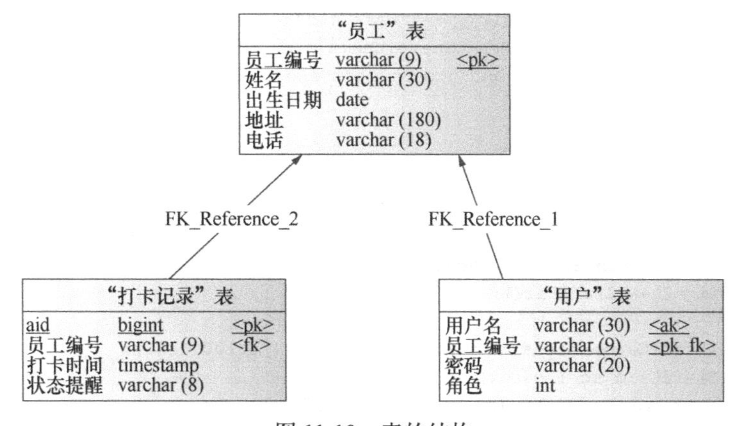

图 11-10 表的结构

1. DTO 方案

数据传输对象（Data Transfer Object，DTO）类似于实体，作用是跨层传输数据。下面我们使用 DTO 作为联合查询的返回结果，传递数据。具体操作如下。

（1）为了实现 3 个表联合查询的结果，新建一个 DTO 类 StaffCard，用于跨层传输。

```
public class StaffCard {
    private  String sno;
    private String name ;   //员工姓名
    private String uname;   //用户名
    private Date ctime;
    private String info;
}
```

（2）定义查询接口。

```
public interface IStaff {
    /**
     * 查询某个员工的打卡信息
    * @return
    */
    public List<StaffCard> getStaffCards(String sno) throws Exception;
}
public interface IStaffDao extends IBaseDao{
    public List<StaffCard> getStaffCards(String sno) throws Exception;
}
public interface IStaffMapper {
    public List<StaffCard> getStaffCards(@Param("sno")String sno)
                                             throws Exception;
}
```

（3）配置 SQL 映射。多表查询的结果通过 resultType 直接映射到 DTO 上。

```
<select id="getStaffCards" resultType="StaffCard">
    select c.ctime,c.info,s.sno,s.name,u.uname from tcard c ,tstaff s,
    tuser u  where c.sno=s.sno and u.sno=s.sno and s.sno=#{sno}
</select>
```

（4）测试。

① 测试代码如下。

```
public static void main(String[] args) {
    IStaff biz = new StaffBiz();
    try {
        List<StaffCard> cards = biz.getStaffCards("101000123");
        for(StaffCard sc : cards) {
            System.out.println(sc.getSno() + ","
                        + sc.getCtime() + "," + sc.getInfo());
        }
    } catch (Exception e) {
```

```
            e.printStackTrace();
        }
    }
```

② 测试结果如下。

```
DEBUG - Opening JDBC Connection
DEBUG - Checked out connection 4792741 from pool.
DEBUG - ==>  Preparing: select c.ctime,c.info,s.sno,s.name,u.uname
  from tcard c ,tstaff s,tuser u where c.sno=s.sno and u.sno=s.sno and s.sno=?
DEBUG - ==> Parameters: 101000123(String)
DEBUG - <==      Total: 3
DEBUG - Closing JDBC Connection
DEBUG - Returned connection 4792741 to pool.
101000123,Tue Jan 14 22:20:38 CST 2020,正常
101000123,Tue Jan 14 22:21:10 CST 2020,正常
101000123,Tue Jan 14 22:21:15 CST 2020,正常
```

2. resultMap 方案

Mapper 层的返回结果为 resultMap，通过 resultMap 映射实体间的对象关系，在查询结果中自动填充多个实体对象。操作步骤如下。

（1）定义 Staff 实体和其他实体对象之间的关系。一个员工对应一个 User 对象，对应多次打卡记录。

```
public class Staff {
    private  String sno;
    private String name ;
    private  Date birthday;
    private  String address;
    private  String tel;
    private User user;
    private List<Card> cards;
}
```

（2）定义查询接口，返回 Staff 集合。

```
public interface IStaff {
    public List<Staff> getStaffCards2(@Param("sno")String sno);
}
public interface IStaffMapper {
    public List<Staff> getStaffCards2(@Param("sno")String sno);
}
```

（3）配置 SQL 映射。使用 resultMap，指明联合查询的结果与实体的对应关系。实体间的关系与表之间的关系类似，也存在一对一、一对多的关系。

```
<resultMap type="Staff" id="staffCardResultMap">
    <id property="sno" column="sno"/>
    <result property="name" column="name"/>
    <association property="user" javaType="User">
```

```
        <result property="uname" column="uname"/>
    </association>
    <collection property="cards" ofType="Card">
        <result property="ctime" column="ctime" />
        <result property="info" column="info"/>
    </collection>
</resultMap>
<select id="getStaffCards2" resultMap="staffCardResultMap">
    select c.ctime,c.info,s.sno,s.name,u.uname from tcard c ,tstaff s,
        tuser u  where c.sno=s.sno and u.sno=s.sno and s.sno=#{sno}
</select>
```

（4）测试。

① 测试代码如下。

```
public static void main(String[] args) {
    IStaff biz = new StaffBiz();
    try {
        List<Staff> staffs =  biz.getStaffCards2("101000123");
        for(Staff sc : staffs) {
            System.out.println(sc.getSno() + ","
                        + sc.getSno() + "," + sc.getName() );
            System.out.println(sc.getUser().getUname() );
            List<Card> cards =  sc.getCards();
            for(Card c : cards) {
                System.out.println(c.getCtime() + "," + c.getInfo());
            }
        }
    } catch (Exception e) {
        e.printStackTrace();
    }
}
```

② 测试结果如下。

```
DEBUG - Checked out connection 4792741 from pool.
DEBUG - ==>  Preparing: select c.ctime,c.info,s.sno,s.name,u.uname
from tcard c ,tstaff s,tuser u where c.sno=s.sno and u.sno=s.sno and s.sno=?
DEBUG - ==> Parameters: 101000123(String)
DEBUG - <==     Total: 2
DEBUG - Closing JDBC Connection [com.mysql.cj.jdbc.ConnectionImpl@4921a5]
DEBUG - Returned connection 4792741 to pool.
101000123,101000123,tom
101000123
Tue Jan 14 22:20:38 CST 2020,正常
Tue Jan 14 22:21:10 CST 2020,正常
```

总之，对于多表联合查询，采用 DTO 方案接收结果，优点是简单明了，缺点是如果系统中存在大量联合查询，则会产生很多 DTO，这增加了程序的复杂度。resultMap 方案是 MyBatis 官方推荐的方案，优点是不会产生多余的类，缺点是需要配置 resultMap 和实体关系等很多内容。

11.5.9 缓存

缓存的作用是在内存中存储常用数据，避免多次查找数据库，从而提高系统性能。默认情况下，基于 MyBatis 的 SqlSession 的本地缓存处于可用状态。缓存分为一级缓存和二级缓存，数据存储在不同类型的缓存中，存续时间和作用域是不一样的。

一级缓存只在 SqlSession 的生命周期内有效。二级缓存的生命周期是 SqlSessionFactory，即它可以跨 SqlSession。二级缓存的使用基于某个 SQL 映射文件。

1. 二级缓存

为了开启或关闭二级缓存，需要在 mybatis.xml 中配置 cacheEnabled。cacheEnabled 的默认值为 true。

```
<settings>
        <setting name="cacheEnabled" value="true"/>
</settings>
```

为了使全局有效的二级缓存可用，还需要在 SQL 映射文件中增加<cache/>配置项。

这个配置项将产生如下默认影响。

- 映射文件中的所有查询结果将被缓存。
- 根据映射文件中的所有插入、更新和删除操作结果，刷新缓存。
- 采用最近最少使用（Least Recently Used，LRU）原则，移除缓存中近期未用的数据。
- 缓存不做定时刷新。
- 缓存最多存储 1024 个对象引用。
- 缓存是可以同时读写的，即并发读写都有安全风险。

另外，还可以不使用默认参数，自己配置二级缓存参数。

在以下代码中，采用先进先出原则移除数据，60s 刷新一次，存储 512 个引用，不允许并发修改。

```
<cache eviction="FIFO" flushInterval="60000"
                      size="512" readOnly="true" />
```

关于数据移除策略，有如下几个选项。

- LRU：近期一直未用的首先移除，这是默认选项。
- FIFO：按照先进先出（First In First Out）的原则移除。

- SOFT：基于垃圾回收器的软引用（Soft Reference）原则。
- WEAK：基于垃圾回收器的弱引用（Weak Reference）原则。

2．定制自己的缓存

可以定制自己的缓存，在使用时声明。

```
<cache type="com.domain.something.MyCustomCache"/>
```

定制自己的缓存的步骤如下。

（1）实现 org.apache.ibatis.cache.Cache 接口。

```
public interface Cache {
    String  getId();
    void    putObject(Object key, Object value);
    Object  getObject(Object key);
    Object  removeObject(Object key);
    void    clear();
    int     getSize();
}
```

（2）在实现类的构造函数中，必须要传入字符型的唯一 id。参考 MyBatis 已实现的缓存类写法。

```
public class BlockingCache implements Cache {
     private long timeout;
    private final Cache delegate;
    private final ConcurrentHashMap<Object, ReentrAntLock> locks;
}
public class FifoCache implements Cache {
    private final Cache delegate;
    private final Deque<Object> keyList;
    private int size;
}
public class LoggingCache implements Cache {
    private final Log log;
    private final Cache delegate;
    protected int requests = 0;
    protected int hits = 0;
}
```

3．cache-ref

在某个命名空间下的映射文件中，可以引用其他映射文件中定义的二级缓存。可以使用 cache-ref 引用其他命名空间中的缓存。

```
<cache-ref namespace="com.someone.application.data.SomeMapper"/>
```

案例 11-3：测试一级缓存

要测试一级缓存，需要完成以下 3 个测试。

测试 1：连续 3 次调用 getAllUser( )方法，观察输入结果。发现 SQL 语句 select * from tuser 只执行了一次，确认后面的两次查询是从缓存中提取的数据。

```
public List<User> getAllUser() throws Exception {
    this.openSession();
    IUserMapper mapper = session.getMapper(IUserMapper.class);
    List<User> allUser = mapper.getAllUser();
                         mapper.getAllUser();
                         mapper.getAllUser();
    return allUser;
}
```

测试结果如下。

```
DEBUG - Checked out connection 4792741 from pool.
DEBUG - ==>  Preparing: select * from tuser
DEBUG - ==> Parameters:
DEBUG - <== Total: 13
```

测试 2：用户登录前，提取所有用户信息，测试发现登录操作并没有使用缓存。

```
public User login(String sno, String pwd) throws Exception {
    this.openSession();
    IUserMapper mapper = this.session.getMapper(IUserMapper.class);
    mapper.getAllUser();
    User user = mapper.login(sno, pwd);
    return user;
}
```

测试结果如下。

```
DEBUG - Checked out connection 4792741 from pool.
DEBUG - ==>  Preparing: select * from tuser
DEBUG - ==> Parameters:
DEBUG - <== Total: 13
DEBUG - ==>  Preparing: select * from tuser where sno = ? and pwd=?
DEBUG - ==> Parameters: 101000123(String), 123(String)
DEBUG - <== Total: 1
DEBUG - Closing JDBC Connection
```

测试 3：多次调用 getUser( )，但是参数不同。

```
public List<User> getAllUser() throws Exception {
    this.openSession();
    IUserMapper mapper = session.getMapper(IUserMapper.class);
    List<User> allUser = mapper.getAllUser();
    mapper.getUser("101000123");
    mapper.getUser("101000124");
    mapper.getUser("101000123");
    return allUser;
}
```

测试结果如下。

```
DEBUG - Checked out connection 4792741 from pool.
DEBUG - ==>  Preparing: select * from tuser
DEBUG - ==> Parameters:
DEBUG - <==      Total: 13
DEBUG - ==>  Preparing: select * from tuser where sno=?
DEBUG - ==> Parameters: 101000123(String)
DEBUG - <==      Total: 1
DEBUG - ==>  Preparing: select * from tuser where sno=?
DEBUG - ==> Parameters: 101000124(String)
DEBUG - <==      Total: 1
```

通过测试发现，对于同一个查询语句，只有参数相同时才会使用一级缓存。

4．ehCache

ehCache 是一个成熟的第三方缓存框架。

Hibernate 也有一级缓存和二级缓存。Hibernate 的一级缓存是基于会话的。基于 HQL 的增删改查操作会影响一级缓存。Hibernate 的二级缓存是基于 HibernateSessionFactory 的，其底层实现是 ehCache 框架。由于 Hibernate 框架的懒加载机制，缓存对 Hibernate 框架非常重要。

为了使 MyBatis 也可以使用 ehCache，需要执行如下操作。

（1）导入 ehcache-core.jar 和 mybatis-ehcache.jar。

（2）在 classpath 下编写 ehcache.xml。

```
<ehcache xmlns:xsi="http://www.w3.org/2001/XMLSchema-instance"
    xsi:noNamespaceSchemaLocation="../config/ehcache.xsd">
    <defaultCache maxElementsInMemory="1000" maxElementsOnDisk="10000000"
       eternal="false"  overflowToDisk="false"  timeToIdleSeconds="120"
       timeToLiveSeconds="120"  diskExpiryThreadIntervalSeconds="120"
       memoryStoreEvictionPolicy="LRU">
    </defaultCache>
</ehcache>
```

（3）在 mapper 文件中配置 ehCache。

```
<cache type="org.mybatis.caches.ehcache.EhcacheCache">
    <property name="timeToIdleSeconds" value="3600" />
    <property name="timeToLiveSeconds" value="3600" />
    <property name="maxEntriesLocalHeap" value="1000" />
    <property name="maxEntriesLocalDisk" value="10000000" />
    <property name="memoryStoreEvictionPolicy" value="LRU" />
</cache>
```

MyBatis 的缓存功能受到很多约束，算不上很实用。二级缓存功能用 Redis 完全可以替代。

11.6 动态 SQL

在映射 SQL 的声明中，使用动态 SQL 语法可以极大地提高 MyBatis 框架的灵活性。

主要使用以下语句实现动态 SQL：

- if；
- choose；
- foreach。

11.6.1 if 语句

if 语句用于在 SQL 声明中进行条件判断。在多条件查询的场景下，经常使用 if 语句进行条件判断。

下面给出一个关于单条件查询的示例。

```
public interface IBlogMapper {
    public Blog findActiveBlogWithTitleLike(@Param("title")String title);
}
<select id="findActiveBlogWithTitleLike" resultType="Blog">
    SELECT * FROM BLOG
    WHERE state = 'ACTIVE'
    <if test="title != null">
        AND title like #{title}
    </if>
</select>
```

关于多条件查询的示例如下。

```
<select id="findActiveBlogLike" resultType="Blog">
    SELECT * FROM BLOG WHERE state = 'ACTIVE'
    <if test="title != null">
        AND title like #{title}
    </if>
    <if test="author != null and author.name != null">
        AND author_name like #{author.name}
    </if>
</select>
```

案例 11-4：模糊查询用户

模糊查询是最常见的查询场景。在下面的示例中，使用模糊查询查找与输入的编号匹配的用户信息。操作步骤如下。

（1）定义 Mapper 接口。

```
public interface IUserMapper {
    /**
     * 模糊查询符合编号条件的所有用户信息
    */
    public List<User> getAllUser(@Param("sno")String sno);
}
```

（2）配置 SQL 映射。为了连接字符串，只能使用 concat 函数，不能使用 JDK 中的加号进行字符命名空间的连接。

```
<select id="getAllUser" resultType="User">
    select * from tuser where 1=1
    <if test="sno != null">
      and sno like concat(concat('%',#{sno}),'%')
    </if>
</select>
```

（3）测试。

① 测试代码如下。

```
public static void main(String[] args) {
    IUser biz = new UserBiz();
    List<User> allUser = biz.getAllUser("101000");
    for(User u : allUser) {
        System.out.println(u.getUname() + "," + u.getPwd() + "," + u.getRole());
    }
}
```

② 测试结果如下。

```
DEBUG - Checked out connection 4792741 from pool.
DEBUG - ==>  Preparing: select * from tuser where 1=1
                     and sno like concat(concat('%',?),'%')
DEBUG - ==> Parameters: 101000(String)
DEBUG - <==      Total: 7
```

11.6.2 choose 语句

如果你不想使用所有条件，而只使用多个条件中的一个，则可以使用 choose 语句。

如下所示，满足多个条件中的一个后，就不再匹配后面的条件。

```
<select id="findActiveBlogLike" resultType="Blog">
    SELECT * FROM BLOG WHERE state = 'ACTIVE'
    <choose>
        <when test="title != null">
            AND title like #{title}
        </when>
        <when test="author != null and author.name != null">
```

```
                AND author_name like #{author.name}
            </when>
            <otherwise>
                AND featured = 1
            </otherwise>
        </choose>
    </select>
```

案例 11-5：根据多个条件模糊查询用户

使用员工系统的员工编号或用户名进行模糊查询。如果员工编号不为空，就使用员工编号进行模糊查询；如果员工编号为空，但是用户名不为空，就使用用户名进行模糊查询。具体操作如下。

（1）定义 Mapper 接口。

```
public interface IUserMapper {
    public List<User> getAllUser(@Param("sno")String sno,
                                @Param("uname")String uname);
}
```

（2）配置 SQL 映射。

```
<select id="getAllUser" resultType="User" >
    select * from tuser where 1=1
    <choose>
        <when test="sno != null and sno!=''">
            and sno like concat(concat('%',#{sno}),'%')
        </when>
        <when test="uname!=null and uname!=''">
            and uname like concat(concat('%',#{uname}),'%')
        </when>
    </choose>
</select>
```

（3）当 sno 不是空值时，进行测试。

① 测试代码如下。

```
public static void main(String[] args) {
    IUser biz = new UserBiz();
    List<User> allUser = biz.getAllUser("101000",null);
    for(User u : allUser) {
        System.out.println(u.getUname() + "," + u.getPwd() + "," + u.getRole());
    }
}
```

② 测试结果如下。

```
DEBUG - Checked out connection 4792741 from pool.
DEBUG - ==>  Preparing: select * from tuser where 1=1
                    and sno like concat(concat('%',?),'%')
```

```
DEBUG - ==> Parameters: 101000(String)
DEBUG - <==     Total: 7
```

（4）当 sno 为空值但用户名不为空值时，继续测试。

① 测试代码如下。

```
public static void main(String[] args) {
    IUser biz = new UserBiz();
    List<User> allUser = biz.getAllUser("","101000");
    for(User u : allUser) {
        System.out.println(u.getUname() + "," + u.getPwd() + "," + u.getRole());
    }
}
```

② 测试结果如下。

```
DEBUG - Opening JDBC Connection
DEBUG - Checked out connection 4792741 from pool.
DEBUG - ==>  Preparing: select * from tuser where 1=1
             and uname like concat(concat('%',?),'%')
DEBUG - ==> Parameters: 101000(String)
DEBUG - <==     Total: 7
```

11.6.3 foreach 语句

foreach 语句中可以使用的属性主要有 item、index、open、separator、close、collection。

- item：集合中元素的别名。该参数必选。
- index：在列表和数组中，index 是元素的序号。在映射中，index 是元素的键。该参数可选。
- open：foreach 语句的开始符号，一般和 close=")"合用。该参数可选。
- separator：元素之间的分隔符。例如，在使用 in( )的时候，separator=","会自动在元素中间用“,”隔开，避免手动输入逗号导致 SQL 错误。
- close: foreach 语句的关闭符号，一般和 open="("合用。该参数可选。
- collection: Mapper 接口中的集合参数，List 对象默认用“list”代替作为键，数组对象用“array”作为键，Map 对象没有默认的键，可以使用@Param("")指定。

示例 11-7：通过 in 语句查询。

```
public List<Post> selectPostIn(Collection<String> list);
<select id="selectPostIn" resultType="domain.blog.Post">
    SELECT * FROM POST P
    WHERE ID in
    <foreach item="item" index="index" collection="list" open="("
        separator="," close=")">
```

```
        #{item}
    </foreach>
</select>
```

注意：以任何 Iterable 对象作为参数均可，如 List、Set、Queue、Collection，也可以使用 Map 或 Array 对象作为参数。

示例 11-8：批量插入数据。

具体代码如下。

```
<insert id="insertAuthor" useGeneratedKeys="true"
    keyProperty="id">
    insert into Author (username, password, email, bio) values
    <foreach item="item" collection="list" separator=",">
        (#{item.username}, #{item.password}, #{item.email}, #{item.bio})
    </foreach>
</insert>
```

案例 11-6：实现用户查询

使用 in 语句，查询用户列表中所有用户的详细信息。具体操作如下。

（1）定义 Mapper 接口。

```
public interface IUserMapper {
    public List<User> getAllUser(List<String> snos);
}
```

（2）配置 SQL 映射。

```
<select id="getAllUser" resultType="User">
    select * from tuser where sno in
    <foreach item="u" index="index" collection="list" open="("
        separator="," close=")">
        #{u}
    </foreach>
</select>
```

（3）测试。

① 测试代码如下。

```
public static void main(String[] args) {
    IUser biz = new UserBiz();
    List<String> snos = new ArrayList<>();
    snos.add("101000124");
    snos.add("101000125");
    List<User> allUser = biz.getAllUser(snos);
```

```
    for(User u : allUser) {
        System.out.println(u.getUname() + "," + u.getPwd() + "," + u.getRole());
    }
}
```

② 测试结果如下。

```
DEBUG - Checked out connection 4792741 from pool.
DEBUG - ==>  Preparing: select * from tuser where sno in ( ? , ? )
DEBUG - ==> Parameters: 101000124(String), 101000125(String)
DEBUG - <==      Total: 2
```

第 12 章
通过 Spring 整合 StaffUser 系统

前面使用 MyBatis 实现了员工系统的持久层代码，在本章中，我们用 Spring 整合 StaffUser 系统。

12.1 下载资源

通过 Spring 整合 MyBatis 的资源可以从 GitHub 下载，操作步骤如下。

（1）进入 GitHub，搜索 MyBatis，在搜索结果中，单击 Spring integration for MyBatis 3，见图 12-1。

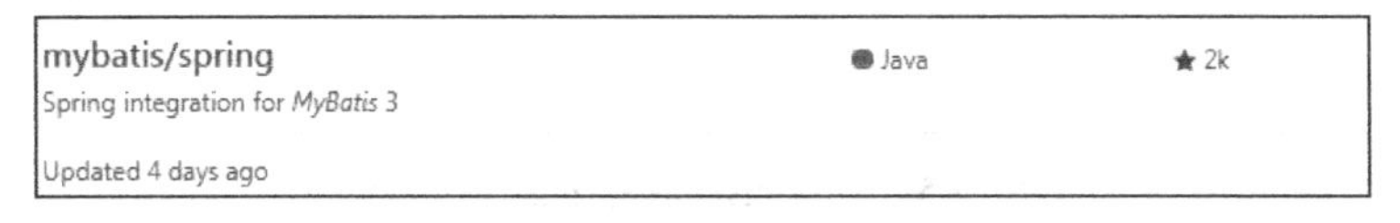

图 12-1　单击 Spring integration for MyBatis 3

（2）如图 12-2 所示，在弹出的页面中，单击“22 releases”，进入图 12-3 所示的页面。其中列出了可下载的 mybatis-spring 资源。

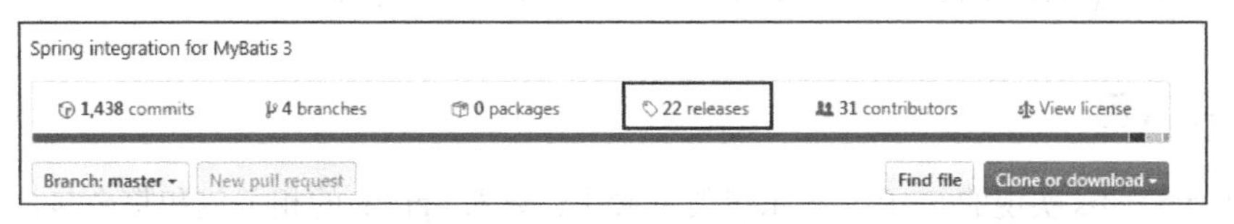

图 12-2　单击“22 releases”

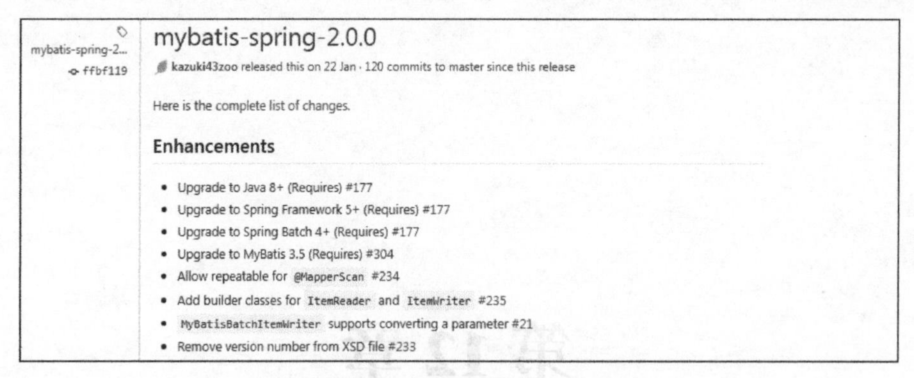

图 12-3　可下载的 mybatis-spring 资源

mybatis-spring-2.*x* 支持的是 Spring Framework 5+，我们采用的是 Spring 4.3.*x*，因此向下滑动页面，找到 mybatis-spring-1.3.2（见图 12-4）。

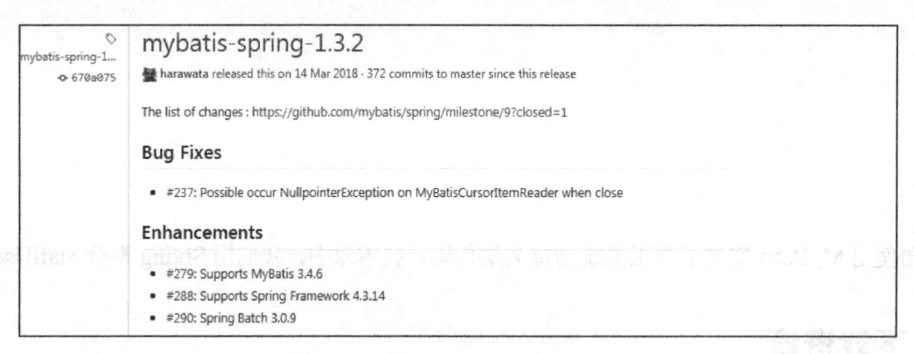

图 12-4　找到 mybatis-spring-1.3.2

（3）下载 mybatis-spring-1.3.2.zip（见图 12-5）。

图 12-5　下载 mybatis-spring-1.3.2.zip

12.2　项目案例：整合 StaffUser 系统

通过 mybatis-spring-1.3.2 整合用 MyBatis 实现的 StaffUser 系统。

12.2.1　导入包

在原来 MyBatis 实现的 StaffUser 系统的基础上，导入 Spring 的相关包和整合包。注意，避免包冲突，如 cglib.jar，Spring 4.3 的核心包已经包含了 cglib 3.2.4，而 MyBatis 3.5.3 依赖 cglib 3.2.10，因此整合后选择 cglib 3.2.10。对于 log4j.jar 与 commons-logging.jar，选择 MyBatis 3.5.3 依赖的高版本包。要导入包，具体操作如下。

（1）导入 Spring 核心包 Spring-core.jar、Spring-context.jar、Spring-beans.jar、Spring-expression.jar。

（2）导入 Spring AOP 包，包括 Spring-AOP.jar 和 Spring-aspects.jar。

（3）导入持久层整合包 Spring-jdbc.jar、Spring-tx.jar 和 Spring-orm.jar。

（4）导入 Spring 整合包 mybatis-spring-1.3.2.jar。

12.2.2 配置 beans.xml 文件

在 classpath 下新建 beans.xml，配置信息如下。

1. 配置 schema

配置 tx 命名空间，用于事务管理。

```
<beans xmlns="http://www.springframework.org/schema/beans"
xmlns:xsi="http://www.w3.org/2001/XMLSchema-instance"
xmlns:tx="http://www.springframework.org/schema/tx"
xmlns:context="http://www.springframework.org/schema/context"
xsi:schemaLocation="http://www.springframework.org/schema/beans
    http://www.springframework.org/schema/beans/spring-beans-4.3.xsd
    http://www.springframework.org/schema/tx
    http://www.springframework.org/schema/tx/spring-tx-4.3.xsd
    http://www.springframework.org/schema/context
   http://www.springframework.org/schema/context/spring-context-4.3.xsd">
</beans>
```

2. 管理数据源

为了通过 Spring 整合 MyBatis，首先要接管 MyBatis 的数据源。

```
<bean id="dataSource"
    class="org.springframework.jdbc.datasource.DriverManagerDataSource">
    <property name="driverClassName"
        value="com.mysql.cj.jdbc.Driver" />
    <property name="url"value="jdbc:mysql://localhost:3306/staff?useSSL=false
            &amp;serverTimezone=UTC&amp;allowPublicKeyRetrieval=true" />
    <property name="username" value="root" />
    <property name="password" value="123456" />
</bean>
```

然后，删除 mybatis.xml 中原来的环境配置。

```
<environments default="developmentMysql">
    <environment id="developmentMysql">
```

```
        <transactionManager type="JDBC" />
        <dataSource type="POOLED">
            <property name="driver" value="${driver}" />
            <property name="url" value="${url}" />
            <property name="username" value="${username}" />
            <property name="password" value="${password}" />
        </dataSource>
    </environment>
</environments>
```

3. 配置 SqlSessionFactoryBean

Spring 使用 SqlSessionFactoryBean 接管 MyBatis 的 SqlSessionFactory 的创建。

```
<bean id="sqlSessionFactory" class="org.mybatis.spring.SqlSessionFactoryBean">
    <property name="configLocation" value="classpath:mybatis.xml" />
    <property name="dataSource" ref="dataSource" />
</bean>
```

4. 配置组件的扫描位置

通过以下代码配置组件的扫描位置。

```
<context:annotation-config />
<context:component-scan base-package="com.icss.biz" />
<context:component-scan base-package="com.icss.dao" />
```

5. 配置事务管理器

通过 Spring 整合 MyBatis 与通过 Spring 整合 JDBC 使用相同的事务管理器。

```
<tx:annotation-driven transaction-manager="txManager" />
<bean id="txManager"
    class="org.springframework.jdbc.datasource.DataSourceTransactionManager">
    <property name="dataSource" ref="dataSource" />
</bean>
```

6. 配置 Mapper 接口

配置 Mapper 接口所在位置，用于扫描注入。

```
<bean class="org.mybatis.spring.mapper.MapperScannerConfigurer">
    <property name="basePackage" value="com.icss.mapper" />
</bean>
```

12.2.3 配置服务层和持久层依赖的对象

服务层依赖持久层对象，持久层依赖 Mapper。有人不使用持久层，让服务层直接依赖 Mapper，这种做法不可取。

通过以下代码使服务层依赖持久层对象。

```
@Service("staffBiz")
public class StaffBiz implements IStaff{
    @Autowired
    private IStaffDao staffDao;
    @Autowired
    private IUserDao userDao;
}
@Service("userBiz")
public class UserBiz implements IUser {
    @Autowired
    private IUserDao userDao;
}
```

通过以下代码使持久层依赖 Mapper。

```
@Repository("staffDao")
public class StaffDaoMysql extends BaseDao implements IStaffDao{
    @Autowired
    private IStaffMapper mapper;
}
@Repository("userDao")
public class UserDaoMysql extends BaseDao implements IUserDao {
    @Autowired
    private IUserMapper mapper;
}
```

12.2.4 管理事务

通过 Spring 整合 MyBatis 使用的事务管理器与通过 Spring 整合 JDBC 使用的事务管理器相同，但是@Transactional 注解的使用方式有所不同。

1. 管理只读事务

在用户登录的过程中，需要数据库操作，但不需要事务。

测试 12-1：使用非事务模式查询，即不配置@Transactional(readOnly=true)。

```
@Service("userBiz")
public class UserBiz implements IUser {
    @Autowired
    private IUserDao userDao;
    public User login(String sno, String pwd) throws Exception {}
}
```

测试代码如下。

```
public static void main(String[] args) {
    IUser u = (IUser)BeanFactory.getBean("userBiz");
    try {
        User user = u.login("101000123", "12345");
        if(user != null) {
```

```
            System.out.println("登录成功，身份是" + user.getRole());
        }else {
            System.out.println("登录失败");
        }
    } catch (Exception e) {
        System.out.println(e.getMessage());
    }
}
```

测试结果如下。

```
DEBUG - Creating a new SqlSession
DEBUG - Creating new JDBC DriverManager Connection
DEBUG - JDBC Connection will not be managed by Spring
DEBUG - ==>  Preparing: select * from tuser where sno = ? and pwd=?
DEBUG - ==> Parameters: 101000123(String), 12345(String)
DEBUG - <==      Total: 1
DEBUG - Closing non transactional SqlSession
DEBUG - Returning JDBC Connection to DataSource
登录成功，身份是 2
```

测试 12-2：使用只读事务查询，即配置@Transactional(readOnly=true)。

```
@Service("userBiz")
public class UserBiz implements IUser {
    @Autowired
    private IUserDao userDao;
    @Transactional(readOnly=true)
    public User login(String sno, String pwd) throws Exception {}
}
```

测试结果如下。

```
DEBUG - Creating new transaction with name [com.icss.biz.impl.UserBiz.login]:
       PROPAGATION_REQUIRED,ISOLATION_DEFAULT,readOnly
DEBUG - Creating a new SqlSession
DEBUG - ==>  Preparing: select * from tuser where sno = ? and pwd=?
DEBUG - ==> Parameters: 101000123(String), 12345(String)
DEBUG - <==      Total: 1
DEBUG - Releasing transactional SqlSession
DEBUG - Committing JDBC transaction on Connection
DEBUG - Resetting read-only flag of JDBC Connection
DEBUG - Releasing JDBC Connection after transaction
DEBUG - Returning JDBC Connection to DataSource
登录成功，身份是 2
```

使用 Spring 整合 MyBatis 与使用 Spring 整合 JDBC 完全不同。对于 login( )方法，若不使用@Transactional(readOnly=true)，不会造成数据库连接不关闭的现象；若使用了@Transactional(readOnly=true)，不会带来好处，反而会影响性能。

2. 管理写操作事务

测试 12-3：使用声明性事务，管理员工与用户的添加。

具体代码如下。

```
@Service("staffBiz")
public class StaffBiz implements IStaff{
    @Autowired
    private IStaffDao staffDao;
    @Autowired
    private IUserDao userDao;
    @Transactional(rollbackFor=Throwable.class)
    public void addStaffUser(Staff staff) throws Exception {
        if(staff == null)
            throw new Exception("staff 入参为空...");
        User user = new User();
        user.setSno(staff.getSno());
        user.setUname(staff.getName());
        user.setRole(2);
        user.setPwd("1234");
        staffDao.addStaff(staff);
        userDao.addUser(user);
    }
}
```

测试结果如下。

```
DEBUG - Creating new transaction with name
                [com.icss.biz.impl.StaffBiz.addStaffUser]
DEBUG - Creating a new SqlSession
DEBUG - ==>  Preparing: insert into tstaff values(?,?,?,?,?)
DEBUG - ==> Parameters: 121000135(String), tom4(String), 1995-10-01,
            北京朝阳区建国门(String), 13522454666(String)
DEBUG - <==    Updates: 1
DEBUG - Releasing transactional SqlSession
DEBUG - Fetched SqlSession from current transaction
DEBUG - ==>  Preparing: insert into tuser values(?,?,?,?)
DEBUG - ==> Parameters: tom4(String), 121000135(String), 1234(String), 2(Integer)
DEBUG - <==    Updates: 1
DEBUG - Releasing transactional SqlSession
DEBUG - Committing JDBC transaction on Connection
DEBUG - Releasing JDBC Connection after transaction
121000135 创建成功....
```

测试 12-4：异常回滚。添加员工成功，添加用户时主动抛出异常，查看员工数据是否回滚了。

```
@Repository("userDao")
public class UserDaoMysql extends BaseDao implements IUserDao {
    @Autowired
    private IUserMapper mapper;
    @Override
    public void addUser(User user) throws Exception {
        mapper.addUser(user);
        throw new RuntimeException("异常测试....");
    }
}
```

测试结果如下。

```
DEBUG - Creating a new SqlSession
DEBUG - JDBC Connection  will be managed by Spring
DEBUG - ==>  Preparing: insert into tstaff values(?,?,?,?,?)
DEBUG - ==> Parameters: 121000136(String), tom4(String), 1995-10-01 ,
       北京朝阳区建国门(String), 13522454666(String)
DEBUG - <==    Updates: 1
DEBUG - Releasing transactional SqlSession
DEBUG - Fetched SqlSession from current transaction
DEBUG - ==>  Preparing: insert into tuser values(?,?,?,?)
DEBUG - ==> Parameters: tom4(String), 121000136(String), 1234(String), 2(Integer)
DEBUG - <==    Updates: 1
DEBUG - Releasing transactional SqlSession
DEBUG - Rolling back JDBC transaction on Connection
DEBUG - Releasing JDBC Connection  after transaction
java.lang.RuntimeException: 异常测试....
    at com.icss.dao.impl.UserDaoMysql.addUser(UserDaoMysql.java:20)
    at com.icss.biz.impl.StaffBiz.addStaffUser(StaffBiz.java:35)
```

通过 Spring 整合 MyBatis 与通过 Spring 整合 JDBC 的事务管理方法一致，都使用服务层方法配置事务管理策略。

第 13 章 通过 SSM 整合书城项目

在前面已经实现的书城项目的基础上，继续整合 MyBatis，即使用 MyBatis 替换原有书城项目中的 JDBC 操作，增加 Spring 对服务层中声明性事务的管理。整合后的项目就是标准的 Spring MVC + Spring + MyBatis 项目了。

13.1 搭建 SSM 整合环境

13.1.1 导入包

在 Spring MVC 实现的原来书城项目的基础上，导入 mybatis-3.5.3.jar 与依赖的包，导入 Spring-jdbc.jar、Spring-orm.jar 和 Spring-tx.jar，导入 mybatis-spring-1.3.2.jar。

13.1.2 配置数据库连接

在 classpath 下的 db.properties 中，配置 MySQL 连接。

```
driver=com.mysql.cj.jdbc.Driver
url=jdbc:mysql://localhost:3306/bk?useSSL=false
            &serverTimezone=UTC&allowPublicKeyRetrieval=true
username=root
password=123456
```

13.1.3 设置 MyBatis 的核心配置文件

在 MyBatis 的核心配置文件中，关于数据库连接部分被 Spring 接管了。其他配置项继续保留。

```
<!DOCTYPE configuration
PUBLIC "-//mybatis.org//DTD Config 3.0//EN"
"http://mybatis.org/dtd/mybatis-3-config.dtd">
<configuration>
    <settings>
        <setting name="jdbcTypeForNull" value="NULL" />
    </settings>
    <typeAliases>
        <typeAlias alias="TUser" type="com.icss.entity.TUser" />
        <typeAlias alias="TBook" type="com.icss.entity.TBook" />
        <typeAlias alias="TBuyDetail" type="com.icss.entity.TBuyDetail" />
        <typeAlias alias="BuyRecord" type="com.icss.dto.BuyRecord" />
        <typeAlias alias="TOrder" type="com.icss.entity.TOrder"/>
    </typeAliases>
    <mappers>
        <mapper resource="com/icss/mapper/UserMapper.xml" />
        <mapper resource="com/icss/mapper/BookMapper.xml" />
    </mappers>
</configuration>
```

注意：读取数据库配置信息的设置将被 Spring 的以下配置取代。

```
<properties resource="db.properties"></properties>
```

13.1.4 设置 Spring 的核心配置文件

书城项目在 Spring MVC 部分使用了配置文件 spring-mvc.xml，服务层和持久层 Bean 可以与 Spring MVC 共用同一个配置文件。另外，也可以把服务层和持久层信息配置在 beans.xml 中，然后导入 spring-mvc.xml。

示例 13-1：共用一个配置文件。

具体实现方式如下。

（1）配置 schema。

```
<beans xmlns="http://www.springframework.org/schema/beans"
xmlns:xsi="http://www.w3.org/2001/XMLSchema-instance"
xmlns:mvc="http://www.springframework.org/schema/mvc"
xmlns:context="http://www.springframework.org/schema/context"
xmlns:tx="http://www.springframework.org/schema/tx"
xsi:schemaLocation="
    http://www.springframework.org/schema/beans
    http://www.springframework.org/schema/beans/spring-beans-4.3.xsd
    http://www.springframework.org/schema/context
    http://www.springframework.org/schema/context/spring-context-4.3.xsd
```

```
    http://www.springframework.org/schema/mvc
    http://www.springframework.org/schema/mvc/spring-mvc-4.3.xsd
    http://www.springframework.org/schema/tx
    http://www.springframework.org/schema/tx/spring-tx-4.3.xsd">
</beans>
```

（2）配置数据库的连接信息，Spring 接管 MyBatis 数据源。

```
<bean id="propertyConfigurer"  class="org.springframework.beans
                .factory.config.PropertyPlaceholderConfigurer">
    <property name="location" value="classpath:db.properties"/>
</bean>
<bean id="dataSource"
    class="org.springframework.jdbc.datasource.DriverManagerDataSource">
    <property name="driverClassName" value="${driver}" />
    <property name="url" value="${url}" />
    <property name="username" value="${username}" />
    <property name="password" value="${password}" />
</bean>
```

（3）Spring 接管 MyBatis 的 SqlSessionFactory。

```
<bean id="sqlSessionFactory" class="org.mybatis.spring.SqlSessionFactoryBean">
    <property name="configLocation" value="classpath:mybatis.xml" />
    <property name="dataSource" ref="dataSource" />
</bean>
```

（4）配置事务管理器。

```
<bean id="txManager"
    class="org.springframework.jdbc.datasource.DataSourceTransactionManager">
    <property name="dataSource" ref="dataSource" />
</bean>
<tx:annotation-driven transaction-manager="txManager" />
```

（5）配置映射的扫描位置。

```
<bean class="org.mybatis.spring.mapper.MapperScannerConfigurer">
    <property name="basePackage" value="com.icss.mapper" />
</bean>
```

（6）配置 Bean 的扫描位置。

```
<context:component-scan base-package="com.icss.action" />
<context:component-scan base-package="com.icss.biz" />
<context:component-scan base-package="com.icss.dao" />
```

（7）完成 Spring MVC 的相关配置。

```
<bean class="org.springframework.web.servlet.view.InternalResourceViewResolver">
    <property name="prefix" value="/WEB-INF/views/" />
</bean>
<mvc:annotation-driven></mvc:annotation-driven>
<bean id="multipartResolver"
```

```
    class="org.springframework.web.multipart.commons.CommonsMultipartResolver">
    <property name="maxUploadSize" value="102400" />
</bean>
<mvc:annotation-driven>
<mvc:message-converters register-defaults="true">
    <bean class="org.springframework.http
                .converter.StringHttpMessageConverter">
            <property name="supportedMediaTypes">
                <list>
                    <value>application/json;charset=gbk</value>
                    <value>text/html;charset=gbk</value>
                </list>
            </property>
    </bean>
    <bean class="org.springframework.http
                    .converter.ByteArrayHttpMessageConverter" />
    <bean  class="org.springframework.http.converter
                    .json.MappingJackson2HttpMessageConverter" />
</mvc:message-converters>
</mvc:annotation-driven>
```

13.2 定义 Mapper 接口和配置 Mapper 文件

1. 定义 Mapper 接口

要定义 Mapper 接口，在包 com.icss.mapper 下，新建接口 IBookMapper 与 IUserMapper。

```
public interface IBookMapper {}
public interface IUserMapper {}
```

2. 配置 Mapper 文件

首先，在包 com.icss.mapper 下，新建 userMapper.xml 和 bookMapper.xml。

然后，在 mybatis.xml 中配置映射文件的路径。

```
<mappers>
    <mapper resource="com/icss/mapper/UserMapper.xml" />
    <mapper resource="com/icss/mapper/BookMapper.xml" />
</mappers>
```

13.3 在持久层配置依赖注入 Mapper

通过以下代码，在持久层配置依赖注入 Mapper。

```
@Repository("bookDao")
public class BookDaoMysql extends BaseDao implements IBookDao{
    @Autowired
    private IBookMapper bookMapper;
```

```
}
    @Repository("userDao")
    public class UserDaoMysql extends BaseDao implements IUserDao{
    @Autowired
    private IUserMapper userMapper;
    @Autowired
    private IBookMapper bookMapper;
}
```

13.4 实现 MyBatis 持久层

13.4.1 显示主页图书列表

对于书城项目的主页，控制层代码在第 8 章已经讲解了，此处只描述持久层和 Mapper 层的实现。具体实现方式如下。

（1）定义 DAO 层的接口。

```
public interface IBookDao{
    public List<TBook> getAllBooks() throws Exception;
}
```

（2）定义 Mapper 接口。

```
public interface IBookMapper {
    public List<TBook> getAllBooks() throws Exception;
}
```

（3）配置 SQL 映射。

```
<mapper namespace="com.icss.mapper.IBookMapper">
    <select id="getAllBooks" resultType="TBook">
        select isbn,bname,press,price,pdate from tbook order by isbn
    </select>
<mapper>
```

13.4.2 显示图片

在书城项目中，为了显示图片，需要根据图书的 ISBN，从数据库中提取 Tbook 表中的 pic 字段。小图片可以使用 BLOB 类型的字段存储于数据库中；大图片建议存储在 Web 项目下，在数据库中记录图片的存储路径。具体操作方式如下。

（1）为了在服务层调用图片，需要直接返回字节流。

```
@Service("bookBiz")
public class BookBiz {
    @Autowired
    private IBookDao bookDao;
```

```
    @Transactional(readOnly=true)
    public byte[] getBookPic(String isbn) throws Exception{
        if(isbn == null || isbn.equals("")) {
            throw new RuntimeException("ISBN不能为空...");
        }
        return bookDao.getBookPic(isbn);
    }
}
```

（2）DAO 层需要从 Mapper 层返回的 Tbook 表中提取图片。

```
@Repository("bookDao")
public class BookDaoMysql extends BaseDao implements IBookDao{
    @Autowired
    private IBookMapper bookMapper;
    public byte[] getBookPic(String isbn) throws Exception{
        TBook bk = bookMapper.getBookPic(isbn);
        return bk.getPic();
    }
}
```

（3）Mapper 层很难直接返回 byte[]，它返回与表对应的实体对象。

```
public interface IBookMapper {
    public TBook getBookPic(@Param("isbn")String isbn) throws Exception;
}
```

（4）要返回 TBook 对象，resultType 不能为 byte。如果把 BLOB 字段直接映射到 byte[]时出现问题，可以使用 resultMap 转换。

```
<mapper namespace="com.icss.mapper.IBookMapper">
    <select id="getBookPic" resultType="TBook">
        select isbn,pic from tbook where isbn = #{isbn}
    </select>
</mapper>
```

13.4.3 显示图书详情

第 8 章介绍过图书详情页，此处只描述如何实现持久层和 Mapper 层。具体步骤如下。

（1）定义 DAO 层的接口。

```
public interface IBookDao{
    public TBook getBookDetail(String isbn) throws Exception;
}
```

（2）定义 Mapper 接口。

```
public interface IBookMapper {
    public TBook getBookDetail(@Param("isbn") String isbn) throws Exception;
}
```

（3）配置 SQL 映射。

```
<select id="getBookDetail" resultType="TBook">
    select isbn,bname,press,price,pdate from tbook where isbn=#{isbn}
</select>
```

13.4.4 管理用户

1. 用户登录与退出

用户登录的过程需要数据库操作。用户在退出时只清除会话，没有数据库操作。

实现用户登录的具体操作如下。

（1）定义 DAO 层的接口。

```
public interface IUserDao{
    public TUser login(String uname,String pwd) throws Exception;
}
```

（2）定义 Mapper 接口。

```
public interface IUserMapper {
    public TUser login(@Param("uname")String uname
                      ,@Param("pwd")String pwd) throws Exception;
}
```

（3）配置 SQL 映射。

```
<select id="login" resultType="TUser">
    select * from tuser where uname=#{uname} and pwd=#{pwd}
</select>
```

实现用户退出的操作与实现用户登录的操作类似，这里不再讨论。

2. 用户注册

通过传入 TUser 实体对象，调用 Mapper 层的 addUser( )，注册用户。具体步骤如下。

（1）定义 DAO 层的接口。

```
public interface IUserDao{
    public void addUser(TUser user) throws Exception;
}
```

（2）定义 Mapper 接口。

```
public interface IUserMapper {
    public void addUser(TUser user) throws Exception;
}
```

（3）配置 SQL 映射。

```
<insert id="addUser" parameterType="TUser">
    insert into tuser value(#{uname},#{pwd},#{account},#{role})
</insert>
```

3. 用户名校验

使用 AJAX 校验输入的用户名是否可用。持久层的接口返回值为 boolean，而 Mapper 接口的返回值应设置为 int。具体操作如下。

（1）定义 DAO 层的接口并实现。

```
public interface IUserDao{
    /**
     * 判断输入的用户名在数据库中是否存在
     */
    public boolean isHaveUserName(String name) throws Exception;
}
public boolean isHaveUserName(String name) throws Exception {
    int num =  userMapper.isHaveUserName(name);
    if(num>0)
        return true;
    else
        return false;
}
```

（2）定义 Mapper 接口。Mapper 层无法直接返回布尔值，可以通过查询到的用户间接判断，也可以使用 select count(*) from tuser where uname=#{uname}来间接判断。

```
public interface IUserMapper {
    public int isHaveUserName(@Param("uname")String name) throws Exception;
}
```

（3）配置 SQL 映射。

```
<select id="isHaveUserName" resultType="int">
    select count(*) from tuser where uname=#{uname}
</select>
```

13.4.5 实现购物车

购物车是存放在会话中的，为了节省会话占用的内存，购物车的数据结构为 Map<String, Integer>，即只存储图书主键 ISBN 与购买数量。为了显示购物车中的图书详情，只能使用一条 SQL 语句从数据库中提取（考虑性能，不能多次提取）。

第 8 章介绍了购物车的样式，此处只描述如何实现持久层和 Mapper 层。具体步骤如下。

（1）定义 DAO 层的接口。读取购物车中的图书信息，不含图片。

```
public interface IBookDao{
    public List<TBook> getBooks(Set<String> isbns) throws Exception;
}
```

（2）定义 Mapper 接口。

```
public interface IBookMapper {
    public List<TBook> getShopCarBooks(@Param("isbns")Set<String> isbns)
                                    throws Exception;
}
```

（3）配置 SQL 映射。

```
<select id="getShopCarBooks" resultType="TBook">
    select isbn,bname,press,price,pdate from tbook where isbn in
    <foreach item="item" index="index" collection="isbns"
        open="(" separator="," close=")">
            #{item}
    </foreach>
</select>
```

13.4.6 用户付款

用户付款功能涉及多张表的操作，分别是账户扣款、生成订单、添加订单明细、更新图书库存数量。

1. 控制付款逻辑

使用声明性事务控制服务层的付款逻辑。

```
@Transactional(rollbackFor=Throwable.class)
public void buyBooks(String uname, double allMoney,
                     Map<String, Integer> shopCar) throws Exception {
    if(uname == null || uname.equals("")) {
        throw new InputNullExcepiton("用户名不能为空");
    }
    if(allMoney < 0) {
        throw new InputNullExcepiton("付款金额异常，请检查");
    }
    if(shopCar == null || shopCar.size()==0) {
        throw new InputNullExcepiton("购物车为空，不能结算");
    }
    userDao.updateUserAccount(uname, -allMoney);
    userDao.addBuyRecord(uname, allMoney, shopCar);
}
```

2. 更新账户金额

调用 IUserMapper 的 updateUserAccount( )更新指定用户的账户金额，实现扣款操作。

考虑代码的重用性，对于用户充值操作，也使用相同的接口。

（1）定义 DAO 层的接口。

```
public interface IUserDao{
    public void updateUserAccount(String uname ,double money) throws Exception;
}
```

（2）定义 Mapper 接口。

```
public interface IUserMapper {
    public void updateUserAccount(@Param("uname")String uname
                    ,@Param("money")double money) throws Exception;
}
```

（3）配置 SQL 映射。

```
<update id="updateUserAccount">
    update tuser set account = account+#{money} where uname=#{uname}
</update>
```

3. 添加订单

在添加订单时，会循环添加多条购买明细。添加的每条购买明细又会触发图书数量的更新。要添加订单，具体操作如下。

（1）定义 DAO 层的接口。

```
public interface IUserDao{
    public void addBuyRecord(String uname,double allMoney,
                    Map<String,Integer> shopCar) throws Exception;
}
@Repository("userDao")
public class UserDaoMysql extends BaseDao implements IUserDao{
    @Autowired
    private IUserMapper userMapper;
    @Autowired
    private IBookMapper bookMapper;
    public void addBuyRecord(String uname,double allMoney,
                Map<String,Integer> shopCar) throws Exception{
        //为订单 id 生成一个有意义的字符串
        String orderNo = OrderUtil.createNewOrderNo();
        TOrder order = new TOrder();
        order.setBuyid(orderNo);
        order.setAllMoney(allMoney);
        order.setBuytime(new Date());
        order.setUname(uname);
        userMapper.addBuyRecord(order);
        Set<String> isbns = shopCar.keySet();
        for(String isbn : isbns){
            TBuyDetail detail = new TBuyDetail();
            detail.setIsbn(isbn);
            detail.setBuycount(shopCar.get(isbn));
            detail.setBuyid(orderNo);
            addBuyDetail(detail);
        }
    }
    private void addBuyDetail(TBuyDetail detail) throws Exception{
        userMapper.addBuyDetail(detail);
        bookMapper.updateBookCount(detail.getIsbn(),-detail.getBuycount());
    }
}
```

（2）定义 Mapper 接口。为了添加购买记录的 DAO 层的接口，会调用 3 个 Mapper 接口方法。

```
public interface IUserMapper {
    public void addBuyRecord(TOrder order) throws Exception;
    public void addBuyDetail(TBuyDetail detail) throws Exception;
}
public interface IBookMapper {
    public void updateBookCount(@Param("isbn")String isbn,
                    @Param("bookCount")int bookCount) throws Exception;
}
```

（3）配置 SQL 映射。

```
<mapper namespace="com.icss.mapper.IUserMapper">
    <insert id="addBuyRecord" parameterType="TOrder">
        insert into tbuyrecord values (#{buyid},#{uname},#{buytime},#{allMoney})
    </insert>
    <insert id="addBuyDetail" parameterType="TBuyDetail">
        insert into tbuydetail values( (select * from (select IFNULL
        (max(autoid),0)+1 from tbuydetail)t1),#{isbn},#{buyid},#{buycount})
    </insert>
    <update id="updateUserAccount">
        update tuser set account = account+#{money} where uname=#{uname}
    </update>
</mapper>
<mapper namespace="com.icss.mapper.IBookMapper">
    <update id="updateBookCount">
        update tbook set bkcount=bkcount+#{bookCount} where isbn = #{isbn}
    </update>
</mapper>
```

13.4.7 上传图书

第 8 章介绍了由后台管理员完成的图书上传功能。对于图书，在数据库中使用的数据类型是 blob，在 Java 中使用的数据类型是 byte[]，Mapper 层的代码在处理字节流写入时没有特殊操作。

在整合后的项目中，实现图书上传的步骤如下。

（1）定义 DAO 层的接口。

```
public interface IBookDao{
    public void addBook(TBook book) throws Exception;
}
```

（2）定义 Mapper 接口。

```
public interface IBookMapper {
    public void addBook(TBook book) throws Exception;
}
```

（3）配置 SQL 映射。

```
<insert id="addBook" parameterType="TBook">
    insert into  tbook(isbn,bname,press,price,pdate,pic,bkcount)
    values(#{isbn},#{bname},#{press},#{price},#{pdate},#{pic},#{bkCount})
</insert>
```

13.4.8 查询用户购买记录

本节介绍在整合后的项目中如何实现复杂的多条件查询，同时考虑翻页操作。

1. DAO 层的接口及其实现

在输入参数中，uname 表示查询条件中的用户名，beginDate 和 endDate 表示查询条件中的开始时间与结束时间。TurnPage 参数为翻页对象，包含页码、每页数量、总记录数、总页数等信息。

```
public class TurnPage {
    public int currentPage =1 ;
    public int rowsOnePage = 10;
    public int allRows;
    public int allPages;
}
public interface IUserDao{
    public List<BuyRecord> getUserBuyRecord(String uname ,
        Date beginDate ,Date endDate,TurnPage tp) throws Exception;
}
@Repository("userDao")
public class UserDaoMysql extends BaseDao implements IUserDao{
    @Autowired
    private IUserMapper userMapper;
    public List<BuyRecord> getUserBuyRecord(String uname ,
            Date beginDate ,Date endDate,TurnPage tp) throws Exception{
        tp.allRows = userMapper.getOrderDetailCount(uname, beginDate, endDate);
        tp.allPages = (tp.allRows-1)/tp.rowsOnePage + 1;
        int iStart,iEnd;
        if(tp.currentPage > tp.allPages)
            tp.currentPage = tp.allPages;
            iStart = (tp.currentPage-1)*tp.rowsOnePage;
            return userMapper.getgetUserBuyRecord(uname,
                            beginDate, endDate, iStart, tp.rowsOnePage);
    }
}
```

2. Mapper 接口

每个 Mapper 接口方法对应一条 SQL 语句。对于翻页操作，需要先提取满足 SQL 条件的记录数量，然后提取指定页码的数据。因此，IUserDao::getUserBuyRecord( )操作对应两个 Mapper 接口。

```
public interface IUserMapper {
/**
 * 满足查询条件的记录数
 * @param uname  用户名
 * @param beginDate  开始时间
 * @param endDate  结束时间
 * @return
 */
public int getOrderDetailCount(@Param("uname")String uname ,
                        @Param("beginDate")Date beginDate ,
                        @Param("endDate")Date endDate);

/**
 * 查询满足条件的用户购买记录
```

```
 * @param uname  用户名
 * @param beginDate  开始时间
 * @param endDate  结束时间
 * @param iStart  起始索引
 * @param rows   返回记录数
 * @return
 */
public List<BuyRecord> getgetUserBuyRecord(@Param("uname")String uname ,
                        @Param("beginDate")Date beginDate ,
                        @Param("endDate")Date endDate,
                        @Param("iStart") int iStart,@Param("rows")int rows);
}
```

3. SQL 映射

对于多条件查询，使用动态 SQL 判断输入的条件是否有效。具体操作如下。

（1）查询满足条件的记录数量。

```
<select id="getOrderDetailCount" resultType="int">
    select count(*) from
         (select d.bcount,bk.bname,bk.isbn,bk.press,bk.price,
            bk.pdate,br.allmoney,br.buytime,br.uname,br.buyid
               from tbuydetail d,tbuyrecord br,tbook bk
               where br.buyid = d.buyid and bk.isbn = d.isbn
          <if test="uname!=null and uname != ''">
                  and br.uname like  concat(concat('%',#{uname}),'%')
          </if>
          <if test="beginDate!=null">
            and br.buytime>=#{beginDate}
          </if>
          <if test="endDate!=null">
                  <![CDATA[ and br.buytime<=#{endDate} ]]>
          </if>
       )tb
</select>
```

（2）使用动态 SQL 拼接多条件查询，返回符合条件的用户购买记录。注意，在 XML 中，小于号必须使用<![CDATA[<]]>。

```
<select id="getgetUserBuyRecord" resultType="BuyRecord">
    select * from (
        select d.bcount,bk.bname,bk.isbn,bk.press,bk.price,
          bk.pdate,br.allmoney,br.buytime,br.uname,br.buyid
          from tbuydetail d,tbuyrecord br,tbook bk
          where br.buyid = d.buyid and bk.isbn = d.isbn
          <if test="uname!=null">
                  and br.uname like  concat(concat('%',#{uname}),'%')
          </if>
          <if test="beginDate!=null">
            and br.buytime>=#{beginDate}
          </if>
          <if test="endDate!=null">
                  <![CDATA[ and br.buytime<=#{endDate} ]]>
          </if>
       ) tb  limit #{iStart},#{rows}
</select>
```

第 14 章 通过 Spring Boot 与 SSM 整合书城项目

14.1 Maven 与环境配置

Spring Boot 的开发需要依赖 Maven 工具，因此本章先介绍一下 Maven 的作用与环境配置。

14.1.1 Maven 的作用

Maven 是一个项目管理工具。Maven 的主要作用如下。

- Maven 中集成了 Ant 工具，可以实现项目的打包和部署（见图 14-1）。

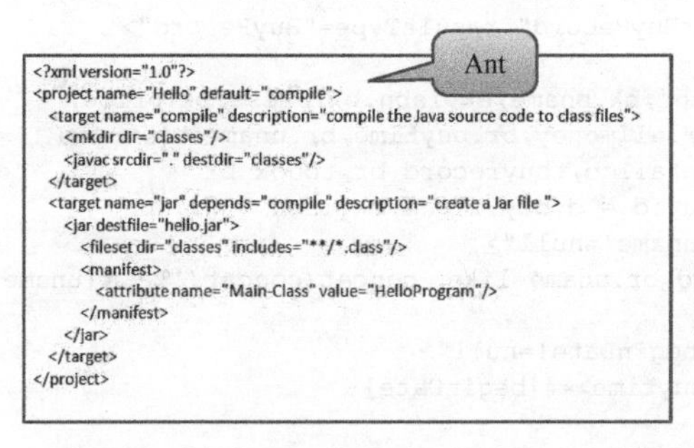

图 14-1　通过 Ant 实现项目的打包和部署

- 可以实现项目资源的共享。Maven 提供了一个中央仓库。一个开发团队统一从中央仓库中某个指定的位置提取资源，不仅加快了资源下载速度，还可以和项目的版本管理工具（如 SVN）协调工作（见图 14-2）。

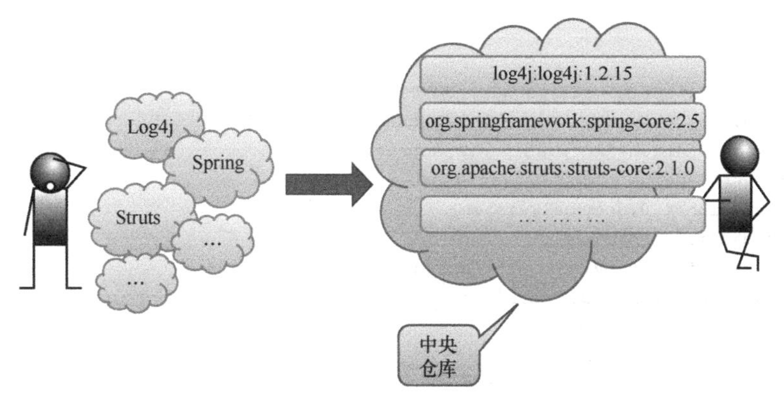

图 14-2　团队开发模式下项目资源的提取

- 管理项目的其他文档资源，如版本控制信息、缺陷跟踪系统信息、开发者信息、许可证信息、javadoc、测试覆盖报告、代码静态分析报告等。

14.1.2　通过 Maven 配置 pom.xml

Maven 依赖配置信息进行资源下载和构建，其配置文件为 pom.xml。

1. Maven 坐标

在 Maven 中，坐标表示每个构件都有自己的一个标识，它由 groupId、artifactId 和 version 等元素组成，依赖这些元素可以精确定位一个资源，如下面的 JAR 包的定位。

```
com/icss/bjetc/1.0/mvc-training-ppt-1.0.jar
```

Maven 坐标包括以下元素。

- groupId：表示组织标识（如项目包名）。
- artifactId：表示项目名称，如 my-app。
- version：表示项目的当前版本。
- packaging：表示项目的打包方式，常见的有.jar 和.war 两种。
- scope：用来控制依赖与编译、测试、运行的 classpath 的关系。
- compile：表示默认编译依赖范围，对编译、测试、运行这 3 种 classpath 都有效。
- test：测试依赖范围，只对测试 classpath 有效。

- classifier: 该元素用来帮助定义构建输出的一些附件，附属构件与主构件对应。

示例 14-1：在 pom.xml 文件的头部，先要显示坐标信息。

具体代码如下。

```
<project xmlns="http://maven.apache.org/POM/4.0.0"
    xmlns:xsi="http://www.w3.org/2001/XMLSchema-instance"
    xsi:schemaLocation="http://maven.apache.org/POM/4.0.0
    http://maven.apache.org/xsd/maven-4.0.0.xsd">
    <modelVersion>4.0.0</modelVersion>
    <groupId>com.icss</groupId>
    <artifactId>firstProvider</artifactId>
    <version>0.0.1-SNAPSHOT</version>
    <packaging>jar</packaging>
    <name>firstProvider</name>
    <url>http://maven.apache.org</url>
</project>
```

2. Maven 仓库

Maven 仓库是一个位置，指项目依赖的第三方库资源所在位置。在 Maven 中，任何一个依赖、插件或者项目构建后输出的 JAR 包，都可以称为构件。Maven 仓库能帮助我们管理构件（主要是 JAR），它就是放置所有 JAR 文件（包括 WAR、ZIP、POM 等文件）的地方。

如图 14-3 所示，Maven 仓库有 3 种类型。

- 本地仓库（一个项目组共同使用一个本地仓库）；
- 中央仓库（唯一内置的远程仓库）；
- 远程仓库（本地仓库从远程仓库下载构件）。

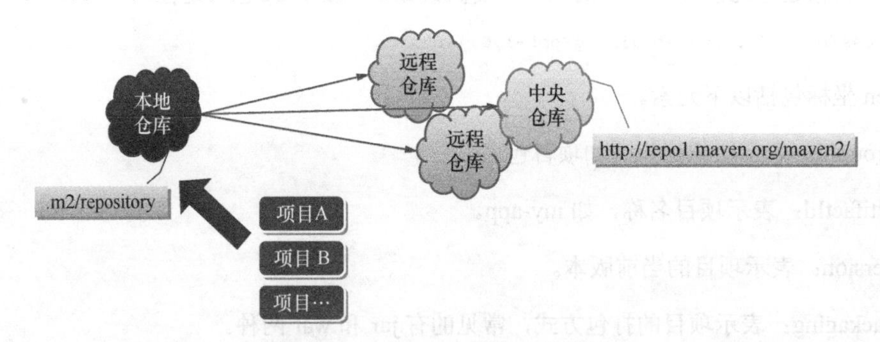

图 14-3 Maven 仓库的 3 种类型

3. 依赖

Maven 的核心特性是依赖管理。当我们处理多模块的项目（包含成百上千个模块或者子项目）

时，模块间的依赖关系（见图 14-4）就非常复杂，难以管理。

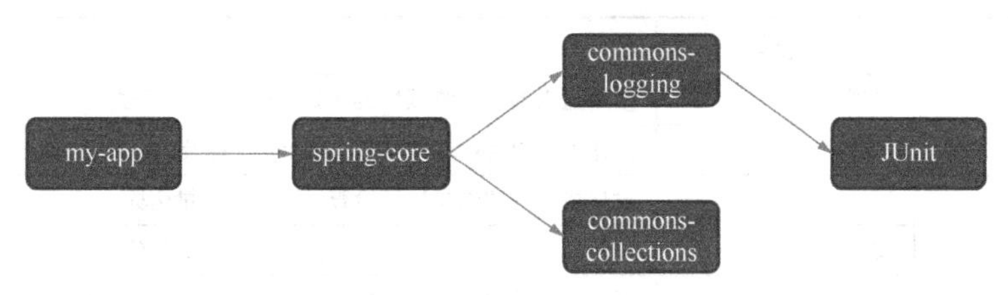

图 14-4　模块间的依赖关系

比如，A 依赖于 B 库，如果一个项目要使用 A 库，那么该项目也需要使用 B 库。Maven 可以避免搜索所有所需库。Maven 通过读取项目配置文件 pom.xml，找出模块之间的依赖关系。我们需要做的只是在每个项目的 pom.xml 文件中定义直接的依赖关系。其他的事情交由 Maven 处理即可。

通常情况下，一个大的项目下有一系列的子项目。在这种情况下，我们可以创建一个公共的 pom.xml 文件，该 pom.xml 文件包含所有的依赖关系（见图 14-5）。

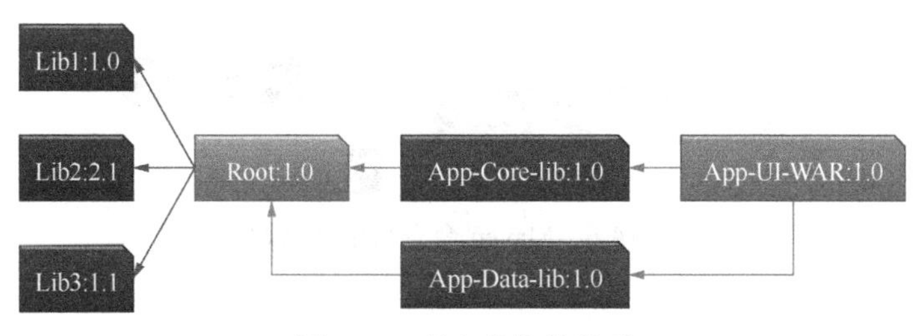

图 14-5　所有的依赖关系

图 14-5 的详情说明了以下几点。

- App-UI-WAR 依赖于 App-Core-lib 和 App-Data-lib。
- Root 是 App-Core-lib 和 App-Data-lib 的父项目。
- Root 依赖于 Lib1、lib2 和 Lib3。

4. Maven 生命周期

Maven 生命周期定义了一个项目的构建、发布过程。Maven 生命周期通常包含几个阶段，见表 14-1。

表 14-1　Maven 生命周期包含的阶段

阶段	处理	描述
验证（validate）	验证项目	验证项目是否正确且所有必需信息是可用的
编译（compile）	执行编译	编译源代码

续表

阶段	处理	描述
测试（test）	测试	使用适当的单元测试框架（如 JUnit）运行测试
包装（package）	打包	创建 JAR/WAR 包，如在 pom.xml 中提及的包
检查（verify）	检查	对集成测试的结果进行检查，以保证质量达标
安装（install）	安装	安装打包的项目到本地仓库，以供其他项目使用
部署（deploy）	部署	复制最终的项目包到远程仓库中，以共享给其他开发人员

如图 14-6 所示，分别执行 mvn clean install、mvn test、mvn site 命令，会按照不同的生命周期进行操作。

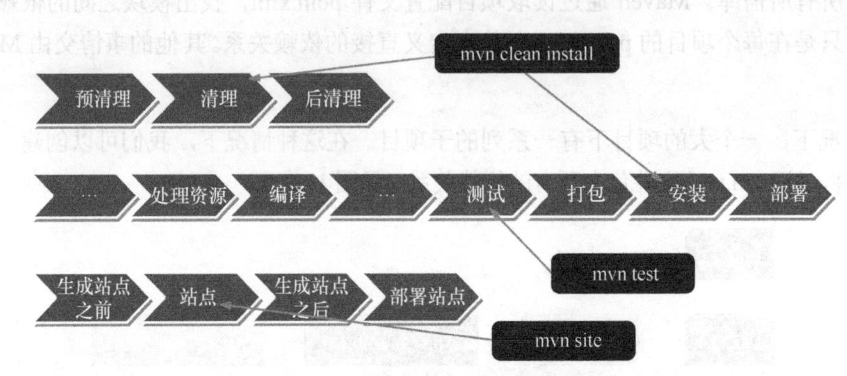

图 14-6　Maven 命令与生命周期

14.1.3　配置 Maven 环境

1．下载 Maven 安装包

Maven 是 Apache 的顶级项目，进入 Apache 官网即可找到 Maven。选择合适的版本下载即可（见图 14-7）。这里选择下载 apache-maven-3.5.0-bin.zip。

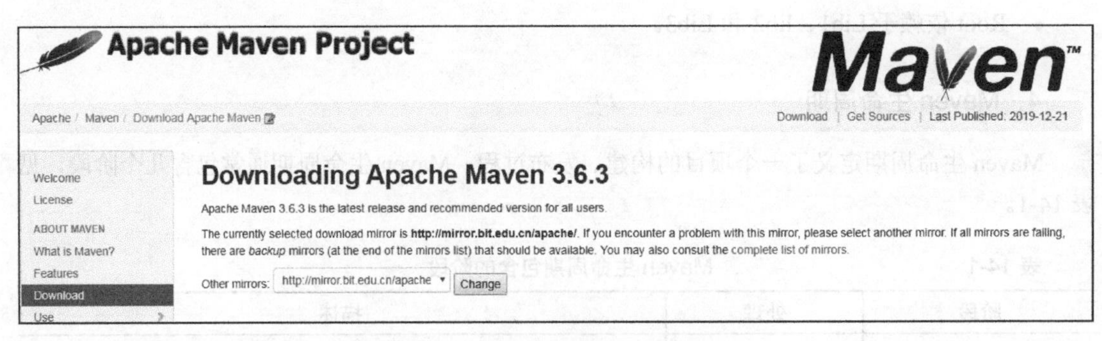

图 14-7　选择合适的 Maven 版本

2. 安装和配置 Maven

安装和配置 Maven 的具体步骤如下。

（1）把下载的 apache-maven-3.5.0-bin.zip 解压到 D 盘根目录下。

（2）安装 JDK 8，并配置环境变量 java_home 与 Path。

（3）配置环境变量 maven_home 与 Path（见图 14-8）。

（4）在命令行窗口中执行命令 mvn -version，如果能正确显示版本信息，则说明配置正确（见图 14-9）。

图 14-8　配置环境变量 maven_home 与 Path

```
D:\Users\Domino>mvn -version
Apache Maven 3.5.0 (ff8f5e7444045639af65f6095c62210b5713f426; 2017-04-04T03:39
6+08:00)
Maven home: D:\apache-maven-3.5.0\bin\..
Java version: 1.8.0_65, vendor: Oracle Corporation
Java home: C:\Java\jdk1.8.0_65\jre
Default locale: zh_CN, platform encoding: GBK
OS name: "windows 7", version: "6.1", arch: "x86", family: "windows"
```

图 14-9　执行 mvn -version 命令

（5）环境配置好后，即可在命令窗口执行 mvn 的构建命令。

3. 配置本地仓库

找到安装 Maven 的本地目录（如 D:\apache-maven-3.5.0），执行如下操作。

（1）在 D:\apache-maven-3.5.0 下，新建文件夹 localRepository。

（2）打开 D:\apache-maven-3.5.0\conf\setting.xml，新增如下配置项。

```
<localRepository>D:/apache-maven-3.5.0/localRepository</localRepository>
```

注意，默认配置项为 Default: ${user.home}/.m2/repository，修改成上面的配置即可。

14.2　Spring Boot 与环境配置

14.2.1　Spring Boot

Spring Boot 是 Spring.io 下的一个独立项目，参见 Spring 官网。Spring Boot 是伴随着 Spring 4.0 而发布的，Boot 是引导的意思，它的主要作用是帮助开发者快速地搭建基于 Spring 框架的开发环境，大幅减少配置信息。

Spring Boot 是一套集成开发环境，具有如下特色。

- 内嵌了 Web 服务器，如 Tomcat、Jetty、Undertow，无须部署 WAR，这给开发和测试都带来了非常大的便利。
- 提供了一系列 starter 依赖，大幅简化了使用 Maven 的依赖配置。
- 自动配置 Spring 和第三方库。
- 有助于进行微服务开发。

14.2.2 Spring Boot 开发环境

STS（Spring Tool Suite）是一个基于 Eclipse 的集成开发环境，用于开发 Spring Boot 项目。换句说话，STS 是一个定制版 Eclipse，由 Spring Framework 官方在 Java EE 版本的 Eclipse 上包装 Spring 插件而来，其核心还是 Java EE 版本的 Eclipse。

STS 的安装方式有 3 种，分别是直接下载 Eclipse 与 STS 的集成开发环境，下载 STS 安装包后离线安装，通过 Eclipse 在线安装。

（1）进入 Spring 官网，可以直接下载 STS 版的 Eclipse（见图 14-10）。

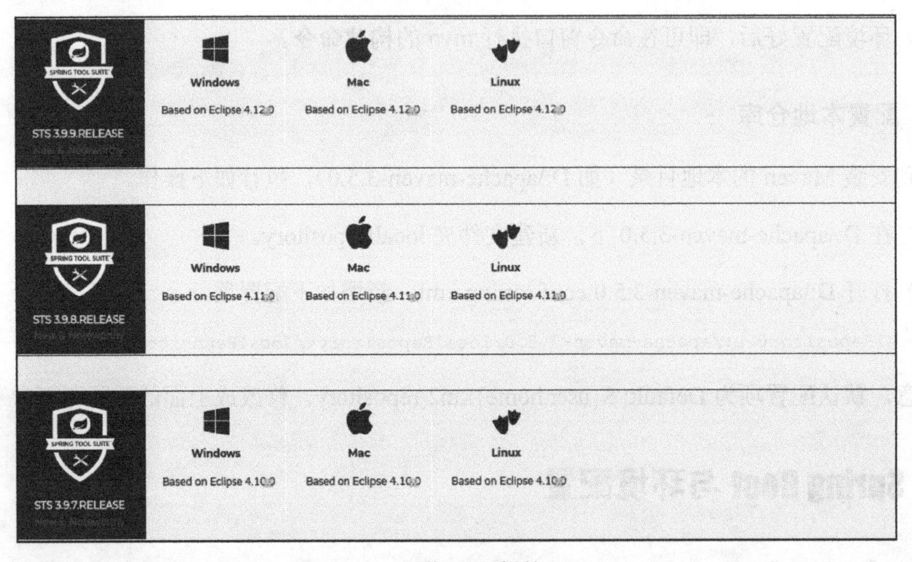

图 14-10　下载 STS 版的 Eclipse

注意：下载的 STS 集成开发环境要与所用机器的 JDK 位数一致，如都是 32 位或 64 位（这是首选方案）。

（2）为了在线安装，在 Eclipse 中，从菜单栏中选择 Help→Eclipse Marketplace，打开 Eclipse

Marketplace 窗口，选择 Search 选项卡并输入 sts 或选择 Popular 选项卡，然后选择 Spring Tool Suite (STS) for Eclipse 3.8.3.RELEASE 插件，并安装（见图 14-11）。

（3）在 Eclipse 中，从菜单栏中选择 Window→Preferences，打开 Preferences 窗口。在左侧面板中，搜索“maven”并选择 User Settings；在右侧面板中，单击 User Settings 文本框后面的 Browse 按钮，找到 settings.xml 的位置并单击 Apply 按钮，完成 Maven 的配置（见图 14-12）。

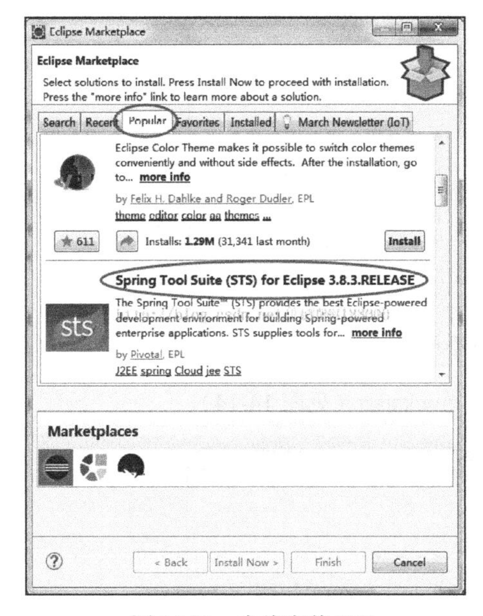

图 14-11　在线安装 STS

图 14-12　在 Eclipse 中配置 Maven

14.3　示例项目

14.3.1　微服务项目

一个微服务项目没有 Web 页面，没有 MVC 转向。所有的 Action 都配置成@RestController 模式，即@RestController = @Controller + @ResponseBody。

本节介绍实现微服务项目的操作步骤。

1．新建 Maven 项目

新建 Maven 项目的具体步骤如下。

（1）在 Maven 中，从菜单栏中选择 File→New→Maven Project，打开 New Maven Project 窗口，新建 Maven 项目，见图 14-13。勾选 User default Workspace location 复选框，单击 Next 按钮。

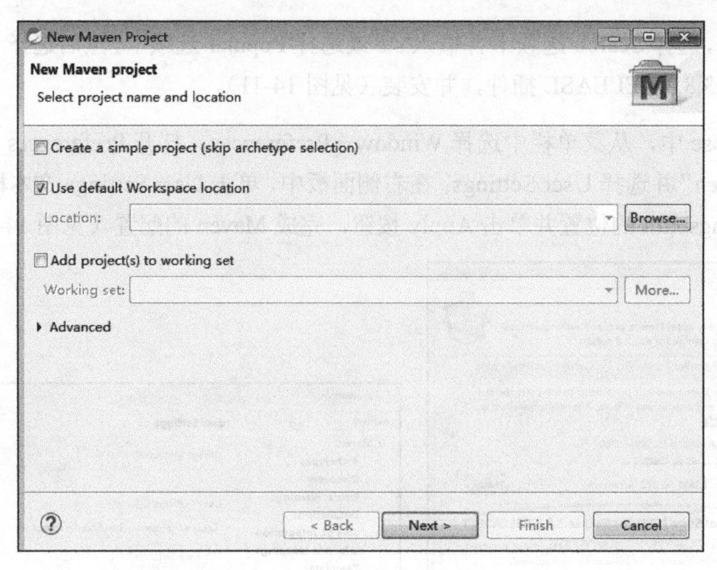

图 14-13　新建 Maven 项目

（2）对于 Archetype，使用默认的 maven-archetype-quickstart（见图 14-14）。

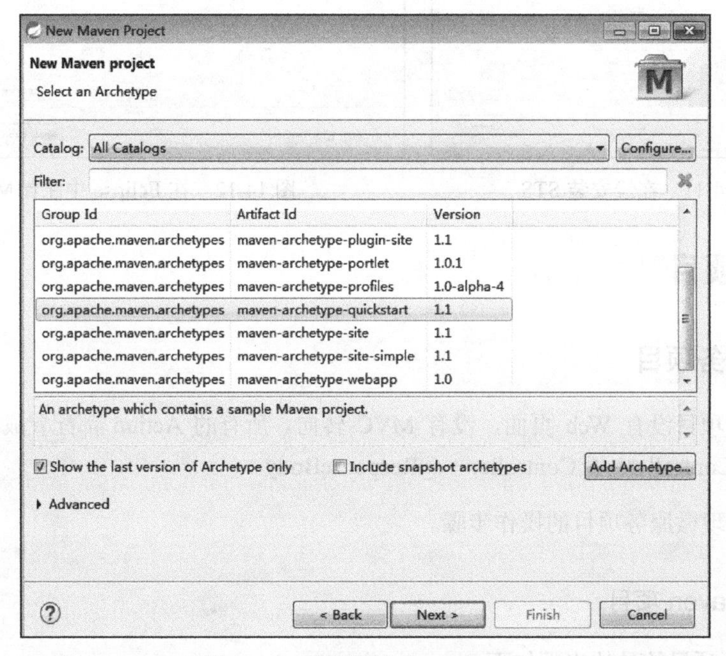

图 14-14　使用默认的 Archetype

（3）设置 Group Id 和 Artifact Id（见图 14-15）。如果集成开发环境配置正确，则单击 Finish 按钮，表示项目创建成功。

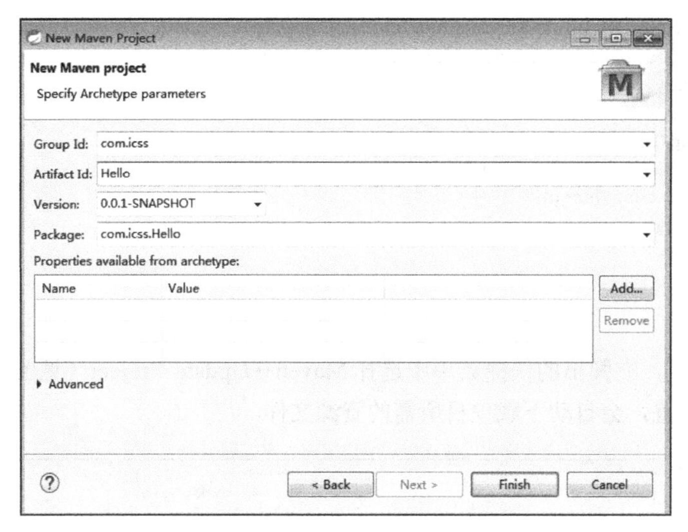

图 14-15 设置 Group Id 和 Artifact Id

（4）Hello 项目的结构如图 14-16 所示。

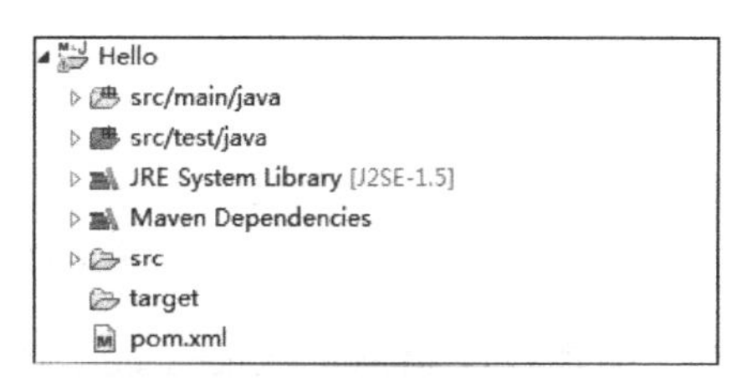

图 14-16 Hello 项目的结构

2. 配置 pom.xml

项目新建成功后，在原来的 pom.xml 文件中增如下配置。

（1）配置 spring-boot-starter-parent。spring-boot-starter-parent 是一个特殊的 Starter，用来提供相关的 Maven 默认依赖。使用它之后，常用的包依赖可以省去 version 标签，即依赖包的版本无须指定，由 Spring Boot 自动匹配。注意，此处的版本使用的是 1.4.1，它对应 Tomcat 8.5、Servlet 3.1 和 Spring Framework 4.3.3。

```
<parent>
    <groupId>org.springframework.boot</groupId>
    <artifactId>spring-boot-starter-parent</artifactId>
    <version>1.4.1.RELEASE</version>
</parent>
```

（2）默认的 Java SE 版本是 1.5，修改成我们需要的 1.8 版本。

```
<properties>
      <project.build.sourceEncoding>UTF-8</project.build.sourceEncoding>
      <java.version>1.8</java.version>
</properties>
```

（3）配置 spring-boot-starter-web。根据这个配置，Maven 会自动下载 Spring 的核心包、Web 包和依赖包。

```
<dependencies>
    <dependency>
```

```
            <groupId>org.springframework.boot</groupId>
            <artifactId>spring-boot-starter-web</artifactId>
        </dependency>
        <dependency>
            <groupId>junit</groupId>
            <artifactId>junit</artifactId>
            <scope>test</scope>
        </dependency>
    </dependencies>
```

3. 下载资源

右击 Hello 项目，在弹出的快捷菜单中选择 Maven→Update Project（见图 14-17），则系统根据前面的 Maven 配置，会自动下载项目所需的资源文件。

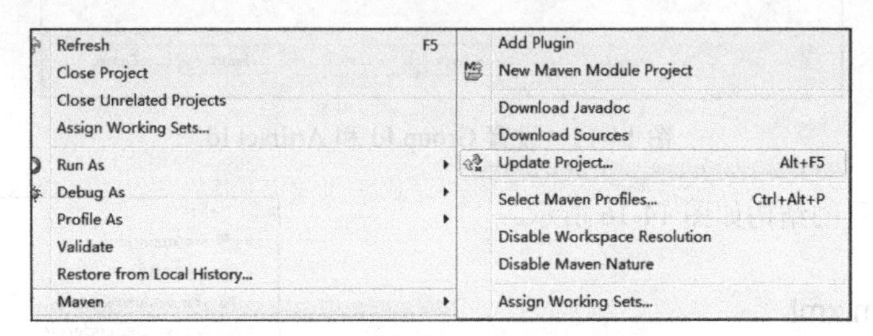

图 14-17 选择 Maven→Update Project

从项目的 Maven Dependencies 中可以看到很多包（见图 14-18），这些包就是根据上面的 pom.xml 的配置下载的。从中可以看出，这比直接使用 Maven 进行依赖配置要简单很多。

图 14-18 Maven Dependencies 中的包

4. 项目配置

在 src/main 下新建 resources 文件夹。在这个文件夹下，新建属性配置文件 application. Properties。在配置文件中填写如下内容。

```
server.port=8088
server.context-path=/Test
```

其中，8080 表示内置的 Tomcat 的启动端口，/Test 表示启动后的站点名字。

application.properties 可以使用的配置项很多，这里就不一一罗列了。

5. 服务代码

编写服务代码。@RestController 表示对外的微服务，注意，此处不是控制器的含义。

```
@RestController
public class HelloController {
    @RequestMapping("/hello")
    public String say() {
        return "hello, xiaohp";
    }
}
```

6. UI 代码

注意，app.java 应该放在所有包的根路径（如 com.icss.app），服务层的包为 com.icss.service，持久层的包为 com.icss.dao。

```
@SpringBootApplication
public class App {
    public static void main( String[] args ){
         SpringApplication.run(App.class, args);
    }
}
```

打开 App.java，右击，在弹出的快捷菜单中选择 Run As→2 Spring Boot App，见图 14-19，启动 Spring Boot。

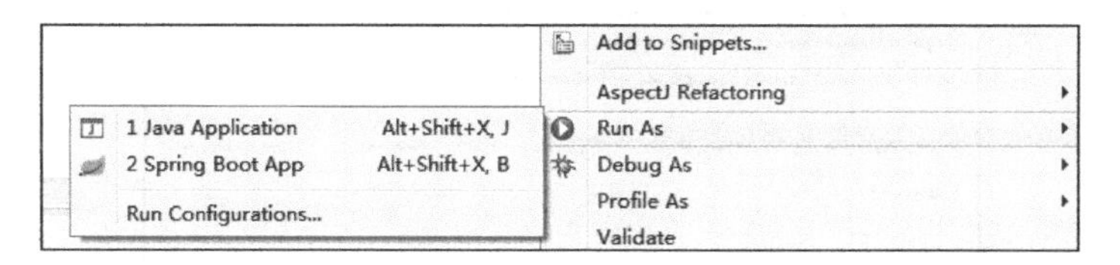

图 14-19　启动 Spring Boot

启动信息见图 14-20。

```
main] s.w.s.m.m.a.RequestMappingHandlerMapping : Mapped "{[/hello]}" onto public java.lang.String com.icss.Hello.H
main] s.w.s.m.m.a.RequestMappingHandlerMapping : Mapped "{[/error]}" onto public org.springframework.http.Response
main] s.w.s.m.m.a.RequestMappingHandlerMapping : Mapped "{[/error],produces=[text/html]}" onto public org.springfr
main] o.s.w.s.handler.SimpleUrlHandlerMapping  : Mapped URL path [/webjars/**] onto handler of type [class org.spr
main] o.s.w.s.handler.SimpleUrlHandlerMapping  : Mapped URL path [/**] onto handler of type [class org.springframe
main] o.s.w.s.handler.SimpleUrlHandlerMapping  : Mapped URL path [/**/favicon.ico] onto handler of type [class org
main] o.s.j.e.a.AnnotationMBeanExporter        : Registering beans for JMX exposure on startup
main] s.b.c.e.t.TomcatEmbeddedServletContainer : Tomcat started on port(s): 8088 (http)
main] com.icss.Hello.App                       : Started App in 8.331 seconds (JVM running for 12.674)
```

图 14-20　启动信息

7. 测试

根据 application.properties 的配置，输入 http://localhost:8088/Test/hello，Web 测试效果见图 14-21。

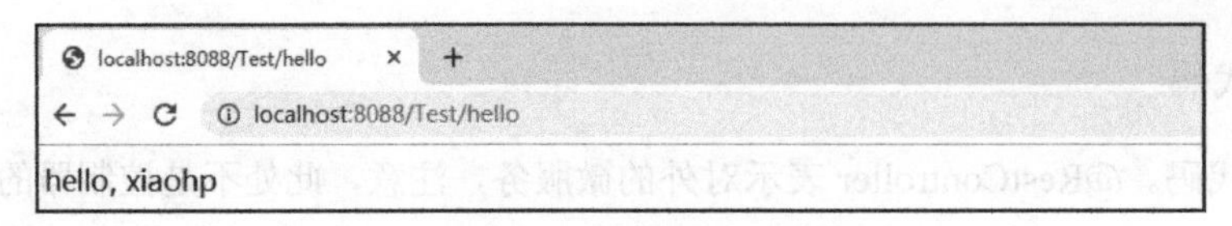

图 14-21　Web 测试效果

14.3.2 Web 项目

前一节中实现的是微服务项目，本节中实现一个 Web 项目。Web 项目需要页面资源，如 JSP、JS、CSS、图片等，这是与微服务项目的主要区别。

1. 新建 Maven 项目

新建 Maven 项目，对于 Archetype，这里选择 maven-archetype-webapp（见图 14-22），并设置 Group Id 和 Artifact Id（见图 14-23）。

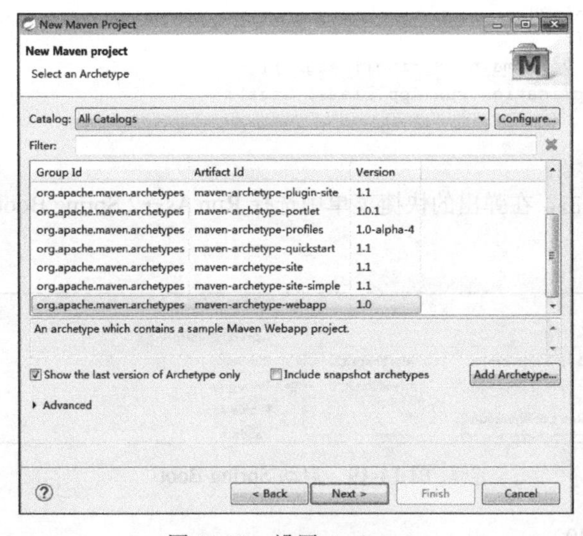

图 14-22　设置 Archetype

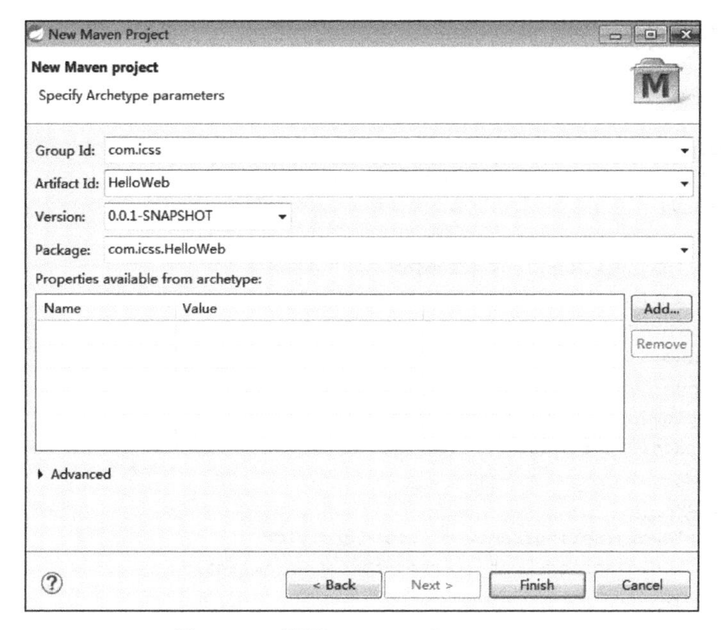

图 14-23　设置 Group Id 和 Artifact Id

HelloWeb 项目的结构如图 14-24 所示。

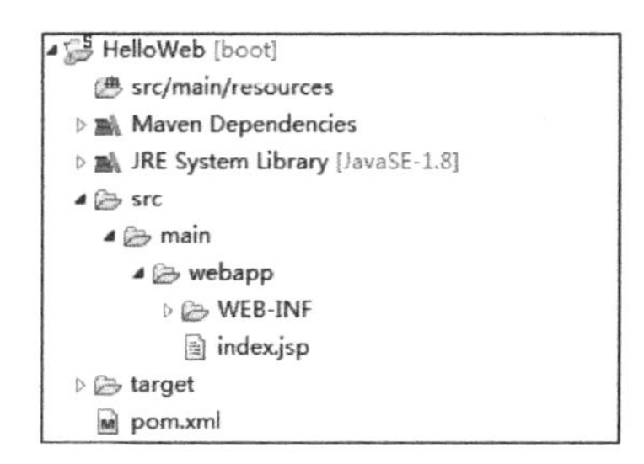

图 14-24　HelloWeb 项目的结构

2. 配置 pom.xml 文件

配置 pom.xml 文件的步骤如下。

（1）如下两个配置仍然需要。

```
<parent>
    <groupId>org.springframework.boot</groupId>
    <artifactId>spring-boot-starter-parent</artifactId>
    <version>1.4.1.RELEASE</version>
</parent>
<properties>
    <project.build.sourceEncoding>UTF-8</project.build.sourceEncoding>
    <java.version>1.8</java.version>
</properties>
```

（2）新增如下依赖项。

```
<dependencies>
    <dependency>
        <groupId>org.springframework.boot</groupId>
        <artifactId>spring-boot-starter-web</artifactId>
    </dependency>
    <dependency>
        <groupId>javax.servlet</groupId>
        <artifactId>javax.servlet-api</artifactId>
        <scope>provided</scope>
    </dependency>
    <dependency>
        <groupId>javax.servlet</groupId>
        <artifactId>jstl</artifactId>
    </dependency>
    <dependency>
        <groupId>org.springframework.boot</groupId>
        <artifactId>spring-boot-starter-tomcat</artifactId>
    </dependency>
    <dependency>
        <groupId>org.apache.tomcat.embed</groupId>
        <artifactId>tomcat-embed-jasper</artifactId>
    </dependency>
    <dependency>
        <groupId>junit</groupId>
        <artifactId>junit</artifactId>
        <scope>test</scope>
    </dependency>
</dependencies>
```

3．编写控制器代码

Web 项目需要控制器代码，在 src\main\java\创建包 com.icss.controller，然后新建控制器 HelloController。

```
@Controller
public class HelloController {
    @RequestMapping("/hello")
    public String say() {
        return "/hello.jsp";
    }
}
```

4．添加视图页面

在 webapp\WEB-INF 下新建文件夹 views，添加页面 hello.jsp（见图 14-25）。

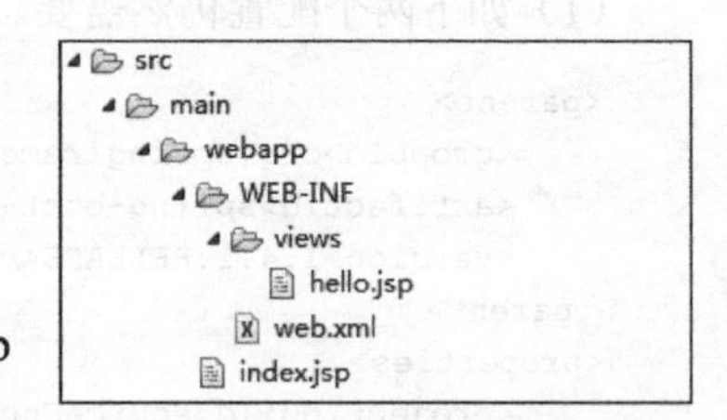

图 14-25　添加页面 hello.jsp

5. 测试

测试的具体步骤如下。

（1）在包 com.icss 下新建启动类 App。启动类应该放在基础包中，控制器为它的子包。

```
@SpringBootApplication
public class App {
    public static void main( String[] args ) {
         SpringApplication.run(App.class, args);
    }
}
```

（2）配置启动信息。

```
server.port=8088
server.context-path=/TestWeb
spring.mvc.view.prefix=/WEB-INF/views/
```

（3）在浏览器地址栏中输入 http://localhost:8088/TestWeb/hello，控制器转向了 hello.jsp 页面，测试结果如图 14-26 所示。

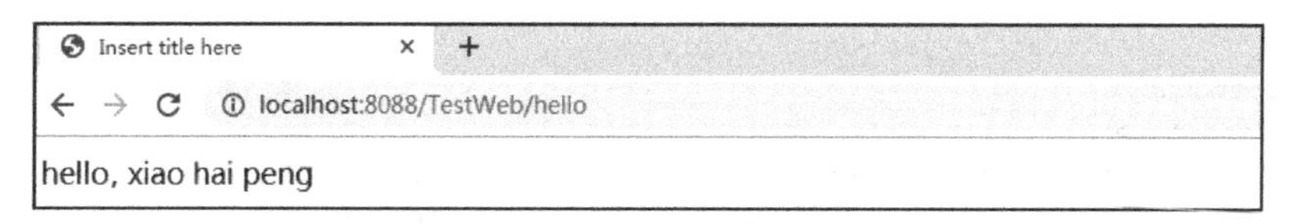

图 14-26　测试结果

14.4　整合书城项目

下面使用 Spring Boot 和 SSM 整合前面的书城项目。Spring Boot+SSM 的编码方式与前面在项目中直接导入包的方式相比有了很多变化，这会给编程带来一些麻烦。

业务代码基本不变，变化的是一些配置方式。

14.4.1　配置书城项目的 Spring Boot 环境

1. 配置 pom.xml 文件

在项目的根路径下，新建 pom.xml 文件，配置信息如下。

（1）配置 Spring Boot 版本。

```
<parent>
      <groupId>org.springframework.boot</groupId>
      <artifactId>spring-boot-starter-parent</artifactId>
```

```
    <version>1.4.1.RELEASE</version>
</parent>
```

（2）配置 Java 版本。

```
<properties>
    <project.build.sourceEncoding>UTF-8</project.build.sourceEncoding>
    <java.version>1.8</java.version>
</properties>
```

（3）配置 Web 依赖。

```
<dependencies>
    <dependency>
        <groupId>org.springframework.boot</groupId>
        <artifactId>spring-boot-starter-web</artifactId>
    </dependency>
    <dependency>
            <groupId>javax.servlet</groupId>
            <artifactId>javax.servlet-api</artifactId>
        <scope>provided</scope>
    </dependency>
    <dependency>
            <groupId>javax.servlet</groupId>
            <artifactId>jstl</artifactId>
    </dependency>
    <dependency>
            <groupId>org.springframework.boot</groupId>
            <artifactId>spring-boot-starter-tomcat</artifactId>
</dependency>
<dependency>
            <groupId>org.apache.tomcat.embed</groupId>
            <artifactId>tomcat-embed-jasper</artifactId>
</dependency>
<dependency>
            <groupId>org.mybatis.spring.boot</groupId>
            <artifactId>mybatis-spring-boot-starter</artifactId>
            <version>1.2.0</version>
</dependency>
</dependencies>
```

（4）配置 MyBatis 依赖。

```
<dependency>
    <groupId>org.mybatis.spring.boot</groupId>
    <artifactId>mybatis-spring-boot-starter</artifactId>
    <version>1.2.0</version>
</dependency>
```

（5）配置数据库依赖（这个项目使用 Oracle Database 11g 中的示例）。

```
<dependency>
    <groupId>com.oracle</groupId>
    <artifactId>ojdbc6</artifactId>
```

```
    <version>11.2.0.4.0-atlassian-hosted</version>
</dependency>
```

2. 配置 application.properties 文件

原来项目的 web.xml 中的信息和 Spring 的核心配置文件中的信息都移到了 application.properties 中，这是 Spring Boot 系统的配置中心。配置信息如下。

```
server.port=8088
server.context-path=/BookStore
spring.mvc.view.prefix=/WEB-INF/views/
spring.datasource.url=jdbc:oracle:thin:@Domino-PC:1521:orcl
spring.datasource.username=aa
spring.datasource.password=123456
spring.datasource.driver-class-name=oracle.jdbc.driver.OracleDriver
mybatis.config-location=classpath:mapper/mybatis.xml
spring.http.multipart.max-file-size=102400
```

（1）按照以下方式配置 Tomcat 启动端口。

```
server.port=8088
```

（2）按照以下方式配置 Web 站点名。

```
server.context-path=/BookStore
```

（3）按照以下方式配置 Spring MVC 的视图默认前缀。

```
spring.mvc.view.prefix=/WEB-INF/views/
```

（4）按照以下方式配置 Oracle 数据源。

```
spring.datasource.url=jdbc:oracle:thin:@Domino-PC:1521:orcl
spring.datasource.username=aa
spring.datasource.password=123456
spring.datasource.driver-class-name=oracle.jdbc.driver.OracleDriver
```

（5）按照以下方式配置 MyBatis 核心配置文件的位置。

```
mybatis.config-location=classpath:mapper/mybatis.xml
```

（6）按照以下方式配置 Spring MVC 中上传文件的最大约束。

```
spring.http.multipart.max-file-size=102400
```

3. 配置 web.xml 文件

配置 web.xml 文件的具体步骤如下。

（1）删除 web.xml 中关于前端控制器的传统配置项。现在 Spring 的配置信息已经移到了 application.properties 中，DispatcherServlet 也无须再进行配置。

```
<servlet>
    <servlet-name>aa</servlet-name>
```

```
        <servlet-class>
            org.springframework.web.servlet.DispatcherServlet
        </servlet-class>
        <init-param>
            <param-name>contextConfigLocation</param-name>
            <param-value>WEB-INF/spring-mvc.xml</param-value>
        </init-param>
        <load-on-startup>1</load-on-startup>
    </servlet>
    <servlet-mapping>
        <servlet-name>aa</servlet-name>
        <url-pattern>*.do</url-pattern>
    </servlet-mapping>
```

（2）删除默认欢迎页的设置。

```
<welcome-file-list>
        <welcome-file>index.jsp</welcome-file>
</welcome-file-list>
```

（3）新建一个 IndexAction，代替欢迎页（注意，把 index.jsp 移到/WEB-INF/views 文件夹下）。

```
@Controller
public class IndexAction {
    @RequestMapping("/")
    public String index() {
        return "index.jsp";
    }
}
```

4. 配置过滤器

在配置过滤器时，为了解决乱码问题，执行以下操作。

（1）在启动的 App 中配置@ServletComponentScan。在 SpringBootApplication 上使用@ServletComponentScan 注解后，Servlet、Filter、Listener 可以直接通过@WebServlet、@WebFilter、@WebListener 注解自动注册，不需要其他代码。于是，系统中用于解决乱码的 CharacterEncodingFilter 就可以生效了。

（2）在 application.properties 中增加字符集配置。

```
spring.http.encoding.force=true
spring.http.encoding.charset=gbk
spring.http.encoding.enabled=true
server.tomcat.uri-encoding=gbk
```

5. 配置异常处理程序

配置异常处理程序的步骤如下。

（1）删除在 spring-mvc.xml 中原来的异常处理程序配置。

```
<bean
   class="org.springframework.web.servlet.handler.SimpleMappingExceptionResolver">
     <property name="exceptionMappings">
         <props>
             <prop key="java.lang.Throwable">/error/error.jsp</prop>
             <prop key="org.springframework
                        .web.multipart.MaxUploadSizeExceededException">
                   /error/OverMaxUploadSize.jsp</prop>
         </props>
     </property>
</bean>
```

（2）使用全局异常@ExceptionHandler 代替 SimpleMappingExceptionResolver。

```
@ControllerAdvice
public class GlobalExceptionHandler {
    public static final String DEFAULT_ERROR_VIEW = "error.jsp";
     @ExceptionHandler(value = Exception.class)
     public ModelAndView defaultErrorHandler(HttpServletRequest req,
                                      Exception e) throws Exception {
         ModelAndView mav = new ModelAndView();
         mav.addObject("exception", e);
         mav.addObject("url", req.getRequestURL());
         mav.setViewName(DEFAULT_ERROR_VIEW);
         return mav;
     }
     @ExceptionHandler(value =
            org.springframework.web.multipart.MultipartException.class)
     public ModelAndView maxUploadSizeExceededException(HttpServletRequest req,
                           Exception e) throws Exception {
         ModelAndView mav = new ModelAndView();
         mav.addObject("exception", e);
         mav.addObject("url", req.getRequestURL());
         mav.setViewName("/error/OverMaxUploadSize.jsp");
         return mav;
    }
}
```

6. 配置自定义拦截器

配置自定义拦截器的步骤如下。

（1）删除原来自定义的拦截器配置。

```
<bean  class="org.springframework.web.servlet
    .mvc.method.annotation.RequestMappingHandlerMapping">
    <property name="interceptors">
        <list>
            <ref bean="officeHoursInterceptor" />
        </list>
    </property>
</bean>
<bean id="officeHoursInterceptor"
```

```
    class="com.icss.intercept.TimeBasedAccessInterceptor">
    <property name="openingTime" value="9" />
    <property name="closingTime" value="20" />
</bean>
```

（2）使用如下配置信息注册拦截器。

```
@Configuration
public class MyWebAppConfig extends WebMvcConfigurerAdapter{
    public void addInterceptors(InterceptorRegistry registry) {
        TimeBasedAccessInterceptor aa = new TimeBasedAccessInterceptor();
        aa.setOpeningTime(8);
        aa.setClosingTime(21);
        registry.addInterceptor(aa);
        super.addInterceptors(registry);
    }
}
```

14.4.2 启动类 App

启动类 App 定义在 com.icss 基础包中，系统启动后会自动扫描子包信息。

```
@SpringBootApplication
@EnableTransactionManagement
@MapperScan("com.icss.bk.dao.batis")
@ServletComponentScan
public class App {
    public static void main( String[] args ){
        SpringApplication.run(App.class, args);
    }
}
```

Spring Boot 与 SSM 项目的集成至此完毕。除了一些配置信息之外，业务代码基本没有变化。使用 Spring Boot 开发项目，在资源依赖、项目发布等很多环节都带来了便利，这种开放模式正在逐渐成为主流。